AF353116

The Impact of Alcoholic Beverages on Human Health

The Impact of Alcoholic Beverages on Human Health

Editor

Peter Anderson

MDPI • Basel • Beijing • Wuhan • Barcelona • Belgrade • Manchester • Tokyo • Cluj • Tianjin

Editor
Peter Anderson
Newcastle University
UK
Maastricht University
The Netherlands

Editorial Office
MDPI
St. Alban-Anlage 66
4052 Basel, Switzerland

This is a reprint of articles from the Special Issue published online in the open access journal *Nutrients* (ISSN 2072-6643) (available at: https://www.mdpi.com/journal/nutrients/special_issues/ Alcoholic_Beverage).

For citation purposes, cite each article independently as indicated on the article page online and as indicated below:

LastName, A.A.; LastName, B.B.; LastName, C.C. Article Title. *Journal Name* **Year**, *Volume Number*, Page Range.

ISBN 978-3-0365-4101-3 (Hbk)
ISBN 978-3-0365-4102-0 (PDF)

Contents

About the Editor

Peter Anderson

Dual trained as specialist in public health medicine and general practice (Oxford University and London School of Hygiene and Tropical Medicine), Dr. Anderson has a background in general practice, public health and academia, and has pioneered research and published on a range of alcohol policy issues. In the European Office of the World Health Organization, he was responsible for the alcohol and tobacco programmes, set up the Organization's first European Alcohol Action Plan, and became Director of the Department of Prevention and Health Promotion. He has brought science to policy across a range of public health issues for the World Economic Forum, several governments, intergovernmental organizations, including European Commission and OECD, and public and private sector think tanks around the world.

 *nutrients*

Editorial

The Impact of Alcoholic Beverages on Human Health

Peter Anderson [1,2]

[1] Population Health Sciences Institute, Newcastle University, Baddiley-Clark Building, Newcastle upon Tyne NE2 4AX, UK; peteranderson.mail@gmail.com
[2] Department of Health Promotion, CAPHRI Care and Public Health Research Institute, Maastricht University, 6211 LK Maastricht, The Netherlands

Citation: Anderson, P. The Impact of Alcoholic Beverages on Human Health. *Nutrients* **2021**, *13*, 4417. https://doi.org/10.3390/nu13124417

Received: 25 November 2021
Accepted: 7 December 2021
Published: 10 December 2021

Publisher's Note: MDPI stays neutral with regard to jurisdictional claims in published maps and institutional affiliations.

As summarized in the World Health Organization's latest Global Status Report on Alcohol, the pleasure of alcohol is indicated by the fact that, worldwide, just over two-fifths of the population aged 15+ years drink alcohol; 2.3 billion people, consuming nearly 35 billion litres of pure ethanol a year, equivalent to just over three drinks (33 g of pure ethanol) a day (see [1]).

The pain of alcohol is indicated in the collection of papers in this special issue: ethanol in alcoholic beverages is toxic to human health, causing 7.1% of all deaths amongst those aged less than 70 years (two million deaths a year), with the three top causes of death being cirrhosis of the liver, road injuries, and tuberculosis [1]. In addition, through a combination of brain damage due to consuming alcohol and genetic predisposition, some 4% of adults experience what is termed alcohol dependence, a complex behavioural syndrome that has at its core the inability to control alcohol consumption despite adverse social, occupational or health consequences [2].

Ethanol:

- Is a teratogen [3];
- Is genotoxic and a carcinogen [4,5];
- Is hepatotoxic [6];
- Is neurotoxic to the brain [2];
- Causes injuries [7];
- Causes a range of cardiovascular diseases [8];
- Increases the risk of a range of communicable diseases, including HIV, TB, pneumonia, and COVID-19 infection [9]; and
- May or may not increase the risk of overweight and obesity [10].

For most conditions, the dose–response curves increase from zero consumption upwards, with many curves being exponential [11]. The exception to this is that some alcohol consumption protects some people against ischaemic diseases to some degree, with potential benefits occurring at about 1 drink every other day [8]; this is an hormetic effect to an environmental agent characterized by a low dose beneficial effect and a high dose toxic effect, that, for ethanol, may be a consequence of human ancestral exposure to naturally occurring low levels of ethanol from ripe fruit [12]. On the other side, though, margins of exposure analysis indicate that the present average daily consumption amongst drinkers across the world (33 g of ethanol a day) exceeds typically accepted thresholds indicating health risks for carcinogens by some 10,000 [5].

The question this supplement raises, is the pleasure worth the pain?

Governments and governmental organizations seem to consider that the pleasure is worth the pain. As Stockwell and colleagues point out, it is possible to provide very specific and detailed advice to governments regarding the public health consequences of policy decisions in such concrete terms as how many people will become ill, injured or die prematurely from alcohol-related reasons if policy X or Y is not introduced [13]. Yet, almost all countries fall far short of implementing effective public health policies to reduce the harm done by alcohol [13]. Further, despite alcohol being a carcinogen, [4,5], at least in Europe, health warning labels are notable by their absence [14]. Additionally, even for people who run into problems, including alcohol dependence, there seems a lack of care and treatment. As Nutt and colleagues point out, it remains the case that, to date, only three pharmacotherapies are licensed for alcohol dependence and only 9% of such individuals receive such treatment [2]; there is simply no moral outrage from non-governmental organizations.

The alcohol industry also seems to consider that the pleasure is worth the pain. Product reformulation of existing products to contain less alcohol, and more extensive market penetration of no and low alcohol products could lead to consumers drinking less ethanol (see [15]). Yet, overall, there seem to be only very limited moves in this direction [15], and the alcohol industry continues to counter the implementation of effective policies that could reduce the harm done by alcohol [13].

Why is there a dissonance between what the science says about alcohol's toxicity, and the failure to prevent two million deaths a year amongst the under seventies and provide adequate treatment to the 4% of adults who experience alcohol dependence? Stockwell et al. [13] mention four reasons:

1. lack of public awareness of both need and the effectiveness of policies;
2. lack of government regulatory mechanisms to implement effective policies;
3. alcohol industry lobbying; and,
4. a failure from the public health community to promote specific and feasible actions as opposed to general principles, e.g., 'increased prices' or 'reduced affordability'.

What would be a litmus test of change that we take the pain of alcohol seriously? The simple specific take of this editorial:
The alcohol industry places a warning label on its products: **ALCOHOL CAUSES CANCER**. Such an (unlikely) action would:

1. Shame governments for failing to protect their citizens against a known carcinogen; and,
2. Demonstrate that the alcohol industry is serious in taking responsibility for its products.

Funding: No funding was received for writing this editorial.

Conflicts of Interest: The author declares no conflict of interest.

References

1. Sohi, I.; Franklin, A.; Chrystoja, B.; Wettlaufer, A.; Rehm, J.; Shield, K. The Global Impact of Alcohol Consumption on Premature Mortality and Health in 2016. *Nutrients* **2021**, *13*, 3145. [CrossRef] [PubMed]
2. Nutt, D.; Hayes, A.; Fonville, L.; Zafar, R.; Palmer, E.; Paterson, L.; Lingford-Hughes, A. Alcohol and the Brain. *Nutrients* **2021**, *13*, 3938. [CrossRef] [PubMed]
3. Popova, S.; Dozet, D.; Shield, K.; Rehm, J.; Burd, L. Alcohol's Impact on the Fetus. *Nutrients* **2021**, *13*, 3452. [CrossRef] [PubMed]
4. Rumgay, H.; Murphy, N.; Ferrari, P.; Soerjomataram, I. Alcohol and Cancer: Epidemiology and Biological Mechanisms. *Nutrients* **2021**, *13*, 3173. [CrossRef] [PubMed]
5. Okaru, A.; Lachenmeier, D. Margin of Exposure Analyses and Overall Toxic Effects of Alcohol with Special Consideration of Carcinogenicity. *Nutrients* **2021**, *13*, 3785. [CrossRef] [PubMed]
6. Pohl, K.; Moodley, P.; Dhanda, A. Alcohol's Impact on the Gut and Liver. *Nutrients* **2021**, *13*, 3170. [CrossRef] [PubMed]
7. Chikritzhs, T.; Livingston, M. Alcohol and the Risk of Injury. *Nutrients* **2021**, *13*, 2777. [CrossRef] [PubMed]
8. Roerecke, M. Alcohol's Impact on the Cardiovascular System. *Nutrients* **2021**, *13*, 3419. [CrossRef]
9. Morojele, N.; Shenoi, S.; Shuper, P.; Braithwaite, R.; Rehm, J. Alcohol Use and the Risk of Communicable Diseases. *Nutrients* **2021**, *13*, 3317. [CrossRef]
10. Fong, M.; Scott, S.; Albani, V.; Adamson, A.; Kaner, E. 'Joining the Dots': Individual, Sociocultural and Environmental Links between Alcohol Consumption, Dietary Intake and Body Weight—A Narrative Review. *Nutrients* **2021**, *13*, 2927. [CrossRef]
11. Rehm, J.; Rovira, P.; Llamosas-Falcón, L.; Shield, K. Dose—Response Relationships between Levels of Alcohol Use and Risks of Mortality or Disease, for All People, by Age, Sex, and Specific Risk Factors. *Nutrients* **2021**, *13*, 2652. [CrossRef] [PubMed]
12. Dudley, R.; Maro, A. Human Evolution and Dietary Ethanol. *Nutrients* **2021**, *13*, 2419. [CrossRef] [PubMed]
13. Stockwell, T.; Giesbrecht, N.; Vallance, K.; Wettlaufer, A. Government Options to Reduce the Impact of Alcohol on Human Health: Obstacles to Effective Policy Implementation. *Nutrients* **2021**, *13*, 2846. [CrossRef] [PubMed]

14. Kokole, D.; Anderson, P.; Jané-Llopis, E. Nature and Potential Impact of Alcohol Health Warning Labels: A Scoping Review. *Nutrients* **2021**, *13*, 3065. [CrossRef] [PubMed]
15. Anderson, P.; Kokole, D.; Llopis, E. Production, Consumption, and Potential Public Health Impact of Low- and No-Alcohol Products: Results of a Scoping Review. *Nutrients* **2021**, *13*, 3153. [CrossRef] [PubMed]

nutrients

MDPI

Review

Human Evolution and Dietary Ethanol

Robert Dudley [1,2,*] and Aleksey Maro [1]

[1] Department of Integrative Biology, University of California, Berkeley, CA 94720, USA;
    alekseymaro@berkeley.edu
[2] Smithsonian Tropical Research Institute, P.O. Box 2072, Balboa, Panama
* Correspondence: wings@berkeley.edu; Tel.: +1-510-642-1555

Abstract: The "drunken monkey" hypothesis posits that attraction to ethanol derives from an evolutionary linkage among the sugars of ripe fruit, associated alcoholic fermentation by yeast, and ensuing consumption by human ancestors. First proposed in 2000, this concept has received increasing attention from the fields of animal sensory biology, primate foraging behavior, and molecular evolution. We undertook a review of English language citations subsequent to publication of the original paper and assessed research trends and future directions relative to natural dietary ethanol exposure in primates and other animals. Two major empirical themes emerge: attraction to and consumption of fermenting fruits (and nectar) by numerous vertebrates and invertebrates (e.g., *Drosophila* flies), and genomic evidence for natural selection consistent with sustained exposure to dietary ethanol in diverse taxa (including hominids and the genus *Homo*) over tens of millions of years. We also describe our current field studies in Uganda of ethanol content within fruits consumed by free-ranging chimpanzees, which suggest chronic low-level exposure to this psychoactive molecule in our closest living relatives.

Keywords: alcoholism; evolution; fermentation; frugivory; *Homo*; primate; yeast

Citation: Dudley, R.; Maro, A. Human Evolution and Dietary Ethanol. *Nutrients* **2021**, *13*, 2419. https://doi.org/10.3390/nu13072419

Academic Editor: Emilio Sacanella

Received: 28 May 2021
Accepted: 13 July 2021
Published: 15 July 2021

Publisher's Note: MDPI stays neutral with regard to jurisdictional claims in published maps and institutional affiliations.

1. Introduction

The argument of the "drunken monkey" hypothesis is that alcohol (and primarily the ethanol molecule) is a low-level but routine component of the diet for all animals that consume fruits and nectar [1,2]. In addition to providing a useful long-distance olfactory cue to localize nutritional resources and to identify ripe and calorically rich fruits up close, ethanol may also act as a feeding stimulant (as in modern humans, via the well-studied aperitif effect; [3]). Humans first began intentional fermentation during the Meso-Neolithic transitional period broadly coincident with the domestication of crops, and ethanol consumption has correspondingly been viewed as a fairly recent phenomenon relative to the origin of our species. However, dietary consumption of ethanol likely characterizes all frugivorous and nectarivorous animals, including primates and the hominoid lineage leading to modern humans. Millions of years of interaction among flowering plants, fermentative yeast, and numerous vertebrate lineages thus suggest a linkage between ethanol ingestion and acquisition of nutritional reward. We also see in diverse animal taxa, as well as in modern humans, substantial genetic variation in the ability to metabolize ethanol that is consistent with natural selection to this end.

The natural role of ethanol in animal nutrition has been largely underestimated in the zoological literature. For example, ethanol in ripe and fermenting fruits has been proposed to be largely aversive to vertebrate consumers [4]. More recently, information from behavioral, ecological, and genomic studies indicates an impressive commonality of behavioral and physiological responses to ethanol, and in taxa ranging from fruit flies to primates. The overarching concept that unites these studies is evolution, which can sometimes provide novel insights into questions of human health and behavior [5,6]. Here, we review advances in the field of comparative ethanol biology since the first publication of the "drunken monkey" hypothesis [1], and describe emerging themes in the 140 English language citations to the original paper (Google Scholar; January 2001–April 2021). We also provide preliminary information on ethanol content of fruits consumed in nature by our nearest living relatives, the chimpanzees. Given that chimpanzees mostly eat ripe fruits (e.g., up to 86% of the time; [7,8]), and that a comparable diet is thought to have pertained to the earliest hominins [9–11], these data suggest that low-level ethanol ingestion was an important feature of human nutrition over evolutionary time. Such ethanol consumption via frugivory could, in turn, have resulted in physiological and sensory adaptations that, today, yield hedonic reward following dietary exposure to this molecule [2]. Predictions of the "drunken monkey" hypothesis and relevant empirical findings since 2000 are provided in Table 1.

Table 1. Predictions of the "drunken monkey" hypothesis and supporting empirical evidence.

Prediction	Supporting Evidence	References
Ethanol occurs naturally at low levels within many fruits and nectars.	A variety of tropical fruits, as well as some nectars, contain ethanol at low concentrations.	[12–15]
Olfaction can be used to localize and preferentially select ethanol-containing nutritional resources.	Fruits consumed by primates produce numerous volatiles, including ethanol. Olfactory abilities are well-developed in primates, but have not been explicitly tested relative to use in fruit localization or selection.	[16,17]
Ethanol at low concentrations is not aversive to frugivores and nectarivores.	Diverse vertebrates consume food items containing low-concentration ethanol.	[18–25]
Ethanol acts as a feeding stimulant.	Modern humans increase caloric ingestion following consumption of an aperitif. Effects of dietary ethanol on ingestion rates for free-ranging primates have not yet been evaluated.	[3]
Genetic variation in the ability to metabolize ethanol is correlated with the extent of dietary exposure.	Substantial variation in ADH tracks dietary inclusion of fruit and nectar among mammals. Ethanol catabolism was up-regulated in African apes ~10 Mya ago, in parallel with terrestrialization.	[26,27]
Hormetic advantage derives from chronic consumption of ethanol.	Mortality is reduced at low levels of ethanol ingestion in modern humans and rodents, and also in *Drosophila* flies exposed to low-concentration ethanol vapor.	[28–33]

2. Vertebrate Responses to Naturally Occurring Ethanol

Sugars within ripe and over-ripe fruits serve as caloric motivation for consumption by animals, primarily mammals and birds, that subsequently disperse the seeds. Ripe fruits must be attractive to these consumers and must also present sufficient nutritional reward so as to elicit consumption. However, the ubiquity of yeasts in natural environments indicates the potential for fermentation prior to consumption by vertebrates [34,35]. Anaerobic fermentation by yeasts and ethanol generation have been dated using molecular methods to coincide with the origin of fleshy and sugar-rich fruits in the Cretaceous period [36] and may specifically have evolved to inhibit activity of bacterial competitors within fruit pulp [37]. Fruit decomposition can thus be viewed as a race in time between microbes and dispersal agents, and correspondingly, there is selection on vertebrate sensory mechanisms to facilitate rapid localization and consumption of these transient resources.

Fermentation of fruit crops is most pronounced in warm, humid environments such as tropical rainforests, the habitat of most frugivorous primates today. For example, ripe palm fruits (*Astrocaryum standleyanum*) contain ~0.6% ethanol within the pulp, but over-ripe fruits have much higher levels, averaging 4.5% [12]. Substantial levels of ethanol within pulp also characterize fruits in Southeast Asia over a range of ripening stages [13]. Animals consuming these fruits will necessarily ingest ethanol at low concentrations. Given that animal frugivores can consume 5%–10% of their body weight daily in ripe fruit, even the aforementioned low concentrations will yield substantial chronic dosage. Floral nectars in the tropics can also ferment and yield substantial ethanol concentrations. Wild tree shrews and slow lorises feed from palm flowers (*Eugeissona tristis*) in Malaysia that contain significant levels of ethanol within the nectar [14]. Although the animals never become overtly inebriated, hair samples contain high levels of a secondary metabolite of ethanol (ethyl glucuronide), consistent with high chronic exposure. Laboratory choice trials with two species of nectar-feeding primates indicate increasing preference for higher-concentration ethanol solutions [18] (see also [19] for analogous experiments with a primate frugivore). Additionally, wild chimpanzees consume anthropogenically sourced fermentations of palm sap within the tree canopy, at least at one site in West Africa [20]. Critically, the assertion that ethanol is toxic and renders fruit unpalatable to vertebrates [4] has been empirically falsified for mammalian dispersal agents [21].

In tropical rainforests, ripe fruit is a transient and spatially heterogeneous resource. Olfactory plumes of ethanol provide, however, an honest signal of caloric availability to potential consumers downwind. The olfactory sensitivity of primates to various alcohols, including ethanol, is well-developed [16,17], but this sensory capacity has not been demonstrated under field conditions. Adult fruit flies, however, use ethanol plumes to locate suitable oviposition sites on ripe fruit. The study of ethanol responses in *Drosophila* now represents a useful model system for understanding molecular pathways of inebriation in humans [38]. Additionally, behavioral preferences by fruit flies for ethanol-containing substrates are correlated with the ability to metabolize ethanol, suggesting a direct link between metabolic capacity and sensory attraction [39]. Similarly, ethanol is not aversive to fruit-feeding birds and bats [22,23] and is sometimes consumed at lethal levels [24,25]. In rodents, ethanol evokes neural hyperactivity in brain-feeding circuits, further supporting evolutionary associations between consumption of fermented substrates and caloric gain [40]. Most importantly, a recent survey of wild primate diets [15] demonstrated the widespread consumption of fruits in the late stages of fermentation (as assessed by human observers). Because ethanol may be present within ripe fruits with no obvious external signs of microbial activity, this study provides a conservative estimate of actual dietary exposure; a quantitative assessment of ethanol concentrations within consumed fruits across the entire spectrum of palatability is clearly now called for.

3. Evolutionary Consequences of Dietary Ethanol

If chronic dietary exposure to ethanol inevitably derives from frugivory (and from nectarivory), then selection will favor the evolution of metabolic adaptations that maximize physiological benefits but minimize costs of exposure. Higher concentrations of ethanol may, by contrast, be stressful and cause harm. Such a nonlinear dose-response curve is termed hormesis and is an evolutionary outcome that increases overall organismal fitness given natural exposure to various compounds at low concentrations [41–43]. A key prediction of the "drunken monkey" hypothesis, therefore, is that hormetic benefits will pertain to animals at low, naturally occurring levels of ethanol exposure.

In support of this claim, longevity (as well as female fecundity) of fruit flies is increased at low atmospheric concentrations of ethanol but decreases at zero exposure and at higher concentrations [28–30]. Laboratory rodents similarly show decreased mortality at intermediate levels of ethanol ingestion [31]. In humans, epidemiological studies suggest a reduction in cardiovascular risk and overall mortality at low levels of ethanol consumption relative either to abstinence or to higher levels of intake [32,33]. Consequences of chronic ethanol ingestion for human reproductive fitness have not been evaluated, but we might expect a similar outcome as with longevity. No current data address the hormetic effects of ethanol on wild animals with variable levels of dietary availability, but logistically, such long-term measurements can be carried out in appropriate contexts (e.g., field-banded tracking of individual hummingbirds through their lifespan at different sites with variable extent of nectar fermentation).

Evolutionary arguments also predict that intra- and interspecific variation in the ability to metabolize ethanol will correspond to its relative dietary inclusion. Alcohol dehydrogenase (ADH) initially converts ethanol to acetaldehyde, which then is acted upon by aldehyde dehydrogenase (ALDH) to yield acetate used for energy yield in oxidative pathways. Both ADH and ALDH exist in a number of different allelic forms characterized by varying catalytic efficiencies, which in *Drosophila* flies are well-known to correlate with natural levels of environmental ethanol exposure [2,39]. Furthermore, in the lineage of great apes that led to modern humans, there is a pronounced genetic signature demonstrating comparable evolutionary responses to chronic dietary exposure to ethanol. Paleogenetic reconstruction of alcohol dehydrogenase genes across the hominid phylogeny indicates a dramatically enhanced catabolic capacity in one particular ADH (ADH4, as encoded by the *ADH7* allele), starting at about 10 Mya [26]. ADH4, although only one of multiple ADH forms present in mammals, is found primarily in the mouth and digestive tract and thus effects the "first pass" at the digestion of ethanol. This enzyme became dramatically better at metabolizing ethanol following the phyletic split between the lineage leading to modern orangutans and to the other great apes, including ourselves. It thus correlates well with increasing terrestrialization among the African apes, possibly yielding greater access to fermenting fruit crops on the ground, and thus resulting in increased ethanol within the diet [26]. The same mutation also characterizes ADH4 of the Madagascan aye-aye, which routinely feeds on nectar from flowers of an endemic palm. Although ethanol content is not characterized for such flowers, studies with captive aye-ayes demonstrate a preference for consumption of low-level ethanol within sugar solutions [18].

Moreover, a recent study [27] evaluated variation in ADH 4 across 79 mammal species; multiple losses of function in *ADH7* (i.e., pseudogenization) and relaxed selection on this allele were found for those taxa with little or no presumed dietary exposure to ethanol (e.g., whales). Contrariwise, natural selection was apparently intensified on *ADH7* for those species specializing on either fruit or nectar [27]. Although the actual extent of dietary ethanol consumption is not known for the study species, clearly the likelihood of its chronic ingestion must be higher for frugivores and nectarivores. Quantitative specification of ethanol exposure, in conjunction with assessment of genetic changes in the other ethanol-metabolizing enzymes (e.g., ADH1, ADH2, and numerous ALDH polymorphisms) is now called for to assess the overall evolutionary response to fermented nutritional substrates.

Hominoids (i.e., the lesser and greater apes) also exhibit an evolutionary loss of uricase as a consequence of accumulating deleterious mutations in the corresponding gene (starting ~20 Mya; [44,45]). Modern humans correspondingly exhibit very high blood levels of uric acid and show amplification of fat accumulation (and of the metabolic syndrome more generally) given chronic fructose ingestion [46–48]. Ethanol consumption also stimulates fructose production by the liver, as well as more widespread production of uric acid, with both effects acting synergistically to increase overall fat storage [49,50]. The psychoactive and hedonic properties of ethanol and fructose are also similar, facilitating addictive responses to these naturally occurring compounds within fruit [51]. Such changes in both the uricase gene and in genes directly involved in ethanol catabolism are consistent with positive selection on dietary preference for fruit sugars and their fermentation products and are possibly linked with sensory mechanisms facilitating their efficient consumption and digestion.

In addition to aforementioned interspecific studies of ethanol metabolism, there is also substantial intraspecific genetic variation in physiological responses to ethanol, at least among modern human populations. In particular, slow-acting ALDH occurs at high frequencies in East Asian humans, and yields toxic acetaldehyde buildup following the con-sumption of ethanol [52,53]. Such variation, in turn, has been correlated with the propensity towards alcoholism for certain populations. Rates of alcoholism, however constructed definitionally, tend to be much lower within East Asian populations, consistent with the deterrent effects of elevated acetaldehyde [54,55]. Although genotype-by-environment interactions are also likely to be involved, the interacting dynamics of ethanol catabolism and accumulation of the acetaldehyde intermediate product are apparently protective against excessive alcohol consumption [56].

Finally, multigenerational exposure to high levels of dietary ethanol can result in significant changes to the gut microbiome, at least in laboratory rodents [57]. This intriguing outcome, mediated either directly by ethanol or by its downstream metabolic products, may also indicate systemic neural regulation of ingestion as influenced by endogenous gut fauna. The role of the microbiome in mediating physiological and behavioral responses to ethanol, either across the lifespan or in evolutionary time, has never been evaluated for free-ranging vertebrates, but clearly is of adaptive relevance. As with aforementioned molecular evolutionary studies of ADH and ALDH, comparative studies of the gut microbiome among frugivorous and nectarivorous species (and including birds as well as mammals) would elucidate correlates of microbical community composition relative to chronic ethanol ingestion and may indicate a role for selection in promoting higher rates of ethanol consumption so as to increase energetic gain while feeding.

4. Natural Ethanol Exposure in Chimpanzees

Recent field studies of chimpanzee-consumed fruits in Uganda suggest a chronic low-level ingestion of ethanol, albeit at sub-inebriating levels that are nonetheless consistent with physiological consequences. The Ngogo population of Eastern chimpanzees (*P. troglodytes schweinfurthii*) in Kibale National Park reside in a forest with a low density of a high-output, asynchronously fruiting fig species (*Ficus mucuso*), which is consumed preferentially more than any other fruit (i.e., 18%–34% of total feeding time; [58,59]). In 2019 and 2020, we determined ethanol concentrations for *F. mucuso* fruits as well as for a diversity of other consumed fruit species. By visiting *F. mucuso* trees with chimpanzees actively foraging in the canopy, we could collect ripe figs either immediately after they fell, following disturbance or by rejection, or within an hour of having fallen (as evidenced by wet latex at the stem). We also collected unripe *F. mucuso* during part of the field season, which the chimpanzees eat during periods of food shortage. Collected figs were frozen at the field station to arrest fermentation. We determined ethanol concentrations within individual fruits using an infrared gas analyzer on homogenized pulp samples, and also via ethanol vapor measurements in the headspace over pulp samples. Prior to these measurements, for each fruit we also assessed its mass, puncture resistance, sugar concentration, surface reflectance, and presence or absence of fig wasps, so as to correlate quantitatively ethanol content with stages of ripeness, and to assess which factors most influence microbial ethanol production. Data obtained to date indicate ethanol levels within ripe figs ranging from negligible amounts to as high as several percent (weight of ethanol/weight of fruit), consistent with values determined for other primate-consumed fruits [12,13].

Levels of ethanol consumption are determined both by ingested food volume and by intrinsic concentration. A typical daily consumption of ~6 kg of fruit at an ethanol content of only 0.23% would correspond to ingestion of one standard drink (i.e., 14 g of ethanol in the USA). Moreover, adult Eastern chimpanzees in the wild weigh substantially less than humans (i.e., only 30–40 kg; [60]), suggesting a much higher body-mass specific exposure. If consumed fruit were to contain 1% ethanol on average, then consuming 6 kg of fruit daily would yield >4 standard drinks daily, and a much higher body-mass specific rate of exposure. These preliminary calculations suggest that ethanol ingestion via frugivory is non-trivial in wild chimpanzees, and can easily approach chronic exposure of physiological relevance, if not of occasional inebriation.

Further assessment of dosage via dietary ingestion would require knowledge of rates of ethanol absorption and catabolism in chimpanzees, which are not necessarily the same as those in humans. Enzymatic activity relative to ethanol degradation is variable among mammalian taxa, and even among modern human populations [56,61,62]. Direct measurement of blood-ethanol levels in free-ranging chimpanzees and other frugivores would be informative in this regard. Nonetheless, chimpanzees may have evolved specific behavioral and physiological responses to ethanol commensurate with its natural occurrence within ripe and over-ripe fruit. Ripening in figs poses particular challenges to frugivores in that overt and substantial color changes otherwise indicating suitability of fruits for consumption are not present in this genus (*Ficus*: Moraceae). Both short-range olfactory and tactile cues are thus more important in identifying ripe fruits, with reduced use of visual cues [63]. Equally relevant to foraging outcomes are features of spatial and temporal heterogeneity in fruit ripening and fermentation. For example, fruits growing within the same tree may be vertically stratified, with fruits higher in the canopy being larger and containing more sugars [64] and possibly higher ethanol content as well.

Moreover, fig wasps are the mutualistic tenants and obligate pollinators of figs and may influence fermentation outcomes via different microbiota that they vector into fruits. Fig wasp behavior and ecology are highly variable among species, as are the chemical and structural features of different figs, which may, in turn, influence outcomes of microbial colonization and growth. Finally, endogenous fermentation of sugars is likely to vary with local climate, and in particular with average ambient temperatures. Lower elevations likely yield fruits with higher ethanol concentrations, given faster yeast growth in hotter climates. Our data for Ngogo (situated at 1400 m above sea level) likely represent conservative values for fruit-ethanol concentrations relative to those within lowland tropical rainforests where most frugivorous animals are found. Nonetheless, these preliminary measurements suggest sustained exposure of our closest living relatives to dietary ethanol and establish a methodological framework for further investigation into the natural consumption of fermented fruits.

5. Conclusions

Behavioral responses to naturally occurring ethanol can be advantageous for many animals, may be ancestral in primates, and have substantial implications for modern humans relative to both benign consumption of alcohol and excessive levels of drinking. A number of empirical questions can be posed to further assess the generality of these evolutionary arguments. In natural ecosystems, how do animals localize fermenting nutritional resources, and what are typical blood-ethanol levels within frugivores and nectarivores? Do the hormetic effects of low-level ethanol consumption extend more generally to all species exposed to this molecule over evolutionary time? Are there particular sensory mechanisms in some species that predispose them to excessive ethanol ingestion under artificial conditions of high supply? For example, ethanol evokes hyperactivity in the brain-feeding circuits of rodents, consistent with a general role as an appetitive stimulant [40]. Fermentation by yeasts of simple carbohydrate substrates is widespread in terrestrial environments, yet the natural background of ethanol availability has been largely ignored by biologists and clinicians alike, nutritional and health implications for modern humans notwithstanding. We therefore encourage further studies of ethanol-seeking activities in the natural world, as such behaviors (and their underlying genetic underpinnings) may yield novel insights into contemporary human consumption and misuse of alcoholic beverages.

Author Contributions: Conceptualization, R.D. and A.M.; methodology, R.D. and A.M.; formal analysis, R.D. and A.M.; investigation, A.M.; resources, A.M.; data curation, R.D. and A.M.; writing—original draft preparation, R.D. and A.M.; writing—review and editing, R.D. and A.M.; supervision, R.D.; project administration, R.D. and A.M.; funding acquisition, R.D. Both authors have read and agreed to the published version of the manuscript.

Funding: This research was funded by the Class of 1933 Chair in the Biological Sciences (University of California, Berkeley) to R.D.

Acknowledgments: We thank John Mitani and Aaron Sandel for ongoing logistical support.

Conflicts of Interest: The authors declare no conflict of interest. The funders had no role in the design of the study; in the collection, analyses, or interpretation of data; in the writing of the manuscript, or in the decision to publish the results.

References

1. Dudley, R. Evolutionary Origins of Human Alcoholism in Primate Frugivory. *Q. Rev. Biol.* **2000**, *75*, 3–15. [CrossRef] [PubMed]
2. Dudley, R. *The Drunken Monkey: Why We Drink and Abuse Alcohol*; University of California Press: Berkeley, CA, USA, 2014.
3. Yeomans, M.R. Effects of alcohol on food and energy intake in human subjects: Evidence for passive and active over-consumption of energy. *Br. J. Nutr.* **2004**, *92*, S31–S34. [CrossRef] [PubMed]
4. Janzen, D.H. Why Fruits Rot, Seeds Mold, and Meat Spoils. *Am. Nat.* **1977**, *111*, 691–713. [CrossRef]
5. Williams, G.C.; Nesse, R.M. *Why We Get Sick: The New Science of Darwinian Medicine*; Times Books: New York, NY, USA, 1994.
6. Nesse, R.M.; Berridge, K.C. Psychoactive Drug Use in Evolutionary Perspective. *Science* **1997**, *278*, 63–66. [CrossRef]
7. McGrew, W.C.; Baldwin, P.J.; Tutin, C.E.G. Diet of wild chimpanzees (*Pan troglodytes verus*) at Mt. Assirik, Senegal: I. Composition. *Am. J. Primatol.* **1998**, *16*, 213–226. [CrossRef]
8. Wrangham, R.W.; Chapman, C.A.; Clark-Arcadi, A.P.; Isabirye-Basuta, G. Social ecology of Kanyawara chimpanzees: Implications for understanding the costs of great ape groups. In *The Great Ape Societies*; McGrew, W.C., Marchant, L.F., Nishida, T., Eds.; Cambridge University Press: Cambridge, UK, 1996; pp. 45–57.
9. Andrews, P.; Martin, L. Hominoid dietary evolution. *Philos. Trans. R. Soc. Lond. B* **1991**, *334*, 199–209.
10. Andrews, P. Palaeoecology and hominoid palaeoenvironments. *Biol. Rev. Cam. Philos. Soc.* **1996**, *71*, 257–300. [CrossRef]
11. Roberts, P.; Boivin, N.; Lee-Thorp, J.; Petraglia, M.; Stock, J. Tropical forests and the genus Homo. *Evol. Anthropol.* **2016**, *25*, 306–317. [CrossRef]
12. Dudley, R. Fermenting fruit and the historical ecology of ethanol ingestion: Is alcoholism in modern humans an evolutionary hangover? *Addiction* **2002**, *97*, 381–388. [CrossRef]
13. Dominy, N.J. Fruits, Fingers, and Fermentation: The Sensory Cues Available to Foraging Primates. *Integr. Comp. Biol.* **2004**, *44*, 295–303. [CrossRef]
14. Wiens, F.; Zitzmann, A.; Lachance, M.-A.; Yegles, M.; Pragst, F.; Wurst, F.M.; von Holst, D.; Guan, S.L.; Spanagel, R. Chronic intake of fermented floral nectar by wild treeshrews. *Proc. Natl. Acad. Sci. USA* **2008**, *105*, 10426–10431. [CrossRef]
15. Schiel, N.; Sanz, C.M.; Schülke, O.; Shanee, S.; Souto, A.; Souza-Alves, J.P.; Stewart, F.; Stewart, K.M.; Stone, A.; Sun, B.; et al. Fermented food consumption in wild nonhuman primates and its ecological drivers. *Am. J. Phys. Anthr.* **2021**, *175*, 513–530. [CrossRef]
16. Simmen, B. Taste discrimination and diet differentiation among New World primates. In *The Digestive System in Mammals: Food, Form, and Function*; Chivers, D.J., Langer, P., Eds.; Cambridge University Press: Cambridge, UK, 1994; pp. 150–165.
17. Laska, M.; Seibt, A. Olfactory sensitivity for aliphatic alcohols in squirrel monkeys and pigtail macaques. *J. Exp. Biol.* **2002**, *205*, 1633–1643. [CrossRef] [PubMed]
18. Gochman, S.R.; Brown, M.B.; Dominy, N.J. Alcohol discrimination and preferences in two species of nectar-feeding primate. *R. Soc. Open Sci.* **2016**, *3*, 160217. [CrossRef] [PubMed]
19. Ibañez, D.D.; Salazar, L.T.H.; Laska, M. Taste Responsiveness of Spider Monkeys to Dietary Ethanol. *Chem. Senses* **2019**, *44*, 631–638. [CrossRef] [PubMed]
20. Hockings, K.J.; Bryson-Morrison, N.; Carvalho, S.; Fujisawa, M.; Humle, T.; McGrew, W.C.; Nakamura, M.; Ohashi, G.; Yamanashi, Y.; Yamakoshi, G.; et al. Tools to tipple: Ethanol ingestion by wild chimpanzees using leaf-sponges. *R. Soc. Open Sci.* **2015**, *2*, 150150. [CrossRef]
21. Peris, J.E.; Rodríguez, A.; Peña, L.; Fedriani, J.M. Fungal infestation boosts fruit aroma and fruit removal by mammals and birds. *Sci. Rep.* **2017**, *7*, 5646. [CrossRef]
22. Sánchez, F.; Korine, C.; Steeghs, M.; Laarhoven, L.-J.; Harren, F.J.M.; Cristescu, S.M.; Dudley, R.; Pinshow, B. Ethanol and Methanol as Possible Odor Cues for Egyptian Fruit Bats (*Rousettus aegyptiacus*). *J. Chem. Ecol.* **2006**, *32*, 1289–1300. [CrossRef]
23. Mazeh, S.; Korine, C.; Pinshow, B.; Dudley, R. Does ethanol in fruit influence feeding in the frugivorous yellow-vented bulbul (*Pycnonotus xanthopygos*)? *Behav. Process.* **2008**, *77*, 369–375. [CrossRef]
24. Fitzgerald, S.D.; Sullivan, J.M.; Everson, R.J. Suspected Ethanol Toxicosis in Two Wild Cedar Waxwings. *Avian Dis.* **1990**, *34*, 488–490. [CrossRef]
25. Kinde, H.; Foate, E.; Beeler, E.; Uzal, F.; Moore, J.; Poppenga, R. Strong circumstantial evidence for ethanol toxicosis in Cedar Waxwings (*Bombycilla cedrorum*). *J. Ornithol.* **2012**, *153*, 995–998. [CrossRef]
26. Carrigan, M.A.; Uryasev, O.; Frye, C.B.; Eckman, B.L.; Myers, C.R.; Hurley, T.D.; Benner, S.A. Hominids adapted to metabolize ethanol long before human-directed fermentation. *Proc. Natl. Acad. Sci. USA* **2015**, *112*, 458–463. [CrossRef]
27. Janiak, M.C.; Pinto, S.L.; Duytschaever, G.; Carrigan, M.A.; Melin, A.D. Genetic evidence of widespread variation in ethanol metabolism among mammals: Revisiting the 'myth' of natural intoxication. *Biol. Lett.* **2020**, *16*, 20200070. [CrossRef]
28. Starmer, W.T.; Heed, W.B.; Rockwood-Sluss, E.S. Extension of longevity in *Drosophila mojavensis* by environmental ethanol: Differences between subraces. *Proc. Natl. Acad. Sci. USA* **1977**, *74*, 387–391. [CrossRef]
29. Etges, W.J.; Klassen, C.S. Influences of atmospheric ethanol on adult *Drosophila mojavensis*: Altered metabolic rates and increases in fitnesses among populations. *Physiol. Zool.* **1989**, *62*, 170–193. [CrossRef]

30. Parsons, P.A. Acetaldehyde utilization in Drosophila: An example of hormesis. *Biol. J. Linn. Soc.* **1989**, *37*, 183–189. [CrossRef]

31. Diao, Y.; Nie, J.; Tan, P.; Zhao, Y.; Tu, J.; Ji, H.; Cao, Y.; Wu, Z.; Liang, H.; Huang, H.; et al. Long-term low-dose ethanol intake improves healthspan and resists high-fat diet-induced obesity in mice. *Aging* **2020**, *12*, 13128–13146. [CrossRef] [PubMed]

32. Poli, A.; Marangoni, F.; Avogaro, A.; Barba, G.; Bellentani, S.; Bucci, M.; Cambieri, R.; Catapano, A.; Costanzo, S.; Cricelli, C.; et al. Moderate alcohol use and health: A consensus document. *Nutr. Metab. Cardiovasc. Dis.* **2013**, *23*, 487–504. [CrossRef] [PubMed]

33. Chiva-Blanch, G.; Badimon, L. Benefits and risks of moderate alcohol consumption on cardiovascular disease: Current findings and controversies. *Nutrients* **2020**, *12*, 108. [CrossRef]

34. Last, F.T.; Price, D. Yeasts associated with living plants and their environs. In *The Yeasts*; Rose, A.H., Harrison, J.S., Eds.; Academic Press: London, UK, 1969; Volume 1, pp. 183–218.

35. Spencer, J.F.T.; Spencer, D.M. Ecology: Where yeasts live. In *Yeasts in Natural and Artificial Habitats*; Spencer, J.F.T., Spencer, D.M., Eds.; Springer: Berlin/Heidelberg, Germany, 1997; pp. 33–58.

36. Benner, S.A.; Caraco, M.D.; Thomson, J.M.; Gaucher, E.A. Planetary biology–paleontological, geological, and molecular his-tories of life. *Science* **2002**, *296*, 864–868. [CrossRef] [PubMed]

37. Ingram, L.O.; Buttke, T.M. Effects of Alcohols on Micro-Organisms. *Adv. Microb. Physiol.* **1984**, *25*, 253–300. [CrossRef]

38. Devineni, A.V.; Heberlein, U. The evolution of Drosophila melanogaster as a model for alcohol research. *Annu. Rev. Neurosci.* **2013**, *36*, 121–138. [CrossRef] [PubMed]

39. Cadieu, N.; Cadieu, J.-C.; El Ghadraoui, L.; Grimal, A.; Lamboeuf, Y. Conditioning to ethanol in the fruit fly—A study using an inhibitor of ADH. *J. Insect Physiol.* **1999**, *45*, 579–586. [CrossRef]

40. Cains, S.; Blomeley, C.; Kollo, M.; Racz, R.; Burdakov, D. Agrp neuron activity is required for alcohol-induced overeating. *Nat. Commun.* **2017**, *8*, 14014. [CrossRef] [PubMed]

41. Gerber, L.M.; Williams, G.C. The nutrient-toxin dosage continuum in human evolution and modern health. *Q. Rev. Biol.* **1999**, *74*, 273–289. [CrossRef] [PubMed]

42. Forbes, V.E. Is hormesis an evolutionary expectation? *Funct. Ecol.* **2000**, *14*, 12–24. [CrossRef]

43. Calabrese, E.J.; Baldwin, L.A. Toxicology rethinks its central belief. *Nature* **2003**, *421*, 691–692. [CrossRef]

44. Oda, M.; Satta, Y.; Takenaka, O.; Takahata, N. Loss of Urate Oxidase Activity in Hominoids and its Evolutionary Implications. *Mol. Biol. Evol.* **2002**, *19*, 640–653. [CrossRef]

45. Kratzer, J.T.; Lanaspa, M.A.; Murphy, M.N.; Cicerchi, C.; Graves, C.L.; Tipton, P.A.; Ortlund, E.A.; Johnson, R.J.; Gaucher, E.A. Evolutionary history and metabolic insights of ancient mammalian uricases. *Proc. Natl. Acad. Sci. USA* **2014**, *111*, 3763–3768. [CrossRef] [PubMed]

46. Johnson, R.J.; Titte, S.; Cade, J.R.; Rideout, B.A.; Oliver, W.J. Uric acid, evolution and primitive cultures. *Semin. Nephrol.* **2005**, *25*, 3–8. [CrossRef]

47. Tapia, E.; Cristóbal, M.; Garcia, F.; Soto, V.; Monroy-Sánchez, F.; Pacheco, U.; Lanaspa, M.A.; Roncal-Jiménez, C.A.; Cruz-Robles, D.; Ishimoto, T.; et al. Synergistic effect of uricase blockade plus physiological amounts of fructose-glucose on glomerular hypertension and oxidative stress in rats. *Am. J. Physiol. Renal. Physiol.* **2013**, *304*, F727–F736. [CrossRef] [PubMed]

48. Softic, S.; Meyer, J.G.; Wang, G.X.; Gupta, M.K.; Batista, T.M.; Lauritzen, H.; Fujisaka, S.; Serra, D.; Herrero, L.; Willoughby, J.; et al. Dietary sugars alter hepatic fatty acid oxidation via transcriptional and post-translational modifications of mitochondrial proteins. *Cell Metab.* **2019**, *30*, 735–753.e734. [CrossRef] [PubMed]

49. Sánchez-Lozada, L.G.; Andres-Hernando, A.; Garcia-Arroyo, F.E.; Cicerchi, C.; Li, N.; Kuwabara, M.; Roncal-Jimenez, C.; Johnson, R.J.; Lanaspa, M. Uric acid activates aldose reductase and the polyol pathway for endogenous fructose production and fat accumulation in the development of fatty liver. *J. Biol. Chem.* **2019**, *294*, 4272–4281. [CrossRef]

50. Wang, M.; Chen, W.Y.; Zhang, J.; Gobejishvili, L.; Barve, S.S.; McClain, C.J.; Joshi-Barve, S. Elevated fructose and uric acid through aldose reductase contribute to experimental and human alcoholic liver disease. *Hepatology* **2020**, *72*, 1617–1637. [CrossRef]

51. Lustig, R.H. Fructose: It's "alcohol without the buzz". *Adv. Nutr.* **2013**, *4*, 226–235. [CrossRef]

52. Shen, Y.-C.; Fan, J.-H.; Edenberg, H.J.; Li, T.-K.; Cui, Y.-H.; Wang, Y.-F.; Tian, C.-H.; Zhou, C.-F.; Zhou, R.-L.; Wang, J.; et al. Polymorphism of ADH and ALDH genes among four ethnic groups in China and effects upon the risk for alcoholism. *Alcohol. Clin. Exp. Res.* **1997**, *21*, 1272–1277. [CrossRef]

53. Osier, M.V.; Pakstis, A.J.; Soodyall, H.; Comas, D.; Goldman, D.; Odunsi, A.; Okonofua, F.; Parnas, J.; Schulz, L.O.; Bertranpetit, J.; et al. A Global Perspective on Genetic Variation at the ADH Genes Reveals Unusual Patterns of Linkage Disequilibrium and Diversity. *Am. J. Hum. Genet.* **2002**, *71*, 84–99. [CrossRef] [PubMed]

54. Agarwal, D.P.; Goedde, H.W. *Alcohol Metabolism, Alcohol Intolerance, and Alcoholism: Biochemical and Pharmacogenetic Approaches*; Springer: Berlin/Heidelberg, Germany, 1990.

55. Helzer, J.E.; Canino, G.J. (Eds.) *Alcoholism in North America, Europe, and Asia*; Oxford University Press: New York, NY, USA, 1992.

56. Li, T.-K. Pharmacogenetics of responses to alcohol and genes that influence alcohol drinking. *J. Stud. Alcohol. Drugs* **2000**, *61*, 5–12. [CrossRef]

57. Ezquer, F.; Quintanilla, M.E.; Moya-Flores, F.; Morales, P.; Munita, J.M.; Olivares, B.; Landskron, G.; Hermoso, M.A.; Ezquer, M.; Herrera-Marschitz, M.; et al. Innate gut microbiota predisposes to high alcohol consumption. *Addict. Biol.* **2021**, *26*, e13018. [CrossRef]

58. Potts, K.B.; Baken, E.; Levang, A.; Watts, D.P. Ecological factors influencing habitat use by chimpanzees at Ngogo, Kibale National Park, Uganda. *Am. J. Primatol.* **2016**, *78*, 432–440. [CrossRef] [PubMed]

59. Watts, D.P.; Potts, K.B.; Lwanga, J.S.; Mitani, J.C. Diet of chimpanzees (*Pan troglodytes schweinfurthii*) at Ngogo, Kibale National Park, Uganda, 2. temporal variation and fallback foods. *Am. J. Primatol.* **2012**, *74*, 130–144. [CrossRef] [PubMed]
60. Pusey, A.E.; Oehlert, G.W.; Williams, J.M.; Goodall, J. Influence of ecological and social factors on body mass of wild chimpanzees. *Int. J. Primatol.* **2005**, *26*, 3–31. [CrossRef]
61. Martínez, C.; Galván, S.; Garcia-Martin, E.; Ramos, M.I.; Gutiérrez-Martín, Y.; Agúndez, J.A. Variability in ethanol biodisposition in whites is modulated by polymorphisms in the ADH1B and ADH1C genes. *Hepatology* **2010**, *51*, 491–500. [CrossRef] [PubMed]
62. Bjerregaard, P.; Mikkelsen, S.S.; Becker, U.; Hansen, T.; Tolstrup, J.S. Genetic variation in alcohol metabolizing enzymes among Inuit and its relation to drinking patterns. *Drug Alcohol Depend.* **2014**, *144*, 239–244. [CrossRef]
63. Dominy, N.J.; Yeakel, J.D.; Bhat, U.; Ramsden, L.; Wrangham, R.W.; Lucas, P.W. How chimpanzees integrate sensory information to select figs. *Interface Focus* **2016**, *6*, 20160001. [CrossRef]
64. Houle, A.; Conklin-Brittain, N.L.; Wrangham, R.W. Vertical stratification of the nutritional value of fruit: Macronutrients and condensed tannins. *Am. J. Primatol.* **2014**, *76*, 1207–1232. [CrossRef]

Article

The Global Impact of Alcohol Consumption on Premature Mortality and Health in 2016

Ivneet Sohi [1,*], Ari Franklin [1], Bethany Chrystoja [1,2,3], Ashley Wettlaufer [1], Jürgen Rehm [1,2,3,4,5,6,7,8] and Kevin Shield [1,2,4]

[1]   Centre for Addiction and Mental Health, Institute for Mental Health Policy Research, 33 Ursula Franklin Street, Toronto, ON M5S 2S1, Canada; Ari.Franklin@camh.ca (A.F.); bethany.chrystoja@ucalgary.ca (B.C.); Ashley.Wettlaufer@camh.ca (A.W.); jtrehm@gmail.com (J.R.); kevin.shield@camh.ca (K.S.)
[2]   Dalla Lana School of Public Health, University of Toronto, 155 College Street, Toronto, ON M5T 3M7, Canada
[3]   Cumming School of Medicine, University of Calgary, 3330 Hospital Drive NW, Calgary, AB T2N 4N1, Canada
[4]   Centre for Addiction and Mental Health, Campbell Family Mental Health Research Institute, 33 Russell Street, Toronto, ON M5S 2S1, Canada
[5]   Department of Psychiatry, University of Toronto, 250 College Street, Toronto, ON M5T 1R8, Canada
[6]   Institute of Medical Science, University of Toronto, 1 King's College Circle, Toronto, ON M5S 1A8, Canada
[7]   Institute of Clinical Psychology and Psychotherapy & Center for Clinical Epidemiology and Longitudinal Studies, Technische Universität Dresden, Chemnitzer Street 46, D-01187 Dresden, Germany
[8]   Department of International Health Projects, Institute for Leadership and Health Management, I.M. Sechenov First Moscow State Medical University, Trubetskaya Street, 8, b. 2, 119992 Moscow, Russia
*   Correspondence: ivneet.sohi@camh.ca

Abstract: This study aimed to estimate the impact of alcohol use on mortality and health among people 69 years of age and younger in 2016. A comparative risk assessment approach was utilized, with population-attributable fractions being estimated by combining alcohol use data from the Global Information System on Alcohol and Health with corresponding relative risk estimates from meta-analyses. The mortality and health data were obtained from the Global Health Observatory. Among people 69 years of age and younger in 2016, 2.0 million deaths and 117.2 million Disability Adjusted Life Years (DALYs) lost were attributable to alcohol consumption, representing 7.1% and 5.5% of all deaths and DALYs lost in that year, respectively. The leading causes of the burden of alcohol-attributable deaths were cirrhosis of the liver (457,000 deaths), road injuries (338,000 deaths), and tuberculosis (190,000 deaths). The numbers of premature deaths per 100,000 people were highest in Eastern Europe (155.8 deaths per 100,000), Central Europe (52.3 deaths per 100,000 people), and Western sub-Saharan Africa (48.7 deaths per 100,000). A large portion of the burden of disease caused by alcohol among people 69 years of age and younger is preventable through the implementation of cost-effective alcohol policies such as increases in taxation.

Keywords: alcohol; burden of disease; death; disability; infectious diseases; non-communicable diseases; injuries; global; policy

Citation: Sohi, I.; Franklin, A.; Chrystoja, B.; Wettlaufer, A.; Rehm, J.; Shield, K. The Global Impact of Alcohol Consumption on Premature Mortality and Health in 2016. *Nutrients* **2021**, *13*, 3145. https://doi.org/10.3390/nu13093145

Academic Editor: Rosa Casas

Received: 2 August 2021
Accepted: 1 September 2021
Published: 9 September 2021

Publisher's Note: MDPI stays neutral with regard to jurisdictional claims in published maps and institutional affiliations.

1. Introduction

Alcohol consumption is a leading risk factor for premature mortality and the burden of disease worldwide. Research in developed and developing countries has found that individuals of younger ages are disproportionately affected by alcohol [1–3]. For instance, alcohol is estimated to be the leading risk factor for the burden of disease among people 15 to 49 years of age, followed by high body mass index, high blood pressure, and dietary risks [3]. This population also has a large proportion of their expected lifespans remaining, contributes relatively more to the economy, and plays important roles in caring for their families [4].

In response to the burden of disease caused by alcohol, the World Health Organization (WHO), through its Global strategy to reduce the harmful use of alcohol and its Global

Action Plan for the prevention and control of NCDs (2013–2020), agreed at the 2010 World Health Assembly to aim for a 10% relative reduction in harmful alcohol use by 2025 [5,6]. Furthermore, the WHO's Sustainable Development Goals (SDG) 3.4 outlines a targeted one-third reduction by 2030 of premature mortality (i.e., deaths among people 69 years of age and younger) due to noncommunicable diseases, with reductions in alcohol-attributable diseases being key to achieving this goal [7,8]. There is a distinct spectrum of alcohol-attributable diseases and injuries which affect people 69 years of age and younger compared to people 70 years of age and older. Therefore, it is necessary to characterize the disease-specific health impacts of alcohol for the purposes of structuring disease-specific health efforts, for example to inform cancer prevention programs [9].

Given the impact of alcohol consumption on premature mortality, the objective of this study was to estimate the alcohol-attributable mortality and burden of disease globally in 2016, and to examine variations in the alcohol-attributable burden over time, by global burden of disease (GBD) region, age, and sex.

2. Materials and Methods

A comparative risk assessment methodology was utilized to estimate the burden of disease attributable to alcohol use in 2016. These estimations were based on the theoretical minimum risk exposure level (TMREL) of lifetime abstention. Lifetime abstention was utilized as a TMREL based on historical precedent, and the observation that lifetime abstainers may have the lowest risk of overall health loss [10]. The population-attributable fraction (PAF) for alcohol use was estimated based on a Levin-based method which combines data on alcohol exposure with corresponding relative risk (RR) estimates [11,12]. Information regarding the methods utilized and the data sources can be found in the Supplemental Material and in the paper by Shield et al. [1].

2.1. Relative Risk Estimates

Alcohol RR estimates for chronic disease outcomes (except from ischemic diseases) were obtained from meta-analyses and based on average drinking (in grams per day) [13]. The lag time between alcohol use and disease occurrence was only modelled for cancer (based on the estimate that there is a 10 year period between exposure and disease outcomes [14]). Heavy episodic drinking (HED) was utilized in the modelling for the RRs for ischemic diseases and injuries [1]. All RR estimates were reviewed and approved by the WHO Technical Advisory Group on Alcohol and Drug Epidemiology. The sources of RR estimates are outlined in Supplemental Material Table S1.

2.2. Mortality, Morbidity, and Population Data

Data on mortality, Years of Life Lost (YLL), morbidity measured using Years Lived with Disability (YLD), age, sex, country, year, and by cause of mortality and/or morbidity were obtained from the WHO's Global Health Estimates [15]. The total burden of disease was measured using Disability Adjusted Life Years (DALYs) lost. Estimates of premature mortality were based on a cut off of deaths which occurred among people 69 years of age and younger [4,16].

Alcoholic cardiomyopathy deaths, YLL, and YLD were estimated using the methods of Manthey and colleagues [17] as they were not directly estimated in the WHO's Global Health Estimates. The WHO's road traffic death database [18] was used to determine the fractions of alcohol-attributable motor vehicle deaths which involved a driver and those traffic deaths which involved people other than the driver.

Population data by age, sex, country, and year were obtained from the UN Population Division [19]. Deaths, YLLs, YLDs, and DALYs lost were aggregated into five-year age groups, beginning at 0 years until 84 years, followed by the category of 85 years and older; alcohol PAFs were applied to these age groupings.

Data were aggregated by GBD region (see: http://ghdx.healthdata.org (accessed on 1 January 2021) for regional groupings) and by Human Development Index (HDI) region (see Figure S1 in the Supplementary Materials for HDI groupings). HDI categories were obtained from the United Nations Development Programme [20]. The HDI is based on having a long and healthy life (i.e., life expectancy at birth), being knowledgeable (i.e., expected years of schooling and mean years of schooling for adults 25 years of age and older), and having a decent standard of living (i.e., Gross National Income *per capita*) [20].

The 95% uncertainty intervals (see Tables S2–S14 in Supplementary Materials) were based on a set of 1000 simulations of all lowest level parameters (i.e., parameters sampled from their respective error distributions). These parameters were then used to estimate 1000 simulated estimates of the alcohol-attributable burden of disease. From these simulations, the 2.5th and 97.5th percentiles were utilized for the 95% uncertainty intervals.

Analyses were performed using the statistical software package R [21].

3. Results

In 2016, there were 2.0 million premature deaths and 117.2 million DALYs lost globally due to alcohol use, representing 7.1% of all premature deaths and 5.5% of all DALYs lost in that year (see Tables 1 and 2). In contrast, 3.2% of all deaths and 3.0% of all DALYs lost among people 70 years of age and older were attributable to alcohol consumption. An estimated 70.7% of all alcohol-attributable deaths and 89.2% of all alcohol-attributable DALYs lost globally in 2016 were premature, i.e., among those 69 years of age and younger. In comparison, 52.0% of all deaths and 81.8% of all DALYs lost globally in 2016 were premature. The alcohol-attributable deaths and DALYs lost among those 69 years of age and younger were greater among men (1.6 million deaths and 90.9 million DALYs lost) compared to women (0.5 million deaths and 26.3 million DALYs lost). The largest proportion of premature deaths that were attributable to alcohol occurred among people 30–39 years of age (13.3%) and 20–29 years of age (13.0%). See Tables S2 to S17 for data on sex-specific alcohol-attributable deaths, YLL, YLD, and DALYs lost.

The leading causes of the burden of premature alcohol-attributable deaths were cirrhosis of the liver (457,000 deaths), road injuries (338,000 deaths), and tuberculosis (190,000 deaths) (see Figure 1). Road injuries and cirrhosis of the liver were the leading causes of alcohol-attributable deaths among those aged 0 to 39 and 40 to 69 years of age, respectively. The proportion of alcohol-attributable deaths due to road injuries decreased with age from 100% of all alcohol-attributable deaths among those aged 0–14 years to 7.0% among those aged 60–69 years. The proportion of alcohol-attributable deaths due to cirrhosis of the liver increased with age, peaking at 26.4% of all alcohol-attributable deaths among those 50–59 years of age. The proportion of alcohol-attributable deaths due to tuberculosis increased with age, peaking at 10.7% of all alcohol-attributable deaths among those 40–49 years of age.

Table 1. Alcohol-attributable deaths globally in 2016, by cause and age.

Cause of Disease or Injury	Alcohol-Attributable Deaths								Population Attributable Fraction (%)							
	0 to 14	15 to 19	20 to 29	30 to 39	40 to 49	50 to 59	60 to 69	≥70	0 to 14	15 to 19	20 to 29	30 to 39	40 to 49	50 to 59	60 to 69	≥70
All Causes	33,939	47,719	248,762	317,457	396,552	510,448	542,012	870,937	0.5	6.9	13.0	13.3	12.0	9.2	5.9	3.2
Communicable, maternal, perinatal and nutritional conditions	0	3073	30,471	50,497	58,869	60,646	55,847	102,481	0.0	1.5	5.4	6.8	8.5	9.1	6.5	4.0
Tuberculosis	0	2299	23,738	34,975	42,291	46,097	40,121	46,762	0.0	8.5	22.0	24.7	25.4	22.0	17.5	12.7
HIV AIDS	0	269	3444	10,566	10,220	4352	1284	302	0.0	1.0	3.0	3.5	3.8	3.5	3.1	2.3
Lower respiratory infections	0	505	3289	4956	6358	10,197	14,442	55,417	0.0	2.2	6.2	8.4	8.5	7.4	5.1	4.1
Noncommunicable diseases	0	5997	42,304	111,176	203,371	330,932	392,643	656,706	0.0	3.2	7.7	11.2	9.8	7.6	5.0	2.8
Malignant neoplasms	0	0	1094	9183	30,653	73,936	102,619	150,215	0.0	0.0	0.9	3.4	4.5	5.2	4.8	3.6
Lip and oral cavity cancer	0	0	224	2542	6678	13,732	14,915	14,085	0.0	0.0	7.9	30.2	32.2	35.4	33.7	27.8
Other pharynx cancers	0	0	63	975	4250	11,026	12,356	9922	0.0	0.0	7.3	28.5	34.6	39.9	38.1	29.6
Oesophagus cancer	0	0	53	712	4519	16,516	27,076	34,070	0.0	0.0	3.5	12.8	18.6	22.4	21.7	17.2
Colon and rectum cancers	0	0	248	1754	4823	12,582	22,287	50,866	0.0	0.0	3.9	9.6	10.8	12.4	12.8	11.4
Liver cancer	0	0	507	3199	10,383	20,080	25,985	41,273	0.0	0.0	5.6	11.5	11.8	12.4	12.6	12.4
Breast cancer	0	0	178	3003	6743	10,445	9238	12,415	0.0	0.0	2.4	7.1	7.4	8.1	7.4	6.5
Larynx cancer	0	0	13	210	1418	4753	6849	7283	0.0	0.0	4.2	17.9	21.7	24.6	24.0	20.3
Diabetes mellitus	0	−41	−362	−681	−1958	−4474	−7816	−19,753	0.0	−1.1	−2.8	−2.7	−2.6	−2.3	−2.0	−2.2
Alcohol use disorders	0	967	7748	21,406	31,222	38,315	28,103	17,804	100.0	100.0	100.0	100.0	100.0	100.0	100.0	100.0
Epilepsy	0	694	3278	3019	2651	2177	1844	2751	0.0	5.9	15.2	17.3	18.7	16.9	14.4	11.7
Cardiovascular diseases	0	680	4328	13,070	30,099	65,934	107,708	347,811	0.0	1.8	3.2	4.2	4.1	3.8	3.1	3.0
Hypertensive heart disease	0	40	457	1260	3204	7079	12,846	41,571	0.0	3.1	9.1	10.2	10.4	9.4	8.2	6.8
Ischaemic heart disease	0	85	139	1195	3027	12,654	32,002	201,657	0.0	0.8	0.3	0.8	0.8	1.3	1.8	3.3
Ischaemic stroke	0	−11	−75	−174	−838	−2823	−10,280	−45,068	0.0	−0.7	−1.3	−1.5	−2.2	−2.1	−2.0	−2.1
Haemorrhagic stroke	0	476	3044	7768	20,968	43,205	68,579	142,988	0.0	5.7	10.4	11.3	11.8	10.8	9.7	9.2
Cardiomyopathy, myocarditis, endocarditis	0	89	763	3021	3738	5819	4562	6662	0.0	1.9	5.0	12.3	11.5	12.3	7.5	3.8
Digestive diseases	0	3697	26,027	61,967	102,541	139,846	144,097	138,180	0.0	12.8	28.4	37.8	39.9	35.6	27.9	13.9
Cirrhosis of the liver	0	3593	24,369	58,145	97,529	134,702	139,105	130,695	0.0	28.8	49.6	54.0	55.3	52.5	47.8	38.9
Pancreatitis	0	104	1658	3822	5012	5144	4993	7485	0.0	11.0	27.3	32.2	31.6	28.3	23.2	18.5
Injuries	33,939	38,649	175,987	155,784	134,313	118,870	93,522	111,750	5.3	12.8	21.9	24.0	25.2	23.2	18.4	12.1
Unintentional injuries	33,939	28,303	116,260	101,421	94,461	87,006	73,167	94,404	5.9	16.0	26.1	27.3	27.6	24.6	19.1	12.0
Road injury	33,939	20,533	78,740	62,508	56,129	48,048	37,797	33,073	23.1	20.2	29.1	30.0	30.0	27.5	24.7	20.8
Poisonings	0	409	2543	2273	2342	1969	1718	1480	0.0	10.0	20.7	22.9	21.8	19.3	13.9	9.5
Falls	0	809	5398	7347	9340	11,524	12,814	32,138	0.0	10.3	22.4	24.6	25.4	20.9	13.5	8.8
Fire, heat and hot substances	0	531	2813	3106	2614	2773	2362	2960	0.0	6.7	14.7	17.5	20.9	20.1	15.5	10.4
Drowning	0	2708	8553	7214	5983	5109	4223	4320	0.0	11.6	23.4	26.0	26.1	22.6	17.3	11.5
Exposure to mechanical forces	0	786	4655	4573	4114	3443	2311	1635	0.0	10.8	23.0	24.9	25.8	23.1	17.7	10.5
Other unintentional injuries	0	2526	13,558	14,400	13,940	14,141	11,942	18,798	0.0	10.4	22.0	24.4	25.1	22.9	16.8	11.3
Intentional injuries	0	10,346	59,726	54,363	39,851	31,863	20,355	17,346	0.0	8.2	16.6	19.7	20.8	20.0	16.2	12.3
Self-harm	0	4491	29,477	31,085	26,207	23,767	16,508	15,484	0.0	8.5	18.4	23.0	23.8	21.9	17.1	12.8
Interpersonal violence	0	5855	30,249	23,278	13,645	8096	3847	1862	0.0	11.1	20.7	22.4	22.0	20.1	16.6	11.1

Table 2. Alcohol-attributable disability adjusted life years lost globally in 2016, by cause for people 0 to 69 years of age.

Cause of Disease or Injury	Alcohol-Attributable DALYs (100,000 s)								Population Attributable Fraction (%)							
	0 to 14	15 to 19	20 to 29	30 to 39	40 to 49	50 to 59	60 to 69	≥70	0 to 14	15 to 19	20 to 29	30 to 39	40 to 49	50 to 59	60 to 69	≥70
All Causes	302.5	451.5	2189.8	2348.6	2349.6	2297.1	1783.2	1419.4	0.5	5.0	9.8	9.9	9.0	7.4	5.1	3.0
Communicable, maternal, perinatal and nutritional conditions	0.0	24.4	215.6	302.8	293.1	238.9	167.2	148.9	0.0	1.1	4.0	5.4	6.7	7.4	5.9	3.9
Tuberculosis	0.0	18.5	168.5	210.4	210.8	182.6	122.0	78.6	0.0	8.5	22.0	24.6	25.4	22.1	18.0	13.0
HIV AIDS	0.0	2.1	24.9	64.0	52.1	18.1	4.2	0.7	0.0	1.0	3.0	3.5	3.8	3.5	3.1	2.2
Lower respiratory infections	0.0	3.8	22.1	28.5	30.2	38.2	41.0	69.6	0.0	2.2	6.2	8.4	8.5	7.4	5.3	4.0
Noncommunicable diseases	6.3	111.5	657.4	975.5	1208.1	1405.3	1191.1	982.8	0.0	2.6	6.0	7.2	6.6	5.6	4.0	2.4
Malignant neoplasms	0.0	0.0	8.4	70.9	184.2	337.5	338.7	263.9	0.0	0.0	1.1	4.6	5.7	6.3	5.5	4.2
Lip and oral cavity cancer	0.0	0.0	1.5	14.5	31.8	52.2	43.1	22.8	0.0	0.0	7.7	30.0	32.2	35.4	33.8	28.7
Other pharynx cancers	0.0	0.0	0.4	5.5	20.0	41.6	35.4	16.6	0.0	0.0	7.1	28.4	34.6	39.9	38.2	30.9
Oesophagus cancer	0.0	0.0	0.3	4.0	21.0	61.6	76.1	53.9	0.0	0.0	3.4	12.8	18.5	22.4	21.7	17.8
Colon and rectum cancers	0.0	0.0	1.6	10.1	23.0	47.9	63.9	75.8	0.0	0.0	3.8	9.6	10.8	12.4	12.8	11.7
Liver cancer	0.0	0.0	3.3	18.2	48.8	75.4	73.3	63.5	0.0	0.0	5.5	11.5	11.8	12.4	12.6	12.4
Breast cancer	0.0	0.0	1.2	17.4	32.9	40.8	27.3	19.2	0.0	0.0	2.4	7.1	7.5	8.2	7.4	6.7
Larynx cancer	0.0	0.0	0.1	1.2	6.7	18.1	19.7	12.0	0.0	0.0	4.1	17.9	21.7	24.6	24.1	20.8
Diabetes mellitus	0.0	−0.7	−7.4	−13.2	−27.2	−37.3	−40.7	−40.2	0.0	−1.3	−3.1	−3.0	−3.1	−2.6	−2.4	−2.2
Alcohol use disorders	6.3	68.0	407.7	451.8	389.3	304.5	157.4	60.6	100.0	100.0	100.0	100.0	100.0	100.0	100.0	100.0
Epilepsy	0.0	9.2	39.3	30.9	24.5	17.7	12.1	9.5	0.0	5.8	14.4	16.0	17.1	15.6	13.9	11.4
Cardiovascular diseases	0.0	5.2	29.2	75.1	142.0	244.7	303.4	462.0	0.0	1.6	2.8	3.9	3.8	3.6	2.9	2.8
Hypertensive heart disease	0.0	0.3	3.0	7.5	16.2	28.1	38.4	58.5	0.0	3.1	9.1	10.0	10.2	9.3	8.3	6.8
Ischaemic heart disease	0.0	0.7	1.1	7.1	14.7	47.5	91.8	252.0	0.0	0.8	0.3	0.8	0.7	1.3	1.8	3.1
Ischaemic stroke	0.0	−0.2	−1.2	−2.4	−8.3	−19.5	−40.0	−71.5	0.0	−1.1	−2.0	−2.3	−3.1	−2.8	−2.3	−2.2
Haemorrhagic stroke	0.0	3.7	21.2	45.5	101.5	166.5	199.6	213.7	0.0	5.8	10.5	11.3	11.9	10.8	9.9	9.0
Cardiomyopathy, myocarditis, endocarditis	0.0	0.7	5.1	17.4	17.9	22.1	13.4	9.3	0.0	1.8	4.7	11.9	11.3	11.9	7.5	3.8
Digestive diseases	0.0	29.8	180.2	360.1	495.3	538.1	420.2	227.0	0.0	12.4	26.6	34.7	36.1	33.2	26.6	14.6
Cirrhosis of the liver	0.0	29.0	169.1	338.1	471.3	518.6	405.8	215.7	0.0	28.9	49.5	53.9	55.3	52.5	48.3	39.5
Pancreatitis	0.0	0.8	11.1	22.0	23.9	19.5	14.4	11.3	0.0	11.1	27.2	32.1	31.5	28.2	23.5	18.8
Injuries	296.2	315.7	1316.8	1070.3	848.4	652.9	425.0	287.7	4.9	12.7	21.8	23.7	24.8	22.8	19.0	13.2
Unintentional injuries	296.2	236.1	904.5	744.6	645.8	522.1	359.6	256.8	5.6	15.6	25.6	26.5	26.8	24.0	19.7	13.4
Road injury	296.2	158.6	557.9	403.0	325.7	241.0	153.7	88.0	23.1	20.3	29.1	30.0	30.2	28.0	25.8	22.4
Poisonings	0.0	3.3	18.2	14.1	12.2	8.2	5.5	2.6	0.0	10.2	20.7	22.9	21.9	19.5	14.6	9.8
Falls	0.0	14.4	76.6	95.2	109.1	113.2	95.6	97.9	0.0	11.9	23.0	24.6	25.2	21.8	16.4	10.7
Fire, heat and hot substances	0.0	5.5	25.9	26.7	21.5	18.1	11.7	7.4	0.0	7.6	15.8	18.4	21.2	20.0	16.7	11.6
Drowning	0.0	20.3	58.0	42.0	29.1	19.8	12.6	7.1	0.0	11.6	23.3	25.9	26.0	22.4	17.6	11.6
Exposure to mechanical forces	0.0	8.1	43.6	43.0	40.1	32.2	20.2	11.0	0.0	11.5	23.4	25.1	26.0	23.6	19.8	13.9
Other unintentional injuries	0.0	25.9	124.3	120.5	108.2	89.6	60.4	42.9	0.0	10.2	21.0	22.5	23.0	20.7	16.5	11.3
Intentional injuries	0.0	79.5	412.3	325.7	202.7	130.8	65.3	30.9	0.0	8.2	16.4	19.1	20.0	19.0	16.0	12.0
Self-harm	0.0	33.7	198.1	180.0	125.9	90.9	48.6	24.7	0.0	8.5	18.4	23.0	23.7	21.9	17.5	12.8
Interpersonal violence	0.0	45.9	214.2	145.7	76.8	39.9	16.7	6.2	0.0	10.9	20.5	22.1	21.8	20.2	17.3	12.0

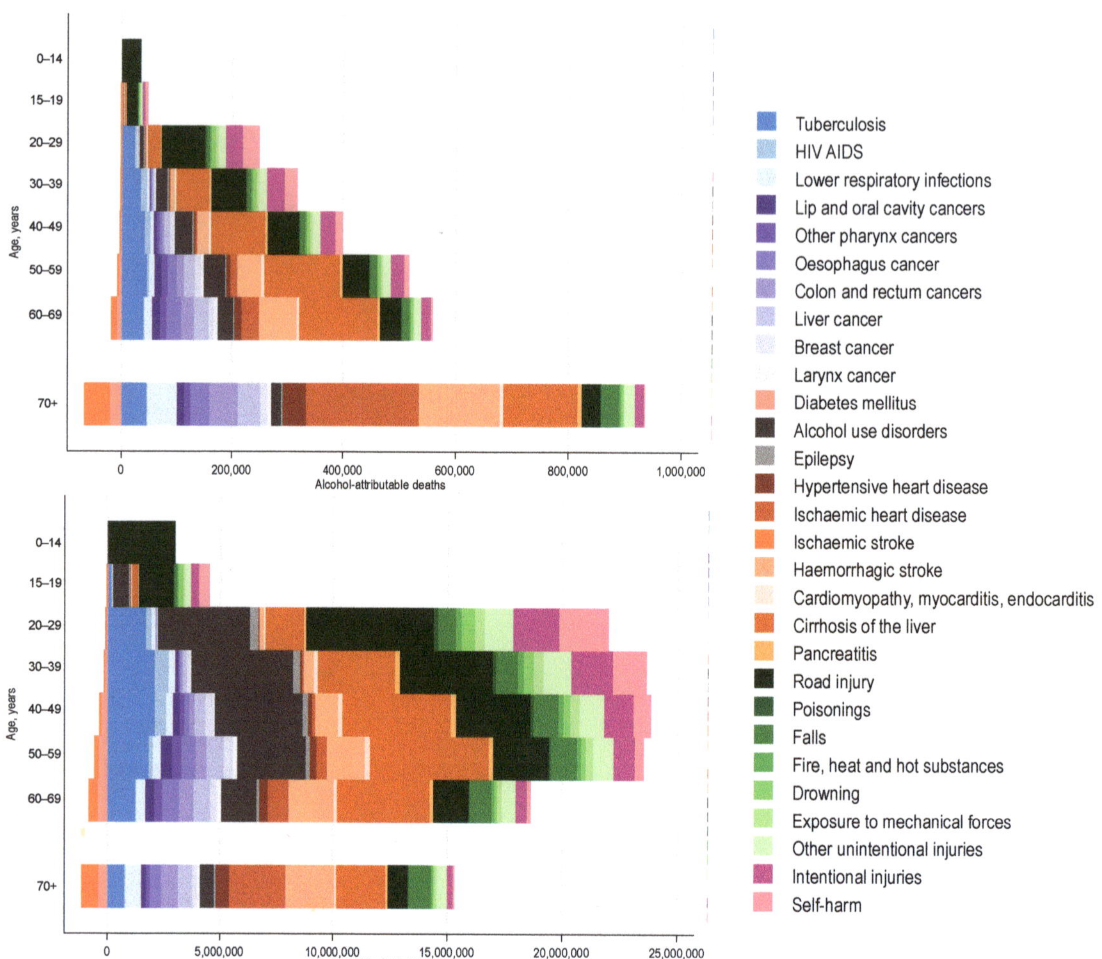

Figure 1. Alcohol-attributable deaths and disability adjusted life years (DALYs) lost by age.

3.1. Alcohol-Attributable Burden of Disease by Region

The numbers of premature alcohol-attributable deaths and DALYs lost per 100,000 people showed large variations globally (see Figures 2 and 3). The numbers of premature alcohol-attributable deaths were highest in Eastern Europe (155.8 deaths per 100,000), Central Europe (52.3 deaths per 100,000 people), and Western sub-Saharan Africa (48.7 deaths per 100,000). In 2016, the two leading contributors to alcohol-attributable deaths among all regions were either cirrhosis of the liver or road injuries, except for three regions: Asia Pacific with self-harm, Southern sub-Saharan Africa with tuberculosis, and Eastern Europe with ischaemic heart disease being the largest contributors, respectively (see Figure 3). The second largest contributors to alcohol-attributable deaths in Southern sub-Saharan Africa and Eastern Europe were HIV/AIDS and alcohol use disorders, respectively. Figure S2 outlines the burden of alcohol-attributable premature YLL and YLD globally. Figure S3 outlines the burden of alcohol-attributable premature YLL and YLD by GBD region.

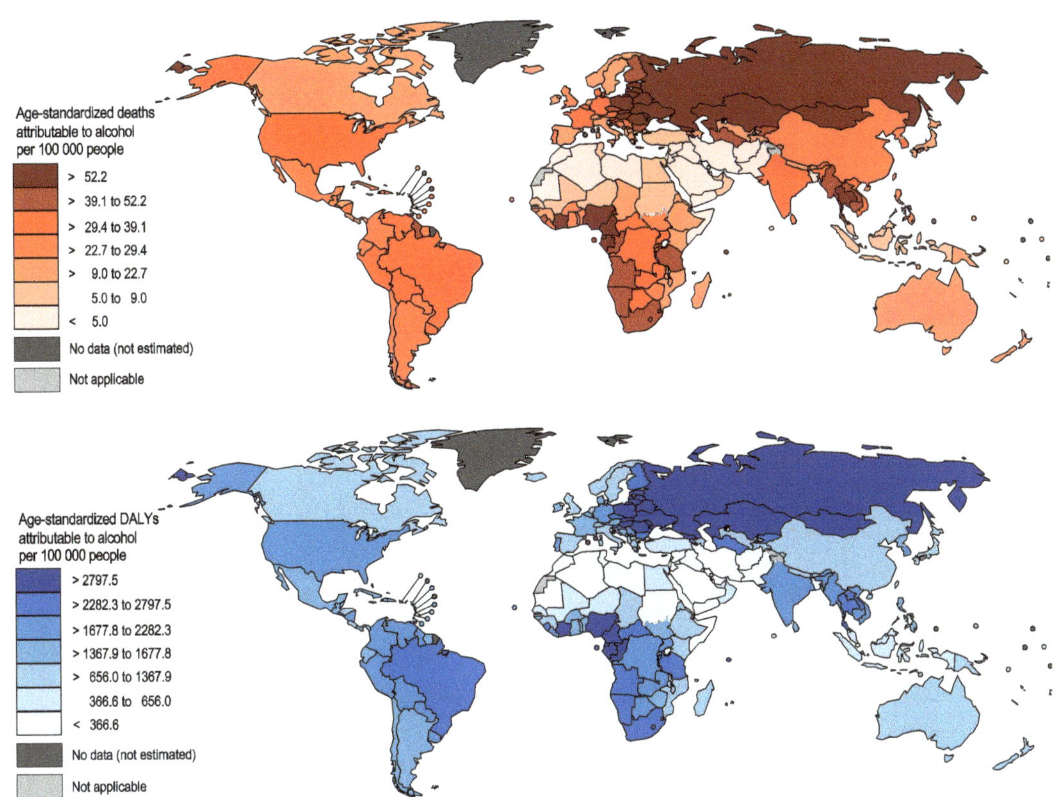

Figure 2. Alcohol-attributable deaths and disability adjusted years of life lost globally in 2016 among people 0 to 69 years of age.

3.2. Alcohol-Attributable Burden of Disease by Human Development Index

The burden of premature alcohol-attributable deaths and DALYs lost varied by HDI region (see Figure 4). The number of alcohol-attributable deaths was highest in countries with a very high HDI (43.3 deaths and 2339.9 DALYs lost per 100,000), followed by low HDI countries (33.7 deaths and 1966.1 DALYs lost per 100,000), high HDI countries (26.8 deaths and 1502.3 DALYs lost per 100,000) and medium HDI countries (24.8 deaths and 1392.5 DALYs lost per 100,000). The leading cause of death was cirrhosis of the liver in very high HDI countries (8.2 deaths per 100,000), low HDI countries (7.4 deaths per 100,000), and medium HDI countries (6.6 deaths per 1,000,000), and was road injuries in high HDI countries (5.9 deaths per 100,000). For DALYs lost, the leading contributor to the premature alcohol-attributable burden of disease was alcohol use disorders for very-high HDI countries (449.3 DALYs lost per 100,000), liver cirrhosis for medium HDI countries (297.3 DALYs lost per 100,000 people), and road injuries for low HDI countries (431.2 DALYs lost per 100,000 people) and high HDI countries (354.4 DALYs lost per 100,000). Figure S4 outlines the burden of premature YLL and YLD by HDI region.

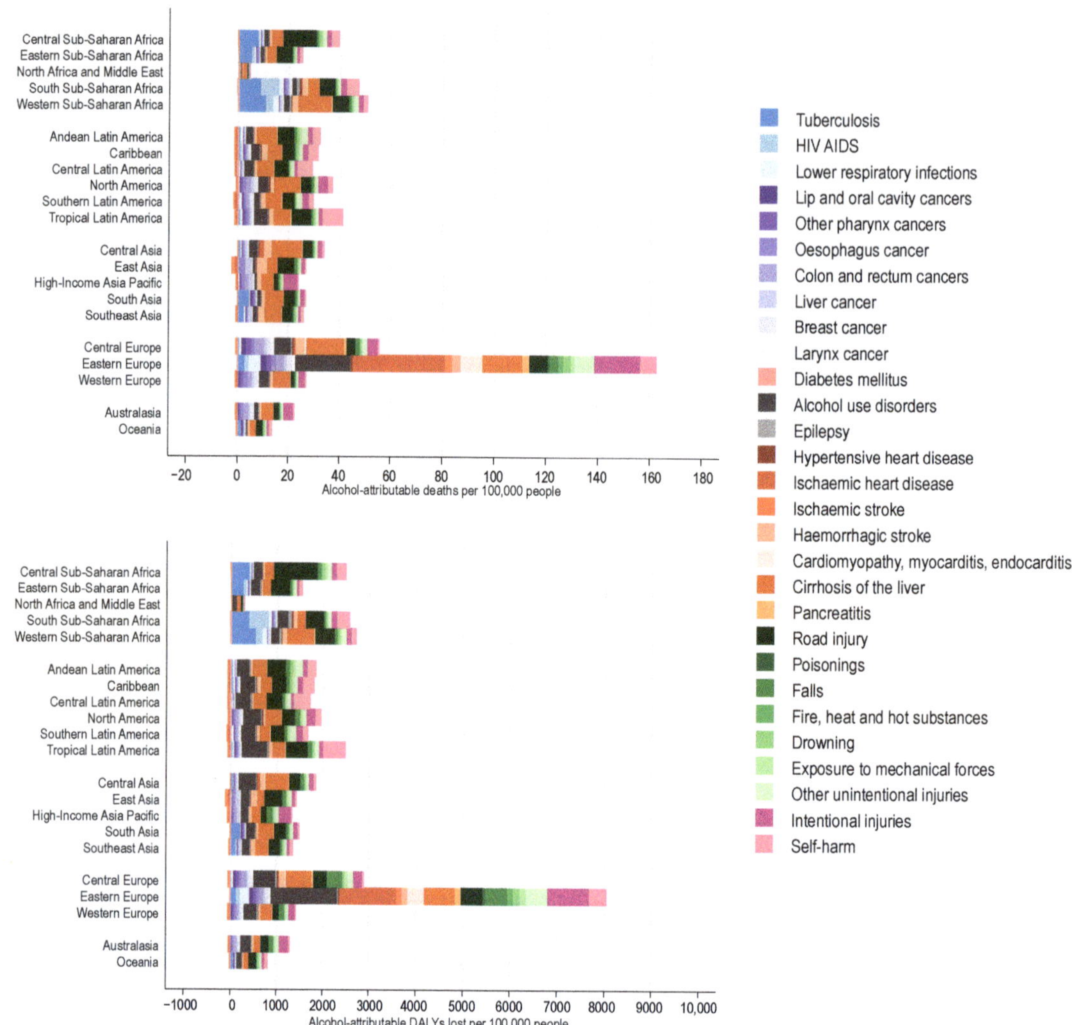

Figure 3. Alcohol-attributable deaths and disability adjusted life years (DALYs) lost among people 0 to 70 years of age by global burden of disease region.

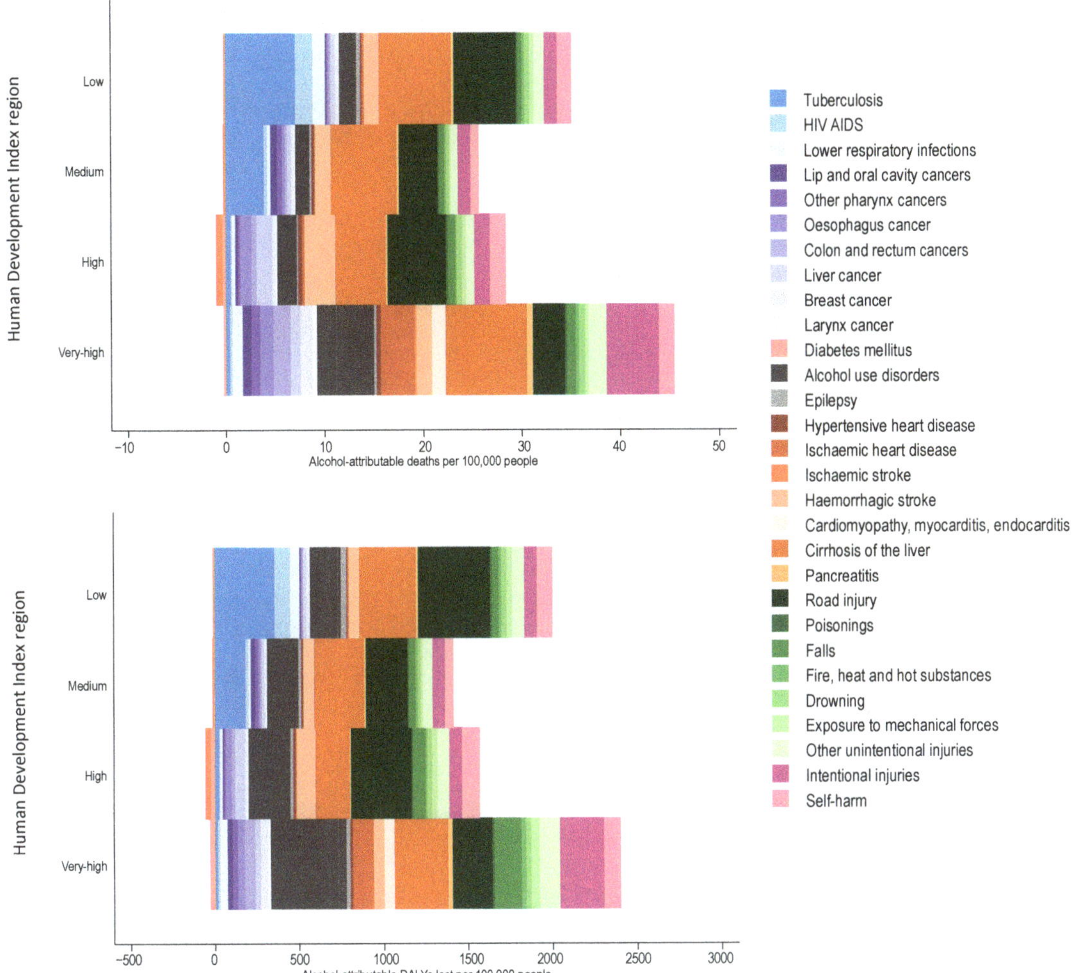

Figure 4. Alcohol-attributable deaths and disability adjusted life years (DALYs) lost among people 0 to 70 years of age by human development index region.

4. Discussion

The results of this study indicate that alcohol-attributable deaths and health loss occurred among people relatively young in age. The proportions of alcohol-attributable deaths and DALYs lost that were premature were greater than the proportions of all-cause deaths and DALYs lost that were considered premature. This indicates that alcohol use disproportionately affects the health of people who are younger in age. The cause composition of the premature alcohol-attributable burden is unique when compared to the burden among people 70 years of age and older, with cirrhosis of the liver, road injuries, and tuberculosis being the primary contributors to this burden. Furthermore, regional and societal development-based variations in the magnitude of the premature alcohol-attributable burden of disease and the cause composition of this burden were observed.

This study modelled both the detrimental and protective effects of alcohol consumption on health. Specifically, alcohol consumed at low amounts, and not on HED occasions has a protective effect on diabetes, ischemic heart disease, and ischemic stroke [13]. This study found that for the premature disease burden, the detrimental effects of alcohol at the population level outweighed the protective effects. At the individual level, the net effect of alcohol consumption on overall health is unknown; however, a recent modelling study found no level of alcohol consumption that provided a net health benefit [10].

The premature burden of disease attributable to alcohol consumption was characterized by tuberculosis, liver cirrhosis, and injuries. Liver cirrhosis is mainly linked to the overall volume of drinking, while injuries attributable to alcohol are mostly related to intoxication (i.e., binge alcohol consumption) [13]. Tuberculosis risk (i.e., the impact of alcohol on the immune system) is related both to the overall volume of alcohol consumed and binge drinking; however, due to the lack of studies, the impact of alcohol use on tuberculosis was modelled based only on the overall volume of alcohol consumed [13]. Therefore, both overall volume of alcohol consumed and drinking to intoxication are factors leading to the premature burden of disease attributable to alcohol consumption.

Tuberculosis remains an enormous public health concern globally, especially in low and medium HDI countries [22]. The treatment of tuberculosis and the interaction between HIV/AIDS and tuberculosis are key public health priorities [23]. Alcohol use is a key risk factor for both diseases (Morojele et al., this issue), which if addressed can substantially reduce the health burden of tuberculosis and HIV/AIDS. The burden of disease due to liver cirrhosis was high in all HDI categories and in most GBD regions. The burden of alcoholic liver cirrhosis is affected by multiple risk factors which interact with alcohol, including hepatitis B and C infections, obesity, and socio-economic status [24]. The burden of alcohol-related injuries is problematic as investment in preventing mortality from injury has fallen behind other causes of death, such as HIV/AIDS and reproductive health [21]. Furthermore, mental health concerns have been overlooked in terms of public health programming, especially in young people [10] where injuries and neuropsychiatric conditions are greatly impacted by alcohol consumption [22].

The burden of premature disease attributable to alcohol consumption was highest in Eastern Europe, Central Europe, and Western sub-Saharan Africa. The Central and Eastern Europe region have a high overall volume of alcohol consumption and a high prevalence of HED [5,6]. Alcohol control policy measures, including increases in alcohol prices and decreases in availability, have been implemented in the Eastern Europe region and have resulted in marked downward shifts in mortality and the burden of disease [25]. The Western sub-Saharan Africa region has a relatively low overall volume of alcohol consumption. The burden of alcohol-attributable premature disease in this region was driven mainly by infectious diseases, liver cirrhosis, and injuries. Cirrhosis-related deaths doubled in the sub-Saharan Africa region between 1980 and 2010, with hepatitis B virus, hepatitis C virus, and alcohol use being contributing factors to this increase [26]. Furthermore, treatment of liver cirrhosis is unavailable in most parts of sub-Saharan Africa, due to a shortage of hepatologists and gastroenterologists, interventional radiologists, hepatobiliary surgeons, and pathologists [27].

4.1. Limitations

The methods used in this paper are limited by several factors. Firstly, estimates of alcohol consumption came from surveys which are susceptible to numerous biases which lead to an underestimation of alcohol use. Per capita consumption of alcohol is utilized to estimate the volume of alcohol use among drinkers to avoid bias; however, no correction exists for the prevalence of HED. This study did not fully account for the interaction between alcohol use and other risk factors, such as smoking (increased risk of cancer [28]), hepatitis B and C (increased risk of liver cirrhosis [29]), and obesity (increased risk of liver cirrhosis [24,30]). Furthermore, the study did not account for the differential alcohol RRs by socio-economic status. Furthermore, although depression has been shown to be causally related to alcohol consumption, it was not included in the estimates of the alcohol-attributable burden of disease due to depression also causally increasing alcohol consumption [13].

This study is also limited as deaths and health loss due to interpersonal harm are based on the alcohol consumption of the person who experiences the harm and not the alcohol consumption of the person who inflicts the harm. This is due to the relative risks of injuries from assault being based on the person who experiences the harm and not the person who is inflicting the harm [13]. Therefore, estimates of intentional harm are likely underestimated for children and women who are often victims of alcohol-related violence [31]. It is important to note that violence against women and children is a major public health, social policy, and human rights concern that spans disciplines and geographical boundaries [32–35]. Globally, domestic violence is one of the largest sources of non-fatal injuries to women and children [36], resulting in avoidable inequities in health status, and increases in the risk of mental health and physical conditions [37].

4.2. Health Policies

The health harms and inequities outlined in the paper should be considered in the context of population-level interventions which can reduce the alcohol-attributable burden of disease and are sustainable, scalable, and politically, economically, and technically feasible [38]. Several alcohol interventions have been designated as "best buys" by the WHO as they are more cost-effective than most other interventions designed for other risk factors [39]. These include increases in taxation and restrictions on availability and marketing. Other policies include WHO cost-effective "very good buys," such as enactment and enforcement of impaired-driving laws and blood-alcohol-concentration limits [39,40].

The need to intervene to reduce the burden of premature health loss attributable to alcohol consumption can be viewed under the framework of utilitarian ageism. The framework of utilitarian ageism, which is often observed in medical practice, states that there should be prioritization of treatments and interventions for health loss among the young (as the old have lived longer) [41,42].

Despite the majority of the burden of disease attributable to alcohol consumption occurring among people 0 to 69 years of age, alcohol leads to a substantial burden of disease among people 70 years of age and older. Interventions such as the WHO best buys and very best buys should also be prioritized to reduce the burden among people 70 years of age and older. Furthermore, these policies should apply equally to beer, wine, spirits, and other alcoholic beverages as the harm caused by alcohol is based on ethanol content regardless of whether the ethanol is consumed in the form of beer, wine, or spirits (with the exception of alcohol poisonings which are caused predominately by the consumption of spirits) [13].

5. Conclusions

Alcohol consumption remains a leading risk factor for the burden of disease, especially among people younger in age. Given the high global alcohol-attributable burden of disease, the development and implementation of cost-effective alcohol control policies can further reduce in the near future the social, economic, and health burdens resulting from the use of alcohol.

Supplementary Materials: The following are available online at https://www.mdpi.com/article/10.3390/nu13093145/s1: the supplemental methodology and supplemental results (Figures S1–S4, and Tables S1–S17).

Author Contributions: Conceptualization, I.S., A.F., J.R. and K.S.; methodology, A.F., J.R. and K.S.; software, K.S.; validation A.F. and J.R.; formal analysis, I.S. and K.S.; investigation, I.S. and K.S.; data curation, A.W., A.F. and K.S.; writing—original draft preparation, I.S.; writing—reviewing and editing, I.S., A.F., J.R., B.C, A.W. and K.S.; visualization, B.C., K.S.; supervision, K.S.; project administration, K.S.; funding acquisition, J.R. and K.S. All authors have read and agreed to the published version of the manuscript.

Funding: This research was funded by the World Health Organization.

Institutional Review Board Statement: Not applicable.

Informed Consent Statement: Not applicable.

Data Availability Statement: All statistical code (i.e., R code) and input files used to produce the results presented in this paper are available to the general public. To obtain the code and input files, please contact the author, Kevin Shield (Kevin.Shield@camh.ca).

Acknowledgments: The authors would like to thank the World Health Organization as well as the regional and country level offices of the World Health Organization for providing data on alcohol consumption and the burden of disease. In particular, the authors thank Alexandra Fleischmann, Vladimir Poznyak, Maristela Monteiro, and Carina Ferreira-Borges for coordinating this data collection.

Conflicts of Interest: The authors declare no conflict of interest. The funders had no role in the design of the study, in the collection, analyses, or interpretation of data, in the writing of the manuscript, or in the decision to publish the results.

References

1. Shield, K.; Manthey, J.; Rylett, M.; Probst, C.; Wettlaufer, A.; Parry, C.D.; Rehm, J. National, regional, and global burdens of disease from 2000 to 2016 attributable to alcohol use: A comparative risk assessment study. *Lancet Public Health* **2020**, *5*, e51–e61. [CrossRef]
2. Shield, K.D.; Rehm, J. Global risk factor rankings: The importance of age-based health loss inequities caused by alcohol and other risk factors. *BMC Res. Notes* **2015**, *8*, 1–4. [CrossRef]
3. Murray, C.J.; Aravkin, A.Y.; Zheng, P.; Abbafati, C.; Abbas, K.M.; Abbasi-Kangevari, M.; Abd-Allah, F.; Abdelalim, A.; Abdollahi, M.; Abdollahpour, I. Global burden of 87 risk factors in 204 countries and territories, 1990–2019: A systematic analysis for the Global Burden of Disease Study 2019. *Lancet* **2020**, *396*, 1223–1249. [CrossRef]
4. Shield, K.; Rehm, J. Substance use and the objectives of current global health frameworks: Measurement matters. *Addiction* **2018**, *114*, 771–773. [CrossRef]
5. Manthey, J.; Shield, K.D.; Rylett, M.; Hasan, O.S.; Probst, C.; Rehm, J. Global alcohol exposure between 1990 and 2017 and forecasts until 2030: A modelling study. *Lancet* **2019**, *393*, 2493–2502. [CrossRef]
6. World Health Organization. *Global Status Report on Alcohol and Health 2018*; World Health Organization: Geneva, Switzerland, 2018.
7. Jiang, H.; Xiang, X.; Room, R.; Hao, W. Alcohol and the Sustainable Development Goals. *Lancet* **2016**, *388*, 1279–1280. [CrossRef]
8. Flor, L.S.; Gakidou, E. The burden of alcohol use: Better data and strong policies towards a sustainable development. *Lancet Public Health* **2020**, *5*, e10–e11. [CrossRef]
9. LoConte, N.K.; Brewster, A.M.; Kaur, J.S.; Merrill, J.K.; Alberg, A.J. Alcohol and cancer: A statement of the American Society of Clinical Oncology. *J. Clin. Oncol.* **2018**, *36*, 83–93. [CrossRef]
10. GBD 2016 Alcohol Collaborators. Alcohol use and burden for 195 countries and territories, 1990–2016: A systematic analysis for the Global Burden of Disease Study 2016. *Lancet* **2018**, *392*, 1015–1035. [CrossRef]
11. Levin, M.L. The occurence of lung cancer in man. *Acta Unio Int. Contra Cancrum* **1953**, *9*, 531–541.
12. Rehm, J.; Irving, H.; Ye, Y.; Kerr, W.C.; Bond, J.; Greenfield, T.K. Are lifetime abstainers the best control group in alcohol epidemiology? On the stability and validity of reported lifetime abstention. *Am. J. Epidemiol.* **2008**, *168*, 866–871. [CrossRef]
13. Rehm, J.; Gmel Sr, G.E.; Gmel, G.; Hasan, O.S.M.; Imtiaz, S.; Popova, S.; Probst, C.; Roerecke, M.; Room, R.; Samokhvalov, A.V.; et al. The relationship between different dimensions of alcohol use and the burden of disease—An update. *Addiction* **2017**, *112*, 968–1001. [CrossRef]
14. Grundy, A.; Poirier, A.E.; Khandwala, F.; McFadden, A.; Friedernreich, C.M.; Brenner, D.R. Cancer inicidence attributable to alcohol consumption in Alberta in 2012. *CMAJ Open* **2016**, *4*, E507. [CrossRef]
15. World Health Organization. *WHO Methods and Data Sources for Country-Level Casues of Death 2000–2016*; WHO: Geneva, Switzerland, 2018.
16. Byass, P. Correlation between noncommunicable disease mortality in people aged 30–69 years and those aged 70–89 years. *Bull. World Health Organ.* **2019**, *97*, 589. [CrossRef]
17. Manthey, J.; Probst, C.; Rylett, M.; Rehm, J. National, regional, and global mortality due to alcoholic cardiomyopathy in 2015. *Heart* **2018**, *104*, 1663–1669. [CrossRef] [PubMed]
18. World Health Organization. *WHO Road Traffic Death Database*; World Health Organization: Geneva, Switzerland, 2018.
19. United Nations. *World Population Prospects: The 2017 Revision*; United Nations: New York, NY, USA, 2017.
20. United Nations Development Programme. Human Development Report 2018. Available online: http://www.webcitation.org/6eYKeeF6z (accessed on 1 January 2021).
21. R Core Team. *R: A Language and Environment for Statistical Computing*; R Core Team: Vienna, Austria, 2013.

22. Kyu, H.H.; Maddison, E.R.; Henry, N.J.; Mumford, J.E.; Barber, R.; Shields, C.; Brown, J.C.; Nguyen, G.; Carter, A.; Wolock, T.M. The global burden of tuberculosis: Results from the Global Burden of Disease Study 2015. *Lancet Infect. Dis.* **2018**, *18*, 261–284. [CrossRef]
23. Corbett, E.L.; Watt, C.J.; Walker, N.; Maher, D.; Williams, B.G.; Raviglione, M.C.; Dye, C. The growing burden of tuberculosis: Global trends and interactions with the HIV epidemic. *Arch. Intern. Med.* **2003**, *163*, 1009–1021. [CrossRef]
24. Rehm, J.; Patra, J.; Brennan, A.; Buckley, C.; Greenfield, T.K.; Kerr, W.C.; Manthey, J.; Purshouse, R.C.; Rovira, P.; Shuper, P.A. The role of alcohol use in the aetiology and progression of liver disease: A narrative review and a quantification. *Drug Alcohol Rev.* **2021**. [CrossRef]
25. NCD Countdown 2030 collaborators. NCD Countdown 2030: Worldwide trends in non-communicable disease mortality and progress towards Sustainable Development Goal target 3.4. *Lancet* **2018**, *392*, 1072–1088. [CrossRef]
26. Mokdad, A.A.; Lopez, A.D.; Shahraz, S.; Lozano, R.; Mokdad, A.H.; Stanaway, J.; Murray, C.J.; Naghavi, M. Liver cirrhosis mortality in 187 countries between 1980 and 2010: A systematic analysis. *BMC Med.* **2014**, *12*, 1–24. [CrossRef]
27. Vento, S.; Dzudzor, B.; Cainelli, F.; Tachi, K. Liver cirrhosis in sub-Saharan Africa: Neglected, yet important. *Lancet Global Health* **2018**, *6*, e1060–e1061. [CrossRef]
28. Voltzke, K.J.; Lee, Y.-C.A.; Zhang, Z.-F.; Zevallos, J.P.; Yu, G.-P.; Winn, D.M.; Vaughan, T.L.; Sturgis, E.M.; Smith, E.; Schwartz, S.M. Racial differences in the relationship between tobacco, alcohol, and the risk of head and neck cancer: Pooled analysis of US studies in the INHANCE Consortium. *Cancer Causes Control* **2018**, *29*, 619–630. [CrossRef]
29. Llamosas-Falcón, L.; Shield, K.D.; Gelovany, M.; Hasan, O.S.; Manthey, J.; Monteiro, M.; Walsh, N.; Rehm, J. Impact of alcohol on the progression of HCV-related liver disease: A systematic review and meta-analysis. *J. Hepatol.* **2021**, *75*, 536–546. [CrossRef]
30. Patra, J.; Buckley, C.; Kerr, W.C.; Brennan, A.; Purshouse, R.C.; Rehm, J. Impact of body mass and alcohol consumption on all-cause and liver mortality in 240 000 adults in the United States. *Drug Alcohol Rev.* **2021**. [CrossRef]
31. Leonard, K. Domestic violence and alcohol: What is known and what do we need to know to encourage environmental interventions? *J. Subst. Use* **2001**, *6*, 235–247. [CrossRef]
32. Watts, C.; Zimmerman, C. Violence against women: Global scope and magnitude. *Lancet* **2002**, *359*, 1232–1237. [CrossRef]
33. Devries, K.M.; Mak, J.Y.; Garcia-Moreno, C.; Petzold, M.; Child, J.C.; Falder, G.; Lim, S.; Bacchus, L.J.; Engell, R.E.; Rosenfeld, L. The global prevalence of intimate partner violence against women. *Science* **2013**, *340*, 1527–1528. [CrossRef]
34. Sinha, M. *Measuring Violence against Women: Statistical Trends. Juristat: Canadian Centre for Justice Statistics*; Minister of Industry: Ottawa, ON, Canada, 2013.
35. World Health Organization. *Violence against Women: Intimate Partner and Sexual Violence against Women*; World Health Organization: Geneva Switzerland, 2016.
36. Kyriacou, D.N.; Anglin, D.; Taliaferro, E.; Stone, S.; Tubb, T.; Linden, J.A.; Muelleman, R.; Barton, E.; Kraus, J.F. Risk factors for injury to women from domestic violence. *N. Engl. J. Med.* **1999**, *341*, 1892–1898. [CrossRef]
37. Coker, A.; Davis, K.; Arias, I.; Desai, S.; Sanderson, M.; Brandt, H.; Smith, P. Physical and mental health effects of intimate partner violence for men and women. *Am. J. Prev. Med.* **2002**, *23*, 260–268. [CrossRef]
38. World Health Organization. *Equity, Social Determinants and Public Health Programmes*; Blas, E., Kurup, A.S., Eds.; World Health Organization,: Geneva, Switzerland, 2010.
39. Chisholm, D.; Moro, D.; Bertram, M.; Pretorius, C.; Gmel, G.; Shield, K.; Rehm, J. Are the "best buys" for alcohol control still valid? An update on the comparative cost-effectiveness of alcohol control strategies at the global level. *JSAD* **2018**, *79*, 514–522. [CrossRef]
40. Pan American Health Organization. *Regional Status Report on Alcohol and Health in the Americas*; PAHO: Washingon, DC, USA, 2015.
41. Harris, J. *The Value of Life: An Introduction to Medical Ethics*; Routledge: London, UK, 1985.
42. Tsuchiya, A. QALYs and ageism: Philosophical theories and age weighting. *Health Econ.* **2000**, *9*, 57–68. [CrossRef]

 *nutrients*

MDPI

Review

Dose–Response Relationships between Levels of Alcohol Use and Risks of Mortality or Disease, for All People, by Age, Sex, and Specific Risk Factors

Jürgen Rehm [1,2,3,4,5,6,7,8,9,*], **Pol Rovira** [9], **Laura Llamosas-Falcón** [1,10] **and Kevin D. Shield** [1,4,6]

1   Centre for Addiction and Mental Health, Institute for Mental Health Policy Research, 33 Ursula Franklin Street, Toronto, ON M5S 2S1, Canada; llamosasfalcon@gmail.com (L.L.-F.); Kevin.Shield@camh.ca (K.D.S.)
2   Institute of Clinical Psychology and Psychotherapy, Technische Universität Dresden, Chemnitzer Str. 46, 01187 Dresden, Germany
3   Center for Interdisciplinary Addiction Research (ZIS), Department of Psychiatry and Psychotherapy, University Medical Center Hamburg-Eppendorf (UKE), Martinistraße 52, 20246 Hamburg, Germany
4   Dalla Lana School of Public Health, University of Toronto, 155 College Street, Toronto, ON M5T 1P8, Canada
5   Faculty of Medicine, Institute of Medical Science, University of Toronto, Medical Sciences Building, 1 King's College Circle, Room 2374, Toronto, ON M5S 1A8, Canada
6   Centre for Addiction and Mental Health, Campbell Family Mental Health Research Institute, 33 Ursula Franklin Street, Toronto, ON M5S 3M1, Canada
7   Department of Psychiatry, University of Toronto, 250 College Street, 8th Floor, Toronto, ON M5T 1R8, Canada
8   Department of International Health Projects, Institute for Leadership and Health Management, I.M. Sechenov First Moscow State Medical University (Sechenov University), Trubetskaya Street 8, b. 2, 119991 Moscow, Russia
9   Program on Substance Abuse, Public Health Agency of Catalonia, 08005 Barcelona, Spain; polrovira26@gmail.com
10 Hospital Universitario 12 de Octubre, Av. de Córdoba, s/n, 28041 Madrid, Spain
*   Correspondence: jtrehm@gmail.com; Tel.: +1-(416)-535-8501 (ext. 36173)

Citation: Rehm, J.; Rovira, P.; Llamosas-Falcón, L.; Shield, K.D. Dose–Response Relationships between Levels of Alcohol Use and Risks of Mortality or Disease, for All People, by Age, Sex, and Specific Risk Factors. *Nutrients* **2021**, *13*, 2652. https://doi.org/10.3390/nu13082652

Academic Editor: Rosa Casas

Received: 2 July 2021
Accepted: 28 July 2021
Published: 30 July 2021

Publisher's Note: MDPI stays neutral with regard to jurisdictional claims in published maps and institutional affiliations.

Abstract: Alcohol use has been causally linked to more than 200 disease and injury conditions, as defined by three-digit ICD-10 codes. The understanding of how alcohol use is related to these conditions is essential to public health and policy research. Accordingly, this study presents a narrative review of different dose–response relationships for alcohol use. Relative-risk (RR) functions were obtained from various comparative risk assessments. Two main dimensions of alcohol consumption are used to assess disease and injury risk: (1) volume of consumption, and (2) patterns of drinking, operationalized via frequency of heavy drinking occasions. Lifetime abstention was used as the reference group. Most dose–response relationships between alcohol and outcomes are monotonic, but for diabetes type 2 and ischemic diseases, there are indications of a curvilinear relationship, where light to moderate drinking is associated with lower risk compared with not drinking (i.e., RR < 1). In general, women experience a greater increase in RR per gram of alcohol consumed than men. The RR per gram of alcohol consumed was lower for people of older ages. RRs indicated that alcohol use may interact synergistically with other risk factors, in particular with socioeconomic status and other behavioural risk factors, such as smoking, obesity, or physical inactivity. The literature on the impact of genetic constitution on dose–response curves is underdeveloped, but certain genetic variants are linked to an increased RR per gram of alcohol consumed for some diseases. When developing alcohol policy measures, including low-risk drinking guidelines, dose–response relationships must be taken into consideration.

Keywords: alcohol; patterns of drinking; disease; mortality; dose response; monotonous; protective effects; curvilinear; alcohol control policy

1. Introduction

Alcohol use has been causally linked to more than 200 disease and injury conditions (based on three-digit ICD-10 codes; [1,2] and Table 1 below), indicating that alcohol use alone is a necessary cause or that it is a component cause (for an epidemiological definition of causality see: [3]). Different dimensions of alcohol use causally lead to a modified risk of disease and injury [4]. Average levels of drinking over time have been linked to many chronic disease categories, such as cancer, gastrointestinal disease, and different categories of heart disease [2]. Heavy drinking occasions are more linked to the more immediate effects of alcohol use, such as unintentional and intentional injuries [5,6], but also affect the risk for ischemic diseases [7], and infectious diseases (e.g., HIV, [2]). Accordingly, these two dimensions of alcohol use are most often used in epidemiological studies to predict disease and injury risk (the average level of consumption over time, and heavy episodic drinking (HED)).

Table 1. Major disease categories causally related to alcohol and modelled in the last comparative risk assessments based on WHO data [8], and their codes in the International Statistical Classification of Diseases and Related Health Problems (ICD).

		Global Health Estimate 2015 Cause Category	ICD-10 Coding
I.		Communicable, maternal, perinatal and nutritional conditions	A00–B99, D50–53, D64.9, E00–02, E40–46, E50–64, G00–04, G14, H65–66, J00–22, N70–73, O00–99, P00–96, U04
	A.	Infectious and parasitic diseases	A00–B99, G00–04, G14, N70–73, P37.3, P37.4
		1 Tuberculosis	A15–19, B90
		3 HIV/AIDS	B20–24
	B.	Respiratory infections	H65–66, J00–22, P23, U04
		1 Lower respiratory infections	J09–22, P23, U04
II.		Noncommunicable diseases	C00–97, D00–48, D55–64 (minus D64.9), D65–89, E03–07, E10–34, E65–88, F01–99, G06–98 (minus G14), H00–61, H68–93, I00–99, J30–98, K00–92, L00–98, M00–99, N00–64, N75–98, Q00–99, X41–42, X44, X45, R95
	A.	Malignant neoplasms	C00–97
		1 Mouth and oropharynx cancers	C00–14
		a. Lip and oral cavity	C00–08
		c. other pharyngeal cancers	C09–10, C12–14
		2 Oesophagus cancer	C15
		4 Colon and rectum cancers	C18–21
		5 Liver cancer	C22
		9 Breast cancer	C50
		19 Larynx cancer	C32
	C.	Diabetes mellitus	E10–14 (minus E10.2–10.29, E11.2–11.29, E12.2, E13.2–13.29, E14.2)
	E.	Mental and substance use disorders	F04–99, G72.1, Q86.0, X41–42, X44, X45
		4 Alcohol use disorders	F10, G72.1, Q86.0, X45
	F.	Neurological conditions	F01–03, G06–98 (minus G14, G72.1)
		3 Epilepsy	G40–41
	H.	Cardiovascular diseases	I00–99
		2 Hypertensive heart disease	I10–15
		3 Ischemic heart disease	I20–25
		4 Stroke	I60–69
		a. Ischemic stroke	G45–46.8, I63–63.9, I65–66.9, I67.2–67.848, I69.3–69.4
		b. Hemorrhagic stroke	I60–62.9, I67.0–67.1, I69.0–69.298
		5 Cardiomyopathy, myocarditis, endocarditis	I30–33, I38, I40, I42
	J.	Digestive diseases	K20–92
		2 Cirrhosis of the liver	K70, K74
		8 Pancreatitis	K85–86
III.		Injuries	V01–Y89 (minus X41–42, X44, X45)
	A.	Unintentional injuries	V01–X40, X43, X46–59, Y40–86, Y88, Y89
		1 Road injury	V01–04, V06, V09–80, V87, V89, V99
		2 Poisonings	X40, X43, X46–48, X49
		3 Falls	W00–19
		4 Fire, heat and hot substances	X00–19
		5 Drowning	W65–74
		6 Exposure to mechanical forces	W20–38, W40–43, W45, W46, W49–52, W75, W76
		8 Other unintentional injuries	Rest of V, W39, W44, W53–64, W77–99, X20–29, X50–59, Y40–86, Y88, Y89
	B.	Intentional injuries	X60–Y09, Y35–36, Y870, Y871
		1 Self-harm	X60–84, Y870
		2 Interpersonal violence	X85–Y09, Y871

The dimensions of average volume of alcohol use and HED are not independent: all people with chronic heavy drinking, such as many of those with alcohol use disorders (but see [9,10]), engage in HED [11]. People who do not engage in chronic heavy drinking can be separated in two categories: those who engage in HED and those who do not, with different impacts on some disease and mortality outcomes [12]. For example, on average, light drinking has been associated with cardio-protectivity compared to lifetime abstention [13]; these effects disappear for light drinkers who also engage in HED [7]. Furthermore, the risk of unintentional injury depends on the blood alcohol concentration and thus is highest in people who have engaged in a heavy drinking occasion prior to sustaining their injury [5,6]. However, this dose–response relationship has been shown to vary based on prior drinking experience. Thus, Gmel and colleagues [14] showed that while people at all average drinking levels are at increased risk for alcohol-related injury, those who normally drink lightly are at higher risk of injury compared to chronic heavy drinkers after consuming the same quantity of alcohol. Similar results have been reported in other studies [15].

It is important to know the exact dose–response relationships between alcohol use and disease outcomes because alcohol control policy measures will in part depend on these relationships [16]. For example, if most dose–response relationships are linear, population measures to lower the population mean of alcohol use (=per capita consumption), such as taxation increases or availability restrictions, are the most effective and appropriate measures (e.g., [17]). If the dose–response relationships are steep and exponential, measures for heavy drinkers could potentially be the more effective and/or cost-effective strategy [16,18]. We will come back to these choices in the Discussion.

Given the numerous diseases which are causally related to alcohol consumption (see Table 1 above), this contribution will systematically examine dose–response curves between different average levels of alcohol use and disease outcomes with a focus on modification of such curves by personal characteristics and/or the drinking context. Thus, this contribution will explore the impact of heavy episodic drinking on dose–response curves, as well as if these risks differ by factors such as sex, age, socio-economic status, genetic constitution, and behavioural risk factors.

2. Materials and Methods

This paper is a narrative review of the relative risk (RR) functions between average level of alcohol use and the occurrence of diseases and injuries [19], mainly based on meta-analytically derived dose–response curves used for comparative risk assessments (e.g., [8,20,21]). We used the meta-analyses reported by the latest WHO comparative risk assessments and from the Global Burden of Disease Study as they are comprehensive, evaluated by special committees, and continuously updated.

These dose–response curves usually compare the relative change in risk from a certain level of average drinking against the risk of a lifetime abstainer (for the rationale, see [22]). Lifetime abstention is selected as a comparison group, rather than abstention, as the people constituting the latter group are comprised of lifetime abstainers and former drinkers, and therefore have different risk levels (for further discussion, see [23,24]). In burden calculations, the theoretical minimum risk exposure level for comparison, rather than abstention, has traditionally been used (e.g., [21]) but, for the topic of this paper, this is not relevant. Whenever possible, the relative-risk curves are sex-specific ([8,20]; for further discussion, see below).

In risk factor epidemiology, in particular in comparative risk assessments, these dose–response curves based on RR for different exposures are then applied to almost all countries, taking into account the respective population distributions for drinking [25]. The only exception is the Russian Federation and surrounding countries, where region-specific dose–response curves are usually applied [26], because the same average level of drinking has been found to be associated with higher risks of mortality and other harms [27]. This procedure assumes that the dose–response curve is fairly stable, an assumption which will be examined below.

3. Results

3.1. Basic Typologies of Dose–Response Relationships

Threshold effects: Most risk curves can be described as continuous, but there is some evidence for threshold effects related to alcohol use. Tuberculosis (TB) provides us with an excellent example of this. In the original meta-analysis, Lönnroth and colleagues [28] examined the relationship between different levels of alcohol consumption, and concluded that all studies with alcohol use below a threshold of 40 g pure alcohol per day found no significant relationship with incidence of active TB, whereas drinking above this threshold resulted in about a three-fold higher risk. In this study, people with alcohol problems, including use disorders, were classified as being above the threshold. A newer study on the alcohol-TB dose–response relationship corroborated these results for people with alcohol problems, but did not identify a threshold when they included only studies involving individuals with an average volume of alcohol consumption [29]. In these studies, a dose–response relationship was found, which can be explained by a monotonic increasing risk. In the linear continuous meta-analysis which was identified as the best-fitting model, the TB risk rose by about 2% per gram pure alcohol intake (95% CI: 0–3%), leading to the following RR: at 25 g/day: 1.57 (95% CI: 1.10–2.23), at 50 g/day: 2.46 (95% CI: 1.21–4.98), at 75 g/day: 3.85 (95% CI: 1.33–11.11), and at 100 g/day: 6.03 (95% CI: 1.47–24.81).

In the current comparative risk analyses [8], there is only one threshold relationship for the risk of HIV/AIDS. This model is not based on meta-analyses but on experimental data, and the threshold of 48 g/day and 60 g/day is modelled based on reaching a minimal blood alcohol level every day [2]. In sum, there is no good evidence for a real threshold effect, but for one dose–response relationship, a conservative threshold was chosen. It was recently decided that the risk for sexual transmitted diseases other than HIV/AIDS would be modelled in the same manner in the next Global Status Report on Alcohol and Health (based on [30]).

Monotonic dose–response relationships: Most dose–response relationships are monotonic [11] if the comparisons are made with lifetime abstainers rather than with the combined group of lifetime abstainers and former drinkers. This means that with increasing average alcohol consumption, the risk for disease or mortality increases. In some instances, a monotonic relationship means that the risk between levels of alcohol use and RR is best modelled linearly, while in others there are exponential or flattening functions [11]. It should be noted that as the RR functions are exponential, an exponential linear function means that the underlying disease or mortality risk is exponential.

Exceptions seem to be ischemic disease and diabetes, which show curvilinear relationships, with light to moderate drinkers showing less risk than lifetime abstainers (for details, see below).

Curvilinear dose–response relationships: A number of important disease and mortality outcomes seem to show curvilinear relationships with the lowest risk at low to moderate drinking levels: ischemic heart disease [7,13], ischemic stroke [31]; diabetes [32,33]. As for ischemic disease, the protective effect seems to be mainly for acute outcomes, especially acute myocardial infarction, and less so for chronic ischemic events [34]. The use of these curvilinear dose–response relationships has received criticism, mainly because of their underlying unclear comparison groups, resulting in either overestimating the protective impact of alcohol or in artificially creating such an impact where there is in reality a monotonous relationship (e.g., [35]). However, at least for ischemic disease categories, there are plausible biological pathways for a protective effect from light to moderate drinking [36,37], so the overall shape of the curve is likely curvilinear with some kind of protective effect before the curve rises up again. However, the protective effect is likely overestimated, especially if meta-analyses based on overall abstention is used; the more former drinkers included, the higher the overestimation (for ischemic heart disease, a meta-analysis estimated the RR of former drinkers for ischemic heart disease mortality; 1.25, 95% CI: 1.15–1.36; and 1.54, 95% CI: 1.17–2.03 for men and women, respectively [38]).

Another controversial dose–response curve involves alcohol use and dementia. Most reviews found a curvilinear relationship with a protective effect for light to moderate drinking [39], even though heavy drinking is clearly detrimentally related to the incidence of dementia, in particular early onset dementia [40].

A final consideration here concerns sex differences in curvilinear relationships: the protective effect and the increase in risk after the nadir are more pronounced in women, both for ischemic disease and diabetes (see risk curves in Appendix 1 of [8]). For more on sex as a modifier, see Section 3.2.

3.2. Modifiers of Dose–Response Relationships

Sex: In all countries, men, on average, consume higher quantities of alcohol, and have more heavy drinking occasions [41]. Accordingly, alcohol-attributable mortality or burden of disease rates are higher in men [8,21]. However, the differences in health harms are attenuated somewhat, as many RRs, especially for chronic diseases, are lower for men for the same level of drinking. In the classical comparative risk assessment of English and colleagues [42], a categorical approach was used with categories of <20 g/day, 20–40 g/day, and >40 g/day for women, and <40 g/day, 40–60 g/day and >60g/day for men, and the RRs were often estimated to be similar for the respective categories. Thus, in the analysis, the RR for the first category of women with a midpoint of 10 g/day was the same as for the first category of men with a midpoint of 30 g/day.

These early quantifications were corroborated by later meta-analyses, and for the following conditions, higher RRs are currently seen for women: HIV, hypertensive heart disease, ischemic heart disease, both stroke types, liver cirrhosis, and pancreatitis. For instance, Figure 1 is based on the most recent comprehensive meta-analysis of Roerecke and colleagues [43] and shows a much higher RR in women for liver cirrhosis at the same level of drinking.

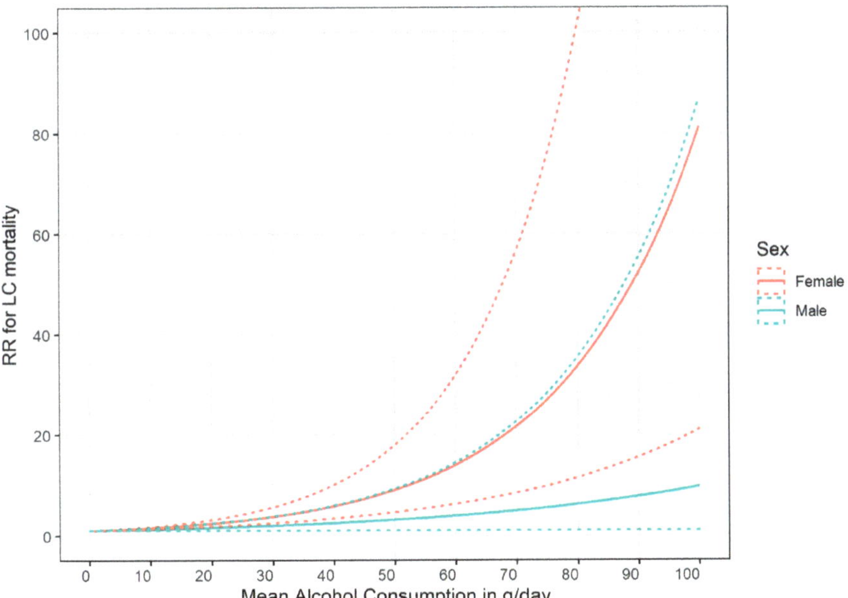

Figure 1. Relative risk for liver cirrhosis mortality as a function of mean alcohol consumption (based on [43]). RR: Relative Risk; LC: liver cirrhosis; solid lines denote point estimates, dashed lines the 95% confidence intervals.

It should be noted that the categorical analyses based on a larger number of studies resulted in fewer exponential dose–response curves, but the women had higher risks for the same amount of drinking. For instance, a consumption of 7 and more drinks per day in women results in a RR of 24.6 (95% CI: 14.8–40.9), whereas the same amount of average daily drinking in men was associated with a RR of 6.9 (95% CI: 1.1–45.0). These differences are based on the fact that for dose–response relationships in the cohorts used in medical epidemiology, there is often not a sufficient number of people included to estimate dose–response curves for higher levels of average drinking, such as those seen in some treatment samples (see Discussion below and [44]).

Before discussing other modifiers, it should be noted that, even now, more than 25 years after the first comparative risk assessment, there is still often not a sufficient number of studies available to adequately separate risk curves between the sexes (see [8,20] for an overview). Data scarcity is even more of a problem for other modifiers (see below and the Discussion).

Age: There are biological and other reasons that dose–response curves for alcohol use should change with age. However, the underlying literature is scarce. Ischemic disease categories are an exception. Klatsky found an attenuation of risk based on age [45]. Based on this, the dose–response curves for ischemic disease were modelled separately for three age groups [46]. Figure 2 gives an example. As these curves also depend on the frequency of HEDs (see above and [21]), different curves are provided. For people without any history of HED, there are potential beneficial effects at up to 30 g/day average consumption, and these effects are most pronounced in younger ages (see the curve in red versus the curves in green or blue). For people with HED, the curve is flat until it reaches 30 g/day average consumption (black line), with no beneficial or detrimental effects, irrespective of age. After that threshold is reached, the curves are the same irrespective of any history of HED [46].

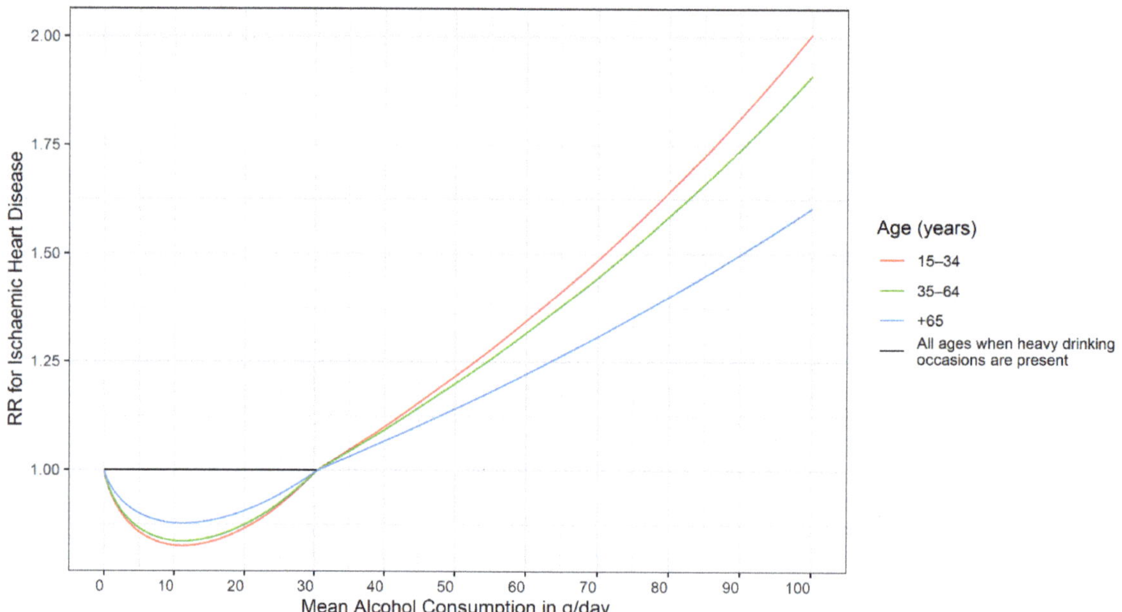

Figure 2. Relative risk for ischemic heart disease mortality as a function of mean alcohol consumption, heavy drinking occasion status, and age for women assuming no heavy drinking occasions for light to moderate drinkers [46]. RR: Relative Risk.

Socioeconomic status and wealth (SES): Epidemiologic studies using a variety of indicators for SES (education, income, professional status) have consistently shown that, for the general population, morbidity and mortality risk increases as SES decreases [47–49]. Behavioural risk factors such as alcohol use and their social patterning have frequently been proposed as factors mediating socioeconomic differences in health [50]. However, there may also be an interaction between such risk factors and SES. Thus, there have been some indications that the dose–response relationships are steeper at lower levels of SES. For instance, the RR of alcohol consumption for those with HIV infections was considerably higher for low SES compared to high SES [51]. In general, systematic reviews and meta-analyses found some indication for an interaction between alcohol use and SES [52]. However, it is not clear if this interaction leading to more harm per litre of ethanol is due to steeper risk curves or due to different drinking patterns or due to differences in reporting (e.g., [53]). The steeper dose–response curves could be due to interactions with other risk factors, such as smoking or BMI (see the discussion on the harm paradox [54]). Overall, exploring this interaction and its possible underlying mechanisms should be a priority in future research endeavours.

The same phenomena can be seen between countries: the harm per litre of alcohol depends strongly on the economic wealth, or on the Human Development Index [55,56]. Analyses stratified by the Human Development Index reveal that a substantial part of the variance for alcohol-attributable all-cause mortality stems from different causes of death between countries. TB once again provides a good example, and has been called the archetypical disease of poverty [57], as it is very much linked to crowding, and other characteristics linked to economically poor environments [58]. Obviously, in rich countries, alcohol consumption and it is effects on the immune system will not lead to TB that often, as there are almost no people with active TB around to contract it from. Thus, the higher risk for the same amount of alcohol can be explained by environmental variables, or interactions with other risk factors, such as crowding. Shield and Rehm [55] provide the list of diseases where various environmental factors play a large role: infectious diseases in general (i.e., all sexually transmitted diseases, including HIV/AIDS, pneumonia) but also liver cirrhosis (via the interaction with hepatitis B and C [59]), and road injuries show the biggest differences in standardized mortality after consuming one litre of pure ethanol.

Genetic constitution: Variants of three genes encoding alcohol-metabolizing enzymes, the aldehyde dehydrogenase gene ALDH2, and the two alcohol dehydrogenase genes, ADH1B and ADH1C, have been associated with different risks for some alcohol-attributable diseases. These variants are more prevalent in Japanese, Chinese, and other Asian populations [60]. Differences in the dose–response curves can be found in disease outcomes for which acetaldehyde plays a role (flushing, most alcohol-attributable cancers [61]) and are especially pronounced for oesophageal cancer [62], where acetaldehyde is one of the most important underlying causal pathways [61,63,64].

4. Discussion

Before we discuss the findings in detail, we want to point out the limitations. Any review is limited by the quality of its underlying literature. In this case, there are four main limitations: First and foremost, the dose–response relationship for levels of alcohol use depends to a considerable degree on the comparison group. Using the non-drinking group for comparison, i.e., not separating between lifetime abstainers and former drinkers will often lead to more pronounced curvilinear relationships which falsely indicate a beneficial impact at light to moderate levels of drinking for conditions where such benefits do not exist. The reason here is due to the inclusion of the so-called 'sick quitters', defined as people who stopped drinking because of health problems [65]. This does not imply that there are no curvilinear relationships between the level of alcohol use and disease and mortality outcomes: as indicated above, there are known biological pathways for the beneficial effects of light to moderate drinking on ischemic disease [36,66,67]. However, the ubiquity of reports on the beneficial health effects from light to moderate drinking [68,69] is mainly due to this effect. Second, most of the risk curves are based on verbal reports of drinkers regarding their consumption levels, thus potentially introducing some biases [44,70,71]. Even seemingly simple questions such as those regarding lifetime abstention may introduce some biases [23]: in a nationally representative US survey with follow-up, more than half (52.9%) of those who reported never having consumed any kind of alcoholic beverage in the 1992 survey had, in fact, reported drinking in previous surveys. Third, most meta-analyses are based on a one-time measurement of alcohol only, with some follow-ups decades later [44]. This assumes that this one-time measurement captures the level of drinking before and after the measurement, and certainly creates regression dilution bias [72] (i.e., underestimation of the true relationship). Finally, in many cases risk curves are based on a few studies from similar cultures. This certainly introduces bias, and also limits the generalizability of our knowledge with respect to groups, defined by sex, age, or other modifiers. Moreover, many of the medical cohorts were selected for their likelihood of returning for follow-up, thus restricting groups with chronic heavy drinking patterns, such as people with alcohol use disorders [73]. For risk curves, this means that the slopes found within the variability of drinking of stable middle-class respondents are simply projected onto slopes for more extreme drinking, where risk acceleration may plateau. This may also create bias, especially for exponentially increasing slopes, and capping the relative risk at values where we have sufficient underlying observations for alcohol exposure may be the answer [74]. Alternatively, RR may be capped at the average risk level for people with alcohol use disorders.

While the above limitations point to the need for more studies to fill in the research gaps, the existing research clearly indicates several implications for alcohol policies, including guidelines: First, as most dose–response curves are monotonous, the lower the level of alcohol consumption overall, the better. While ischemic diseases and diabetes may constitute exceptions, even based on current meta-analyses, it seems clear that less consumption is better (i.e., between 10 and 20 g/day [34,75]), and the risk is sex-specific. This means that most current low-risk drinking guidelines have thresholds which are too high [75]. Second, as many of the dose–response curves are exponential, risk reduction is greater for heavier drinkers compared to moderate drinkers, if both reduce their drinking by the same number of drinks per day [18]. Empirical evidence suggests that this is best achieved by moving the overall population mean downwards [56,76].

5. Conclusions

Dose–response relationships are crucial for determining the best medical recommendations (such as low-risk drinking guidelines; [77]) and for creating effective alcohol control policy measures. More research is necessary to better understand their variability and determinants.

Author Contributions: Conceptualization, J.R.; methodology, J.R.; validation, J.R., P.R., L.L.-F., K.D.S.; formal analysis, P.R. and K.D.S.; investigation, J.R. and K.D.S.; resources, J.R.; data curation, J.R.; writing—original draft preparation, J.R.; writing—review and editing, J.R., P.R., L.L.-F., K.D.S.; visualization, P.R.; supervision, J.R.; project administration, J.R.; funding acquisition, J.R. All authors have read and agreed to the published version of the manuscript.

Funding: This research was supported by funding from the US National Institute on Alcohol Abuse and Alcoholism (1R01AA028009-01A1) and the Canadian Institutes of Health Research, Institute of Neurosciences, and Mental Health and Addiction (CRISM Ontario Node grant no. SMN-13950) to the first author.

Institutional Review Board Statement: Not applicable, as the study did not involve humans or animals.

Informed Consent Statement: Not applicable.

Data Availability Statement: Not applicable.

Acknowledgments: We would like to acknowledge Astrid Otto for referencing and copy-editing the manuscript.

Conflicts of Interest: The authors declare no conflict of interest. The funders had no role in the design of the study, in the collection, analyses, or interpretation of data, in the writing of the manuscript, or in the decision to publish the results.

References

1. Rehm, J.; Mathers, C.; Popova, S.; Thavorncharoensap, M.; Teerawattananon, Y.; Patra, J. Global burden of disease and injury and economic cost attributable to alcohol use and alcohol-use disorders. *Lancet* **2009**, *373*, 2223–2233. [CrossRef]
2. Rehm, J.; Probst, C.; Shield, K.D.; Shuper, P.A. Does alcohol use have a causal effect on HIV incidence and disease progression? A review of the literature and a modeling strategy for quantifying the effect. *Popul. Health Metr.* **2017**, *15*, 4. [CrossRef] [PubMed]
3. Rothman, K.J.; Greenland, S.; Lash, T.L. *Modern Epidemiology*, 3rd ed.; Lippincott Williams & Wilkins: Philadelphia, PA, USA, 2008.
4. Rehm, J.; Ashley, M.J.; Room, R.; Single, E.; Bondy, S.; Ferrence, R.; Giesbrecht, N. On the emerging paradigm of drinking patterns and their social and health consequences. *Addiction* **1996**, *91*, 1615–1621. [CrossRef]
5. Cherpitel, C.J.; Ye, Y.; Bond, J.; Borges, G.; Monteiro, M. Relative risk of injury from acute alcohol consumption: Modeling the dose-response relationship in emergency department data from 18 countries. *Addiction* **2015**, *110*, 279–288. [CrossRef] [PubMed]
6. Taylor, B.; Irving, H.; Kanteres, F.; Room, R.; Borges, G.; Cherpitel, C.; Greenfield, T.; Rehm, J. The more you drink, the harder you fall: A systematic review and meta-analysis of how acute alcohol consumption and injury or collision risk increase together. *Drug Alcohol Depend.* **2010**, *110*, 108–116. [CrossRef]
7. Roerecke, M.; Rehm, J. Alcohol consumption, drinking patterns, and ischemic heart disease: A narrative review of meta-analyses and a systematic review and meta-analysis of the impact of heavy drinking occasions on risk for moderate drinkers. *BMC Med.* **2014**, *12*, 182. [CrossRef] [PubMed]
8. Shield, K.; Manthey, J.; Rylett, M.; Probst, C.; Wettlaufer, A.; Parry, C.D.H.; Rehm, J. National, regional, and global burdens of disease from 2000 to 2016 attributable to alcohol use: A comparative risk assessment study. *Lancet Public Health* **2020**, *5*, e51–e61. [CrossRef]
9. Heilig, M.; MacKillop, J.; Martinez, D.; Rehm, J.; Leggio, L.; Vanderschuren, L.J. Addiction as a brain disease revised: Why it still matters, and the need for consilience. *Neuropsychopharmacology* **2021**, 1–9. [CrossRef]
10. Jiang, H.; Lange, S.; Tran, A.; Imtiaz, S.; Rehm, J. Determining the sex-specific distributions of average daily alcohol consumption using cluster analysis: Is there a separate distribution for people with alcohol dependence? *Popul. Health Metr.* **2021**, *19*, 1–11. [CrossRef]
11. Rehm, J.; Gmel, G.E., Sr.; Gmel, G.; Hasan, O.S.M.; Imtiaz, S.; Popova, S.; Probst, C.; Roerecke, M.; Room, R.; Samokhvalov, A.V.; et al. The relationship between different dimensions of alcohol use and the burden of disease-an update. *Addiction* **2017**, *112*, 968–1001. [CrossRef] [PubMed]
12. Gmel, G.; Kuntsche, E.; Rehm, J. Risky single—Occasion drinking: Bingeing is not bingeing. *Addiction* **2011**, *106*, 1037–1045. [CrossRef]
13. Roerecke, M.; Rehm, J. The cardioprotective association of average alcohol consumption and ischaemic heart disease: A systematic review and meta—Analysis. *Addiction* **2012**, *107*, 1246–1260. [CrossRef]
14. Gmel, G.; Bissery, A.; Gammeter, R.; Givel, J.C.; Calmes, J.M.; Yersin, B.; Daeppen, J.B. Alcohol—Attributable injuries in admissions to a Swiss emergency room—An analysis of the link between volume of drinking, drinking patterns, and preattendance drinking. *Alcohol. Clin. Exp. Res.* **2006**, *30*, 501–509. [CrossRef]
15. Krüger, H.P.; Kazenwadel, J.; Vollrath, M. Grand Rapids effects revisited: Accidents, alcohol and risk. In *Alcohol, Drugs, and Traffic Safety, Proceedings of the 13th International Conference on Alcohol, Drugs and Traffic Safety, Adelaide, Australia, 13–18 August 1995*; Kloeden, C.N., McLean, A.J., Eds.; NHMRC Road Accident Research Unit, University of Adelaide: Adelaide, Australia, 1995; pp. 222–230.
16. Keyes, K.M.; Galea, S. *Population Health Science*; Oxford University Press: Oxford, UK, 2016.
17. Rose, G. Sick individuals and sick populations. *Int. J. Epidemiol.* **1985**, *14*, 32–38. [CrossRef]
18. Nutt, D.J.; Rehm, J. Doing it by numbers: A simple approach to reducing the harms of alcohol. *J. Psychopharmacol.* **2014**, *28*, 3–7. [CrossRef] [PubMed]
19. Cook, D.J.; Mulrow, C.D.; Haynes, R.B. Systematic reviews: Synthesis of best evidence for clinical decisions. *Ann. Intern. Med.* **1997**, *126*, 376–380. [CrossRef]
20. Rehm, J.; Sherk, A.; Shield, K.D.; Gmel, G. Risk Relations between Alcohol Use and Non-Injury Causes of Death Toronto, ON: Centre for Addiction and Mental Health. 2017. Available online: https://www.camh.ca/-/media/files/pdfs---reports-and-books---research/camh-risk-relations-between-alcohol-use-and-non-injury-causes-of-death-sept2017-pdf.pdf?la=en&hash=393C0B60D26A218F2512C39C1B61EFCE0FCC5128 (accessed on 29 July 2021).
21. GBD. 2019 Risk Factors Collaborators. Global burden of 87 risk factors in 204 countries and territories, 1990–2019: A systematic analysis for the Global Burden of Disease Study 2019. *Lancet* **2020**, *396*, 1223–1249. [CrossRef]
22. Rehm, J.; Monteiro, M.; Room, R.; Gmel, G.; Jernigan, D.; Frick, U.; Graham, K. Steps towards constructing a global comparative risk analysis for alcohol consumption: Determining indicators and empirical weights for patterns of drinking, deciding about theoretical minimum, and dealing with different consequences. *Eur. Addict. Res.* **2001**, *7*, 138–147. [CrossRef]

23. Rehm, J.; Irving, H.; Ye, Y.; Kerr, W.C.; Bond, J.; Greenfield, T.K. Are lifetime abstainers the best control group in alcohol epidemiology? On the stability and validity of reported lifetime abstention. *Am. J. Epidemiol.* **2008**, *168*, 866–871. [CrossRef] [PubMed]
24. Klatsky, A.L. Invited commentary: Never, or hardly ever? It could make a difference. *Am. J. Epidemiol.* **2008**, *168*, 872–875. [CrossRef] [PubMed]
25. Rehm, J.; Kehoe, T.; Gmel, G.; Stinson, F.; Grant, B.; Gmel, G. Statistical modeling of volume of alcohol exposure for epidemiological studies of population health: The US example. *Popul. Health Metr.* **2010**, *8*, 3. [CrossRef]
26. Shield, K.; Rehm, J. Russia-specific relative risks and their effects on the estimated alcohol-attributable burden of disease. *BMC Public Health* **2015**, *15*, 482. [CrossRef] [PubMed]
27. Zaridze, D.; Lewington, S.; Boroda, A.; Scélo, G.; Karpov, R.; Lazarev, A.; Konobeevskaya, I.; Igitov, V.; Terechova, T.; Boffetta, P. Alcohol and mortality in Russia: Prospective observational study of 151,000 adults. *Lancet* **2014**, *383*, 1465–1473. [CrossRef]
28. Lönnroth, K.; Williams, B.; Stadlin, S.; Jaramillo, E.; Dye, C. Alcohol use as a risk factor for tuberculosis—A systematic review. *BMC Public Health* **2008**, *8*, 289. [CrossRef] [PubMed]
29. Imtiaz, S.; Shield, K.D.; Roerecke, M.; Samokhvalov, A.V.; Lönnroth, K.; Rehm, J. Alcohol consumption as a risk factor for tuberculosis: Meta-analyses and burden of disease. *Eur. Respir. J.* **2017**, *50*, 1700216. [CrossRef] [PubMed]
30. Llamosas-Falcón, L. *Proposal: To Include Mortality and Morbidity for Sexually Transmitted Diseases (STDs) as Alcohol-Attributable Disease Conditions in Comparative Risk Assessments—Rationale and Procedures*; Centre for Addiction and Mental Health: Toronto, ON, Canada, 2021.
31. Larsson, S.C.; Wallin, A.; Wolk, A.; Markus, H.S. Differing association of alcohol consumption with different stroke types: A systematic review and meta-analysis. *BMC Med.* **2016**, *14*, 178. [CrossRef]
32. Baliunas, D.; Taylor, B.; Irving, H.; Roerecke, M.; Patra, J.; Mohapatra, S.; Rehm, J. Alcohol as a risk factor for type 2 diabetes—A systematic review and meta-analysis. *Diabetes Care* **2009**, *32*, 2123–2132. [CrossRef]
33. Knott, C.; Bell, S.; Britton, A. Alcohol consumption and the risk of type 2 diabetes: A systematic review and Dose-Response Meta-analysis of more than 1.9 million individuals from 38 observational studies. *Diabetes Care* **2015**, *38*, 1804–1812. [CrossRef]
34. Wood, A.M.; Kaptoge, S.; Butterworth, A.S.; Willeit, P.; Warnakula, S.; Bolton, T.; Paige, E.; Paul, D.S.; Sweeting, M.; Burgess, S.; et al. Risk thresholds for alcohol consumption: Combined analysis of individual-participant data for 599,912 current drinkers in 83 prospective studies. *Lancet* **2018**, *391*, 1513–1523. [CrossRef]
35. Zhao, J.; Stockwell, T.; Roemer, A.; Naimi, T.; Chikritzhs, T. Alcohol consumption and mortality from coronary heart disease: An updated meta-analysis of cohort studies. *J. Stud. Alcohol Drugs* **2017**, *78*, 375–386. [CrossRef]
36. Brien, S.E.; Ronksley, P.E.; Turner, B.J.; Mukamal, K.J.; Ghali, W.A. Effect of alcohol consumption on biological markers associated with risk of coronary heart disease: Systematic review and meta-analysis of interventional studies. *Br. Med. J.* **2011**, *342*, d636. [CrossRef]
37. Rehm, J. On the limitations of observational studies. *Addict. Res. Theory* **2007**, *15*, 20–22.
38. Roerecke, M.; Rehm, J. Ischemic heart disease mortality and morbidity in former drinkers: A meta-analysis. *Am. J. Epidemiol.* **2011**, *73*, 245–258. [CrossRef]
39. Rehm, J.; Hasan, O.S.M.; Black, S.E.; Shield, K.D.; Schwarzinger, M. Alcohol use and dementia: A systematic scoping review. *Alzheimers Res. Ther.* **2019**, *11*, 1. [CrossRef] [PubMed]
40. Schwarzinger, M.; Pollock, B.G.; Hasan, O.S.M.; Dufouil, C.; Rehm, J.; QalyDays Study Group. Contribution of alcohol use disorders to the burden of dementia in France 2008-13: A nationwide retrospective cohort study. *Lancet Public Health* **2018**, *3*, e124–e132. [CrossRef]
41. Manthey, J.; Shield, K.D.; Rylett, M.; Hasan, O.S.M.; Probst, C.; Rehm, J. Global alcohol exposure between 1990 and 2017 and forecasts until 2030: A modelling study. *Lancet* **2019**, *393*, 2493–2502. [CrossRef]
42. English, D.; Holman, C.; Milne, E.; Winter, M.; Hulse, G.; Codde, J.; Bower, C.; Corti, B.; De Klerk, N.; Knuiman, M.; et al. *The Quantification of Drug caused Morbidity and Mortality in Australia 1995*; Australian Government Publishing Service: Canberra, Australia, 1995.
43. Roerecke, M.; Vafaei, A.; Hasan, O.S.M.; Chrystoja, B.R.; Cruz, M.; Lee, R.; Neuman, M.R.; Rehm, J. Alcohol Consumption and Risk of Liver Cirrhosis: A Systematic Review and Meta-Analysis. *Am. J. Gastroenterol.* **2019**, *114*, 1574–1586. [CrossRef] [PubMed]
44. Rehm, J.; Gmel, G.; Sempos, C.T.; Trevisan, M. Alcohol-related morbidity and mortality. *Alcohol Res. Health* **2003**, *27*, 39.
45. Klatsky, A.L.; Udaltsova, N. Alcohol drinking and total mortality risk. *Ann. Epidemiol.* **2007**, *17*, S63–S67. [CrossRef]
46. Rehm, J.; Shield, K.D.; Roerecke, M.; Gmel, G. Modelling the impact of alcohol consumption on cardiovascular disease mortality for comparative risk assessments: An overview. *BMC Public Health* **2016**, *16*, 363. [CrossRef]
47. Bond Huie, S.A.; Krueger, P.M.; Rogers, R.G.; Hummer, R.A. Wealth, race, and mortality. *Soc. Sci. Q.* **2003**, *84*, 667–684. [CrossRef]
48. Feinstein, J.S. The relationship between socioeconomic status and health: A review of the literature. *Milbank Q.* **1993**, *71*, 279–322. [CrossRef]
49. Mackenbach, J.P.; Stirbu, I.; Roskam, A.-J.R.; Schaap, M.M.; Menvielle, G.; Leinsalu, M.; Kunst, A.E. Socioeconomic inequalities in health in 22 European countries. *N. Engl. J. Med.* **2008**, *358*, 2468–2481. [CrossRef]
50. Petrovic, D.; de Mestral, C.; Bochud, M.; Bartley, M.; Kivimäki, M.; Vineis, P.; Mackenbach, J.; Stringhini, S. The contribution of health behaviors to socioeconomic inequalities in health: A systematic review. *Prev. Med.* **2018**, *113*, 15–31. [CrossRef] [PubMed]

51. Probst, C.; Simbayi, L.C.; Parry, C.D.; Shuper, P.A.; Rehm, J. Alcohol use, socioeconomic status and risk of HIV infections. *AIDS Behav.* **2017**, *21*, 1926–1937. [CrossRef]

52. Probst, C.; Roerecke, M.; Behrendt, S.; Rehm, J. Socioeconomic differences in alcohol-attributable mortality compared with all-cause mortality: A systematic review and meta-analysis. *Int. J. Epidemiol.* **2014**, *43*, 1314–1327. [CrossRef] [PubMed]

53. Probst, C.; Kilian, C.; Sanchez, S.; Lange, S.; Rehm, J. The role of alcohol use and drinking patterns in socioeconomic inequalities in mortality: A systematic review. *Lancet Public Health* **2020**, *5*, e324–e332. [CrossRef]

54. Boyd, J.; Sexton, O.; Angus, C.; Meier, P.; Purshouse, R.C.; Holmes, J. Causal mechanisms proposed for the alcohol harm paradox—A systematic review. *Addiction* **2021**. [CrossRef]

55. Shield, K.D.; Rehm, J. Societal development and the alcohol-attributable burden of disease. *Addiction* **2021**. [CrossRef]

56. Babor, T.F.; Casswell, S.; Graham, K.; Huckle, T.; Livingston, M.; Österberg, E.; Rehm, J.; Room, R.; Rossow, I.; Sornpaisarn, B. *Alcohol: No Ordinary Commodity. Research and Public Policy*, 3rd ed.; Oxford University Press: Oxford, UK, 2021, (forthcoming).

57. Anonymous. Tackling poverty in tuberculosis control. *Lancet* **2005**, *366*, 2063. [CrossRef]

58. Rehm, J.; Samokhvalov, A.V.; Neuman, M.G.; Room, R.; Parry, C.; Lönnroth, K.; Patra, J.; Poznyak, V.; Popova, S. The association between alcohol use, alcohol use disorders and tuberculosis (TB). A systematic review. *BMC Public Health* **2009**, *9*, 1–12. [CrossRef]

59. Llamosas-Falcón, L.; Shield, K.D.; Gelovany, M.; Hasan, O.S.; Manthey, J.; Monteiro, M.; Walsh, N.; Rehm, J. Impact of alcohol on the progression of HCV-related liver disease: A systematic review and meta-analysis. *J. Hepatol.* **2021**. [CrossRef]

60. Eng, M.Y.; Luczak, S.E.; Wall, T.L. ALDH2, ADH1B, and ADH1C genotypes in Asians: A literature review. *Alcohol Res. Health* **2007**, *30*, 22.

61. International Agency for Research on Cancer. *International Agency for Research on Cancer. IARC Monographs on the Evaluation of Carcinogenic Risks to Humans: Volume 96—Alcohol Consumption and Ethyl Carbamate*; International Agency for Research on Cancer: Lyon, France, 2010.

62. Roerecke, M.; Shield, K.D.; Higuchi, S.; Yoshimura, A.; Larsen, E.; Rehm, M.X.; Rehm, J. Estimates of alcohol-related oesophageal cancer burden in Japan: Systematic review and meta-analyses. *Bull. World Health Organ.* **2015**, *93*, 329–338. [CrossRef]

63. Baan, R.; Straif, K.; Grosse, Y.; Secretan, B.; El Ghissassi, F.; Bouvard, V.; Altieri, A.; Cogliano, V.; WHO International Agency for Research on Cancer Monograph Working Group. Carcinogenicity of alcoholic beverages. *Lancet Oncol.* **2007**, *8*, 292–293. [CrossRef]

64. Brooks, P.J.; Enoch, M.-A.; Goldman, D.; Li, T.-K.; Yokoyama, A. The alcohol flushing response: An unrecognized risk factor for esophageal cancer from alcohol consumption. *PLoS Med.* **2009**, *6*, e1000050. [CrossRef] [PubMed]

65. Sarich, P.; Canfell, K.; Banks, E.; Paige, E.; Egger, S.; Joshy, G.; Korda, R.; Weber, M. A prospective study of health conditions related to alcohol consumption cessation among 97,852 drinkers aged 45 and over in Australia. *Alcohol. Clin. Exp. Res.* **2019**, *43*, 710–721. [CrossRef]

66. Puddey, I.B.; Rakic, V.; Dimmitt, S.B.; Beilin, L.J. Influence of pattern of drinking on cardiovascular disease and cardiovascular risk factors–a review. *Addiction* **1999**, *94*, 649–663. [CrossRef]

67. Rehm, J.; Roerecke, M. Cardiovascular effects of alcohol consumption. *Trends Cardiovasc. Med.* **2017**, *27*, 534–538. [CrossRef] [PubMed]

68. Fekjær, H.O. Alcohol—A universal preventive agent? A critical analysis. *Addiction* **2013**, *108*, 2051–2057. [CrossRef] [PubMed]

69. Rehm, J. The role of the comparison group in epidemiology and general limitations. *Addiction* **2013**, *108*, 2058–2059. [CrossRef] [PubMed]

70. Del Boca, F.K.; Darkes, J. The validity of self—Reports of alcohol consumption: State of the science and challenges for research. *Addiction* **2003**, *98*, 1–12. [CrossRef]

71. Rehm, J.; Kilian, C.; Rovira, P.; Shield, K.D.; Manthey, J. The elusiveness of representativeness in general population surveys for alcohol. *Drug Alcohol Rev.* **2021**, *40*, 161–165. [CrossRef] [PubMed]

72. Clarke, R.; Shipley, M.; Lewington, S.; Youngman, L.; Collins, R.; Marmot, M.; Peto, R. Underestimation of risk associations due to regression dilution in long-term follow-up of prospective studies. *Am. J. Epidemiol.* **1999**, *150*, 341–353. [CrossRef] [PubMed]

73. Rehm, J.; Marmet, S.; Anderson, P.; Gual, A.; Kraus, L.; Nutt, D.; Room, R.; Samokhvalov, A.V.; Scafato, E.; Trapencieris, M.; et al. Defining substance use disorders: Do we really need more than heavy use? *Alcohol Alcohol.* **2013**, *48*, 633–640. [CrossRef]

74. Gmel, G.; Shield, K.D.; Kehoe-Chan, T.A.; Rehm, J. The effects of capping the alcohol consumption distribution and relative risk functions on the estimated number of deaths attributable to alcohol consumption in the European Union in 2004. *BMC Med. Res. Methodol.* **2013**, *13*, 1–9. [CrossRef]

75. Shield, K.D.; Gmel, G.; Mäkelä, P.; Probst, C.; Room, R.; Gmel, G.; Rehm, J. Risk, individual perception of risk and population health. *Addiction* **2017**, *112*, 2272–2273. [CrossRef]

76. Rossow, I.; Mäkelä, P. Public health thinking around alcohol-related harm: Why does per capita consumption matter? *J. Stud. Alcohol Drugs* **2021**, *82*, 9–17. [CrossRef]

77. Bondy, S.J.; Rehm, J.; Ashley, M.J.; Walsh, G.; Single, E.; Room, R. Low-risk drinking guidelines: The scientific evidence. *Can. J. Public Health* **1999**, *90*, 264–270. [CrossRef]

Review

Alcohol and Cancer: Epidemiology and Biological Mechanisms

Harriet Rumgay [1,*], Neil Murphy [2], Pietro Ferrari [2] and Isabelle Soerjomataram [1]

[1]   Cancer Surveillance Branch, International Agency for Research on Cancer, CEDEX 08, 69372 Lyon, France; SoerjomataramI@iarc.fr

[2]   Nutrition and Metabolism Branch, International Agency for Research on Cancer, CEDEX 08, 69372 Lyon, France; murphyn@iarc.fr (N.M.); ferrarip@iarc.fr (P.F.)

*   Correspondence: RumgayH@students.iarc.fr

Abstract: Approximately 4% of cancers worldwide are caused by alcohol consumption. Drinking alcohol increases the risk of several cancer types, including cancers of the upper aerodigestive tract, liver, colorectum, and breast. In this review, we summarise the epidemiological evidence on alcohol and cancer risk and the mechanistic evidence of alcohol-mediated carcinogenesis. There are several mechanistic pathways by which the consumption of alcohol, as ethanol, is known to cause cancer, though some are still not fully understood. Ethanol's metabolite acetaldehyde can cause DNA damage and block DNA synthesis and repair, whilst both ethanol and acetaldehyde can disrupt DNA methylation. Ethanol can also induce inflammation and oxidative stress leading to lipid peroxidation and further DNA damage. One-carbon metabolism and folate levels are also impaired by ethanol. Other known mechanisms are discussed. Further understanding of the carcinogenic properties of alcohol and its metabolites will inform future research, but there is already a need for comprehensive alcohol control and cancer prevention strategies to reduce the burden of cancer attributable to alcohol.

Keywords: alcohol; acetaldehyde; oxidative stress; inflammation; one carbon metabolism; lipid metabolism; DNA damage; cancer; carcinogenesis

Citation: Rumgay, H.; Murphy, N.; Ferrari, P.; Soerjomataram, I. Alcohol and Cancer: Epidemiology and Biological Mechanisms. *Nutrients* **2021**, *13*, 3173. https://doi.org/10.3390/nu13093173

Academic Editor: Peter Anderson

Received: 23 July 2021
Accepted: 9 September 2021
Published: 11 September 2021

Publisher's Note: MDPI stays neutral with regard to jurisdictional claims in published maps and institutional affiliations.

1. Introduction

Approximately 4% of cancers worldwide are caused by alcohol consumption, equating to more than 740,000 cases of cancer globally in 2020 [1]. The impact of alcohol consumption on cancer burden differs by cancer type, and cancers of the oesophagus, liver, and breast represent the most alcohol-attributable cases of cancer globally (Figure 1). Drinking alcohol even at lower levels of intake can increase the risk of cancer and we previously estimated that over 100,000 cases of cancer in 2020 were caused by light and moderate drinking of the equivalent of around one or two alcoholic drinks per day [1]. Despite this, there is low public awareness of the causal link between alcohol and cancer and alcohol use is growing in several regions of the world [2,3].

More than 30 years ago, in 1988, the International Agency for Research on Cancer (IARC) classified alcoholic beverages as a group 1 carcinogen, the most severe classification [4]. The IARC Monographs program aims to classify cancerous agents according to the strength of the available epidemiological and experimental evidence. Cancers of the oral cavity, pharynx, larynx, oesophagus, and liver were first classified as being causally related to the consumption of alcoholic beverages, and this was expanded to include cancers of the colorectum and female breast in the later monographs on alcoholic beverages in 2010 and 2012, with a positive association observed for cancer of the pancreas [5,6].

The World Cancer Research Fund (WCRF) also conducts classification of physical and dietary components and their potential cancerous effects as part of their Continuous Update Project. The WCRF base their conclusions on the quality of epidemiological evidence and carry out meta-analyses of the association with cancer risk. In the most recent report on Diet, Nutrition, Physical Activity and Cancer, WCRF concluded that there was strong evidence that alcohol consumption increased the risk of cancers of the mouth, pharynx and larynx, oesophagus (squamous cell carcinoma), liver, colorectum, and breast (postmenopausal), with a probable increased risk of stomach cancer and premenopausal breast cancer [7].

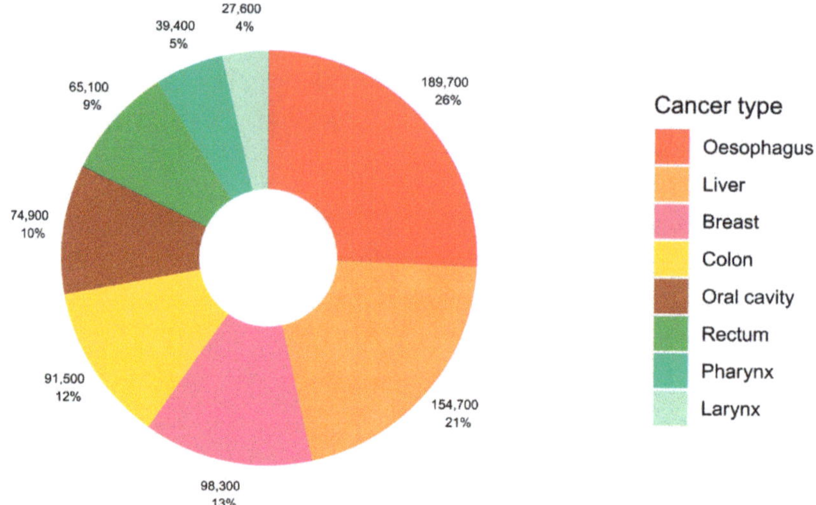

Figure 1. Global number and proportion of cancer cases attributable to alcohol consumption according to cancer type. Source of alcohol-attributable cases: Rumgay and colleagues [1].

In addition to associations from epidemiological studies, multiple mechanistic pathways through which alcohol can cause cancer have been proposed. In this review, we aim to summarise the epidemiological evidence on alcohol and cancer risk and the mechanistic evidence of alcohol-driven carcinogenesis. We searched the PubMed and Cochrane databases for reviews, umbrella reviews, meta-analyses, and Mendelian randomisation studies on total alcohol use and cancer risk and mechanisms of alcohol-related carcinogenesis published up until June 2021. We also searched the WCRF's Continuous Update Project reports for meta-analyses on alcohol consumption and cancer risk.

2. Alcohol and Cancer Risk

The effects of alcohol consumption on cancer risk have been studied for many decades and an association with alcohol has been observed for multiple cancer sites. Here, we discuss evidence from large meta-analyses of observational studies and emerging evidence from Mendelian randomisation studies. Figures 2 and 3 present the dose-response relationships for the risk of cancer at several sites per 10 g/day increase in alcohol consumption from the meta-analyses carried out in the WCRF Continuous Update Project [7], and the risk of cancer at several sites according to three levels of alcohol intake [light (up to 12.5 g/day), moderate (12.5 to 50 g/day), and heavy (more than 50 g/day)] from a meta-analysis conducted by Bagnardi and colleagues [8], both with respect to the reference category of alcohol non-drinkers.

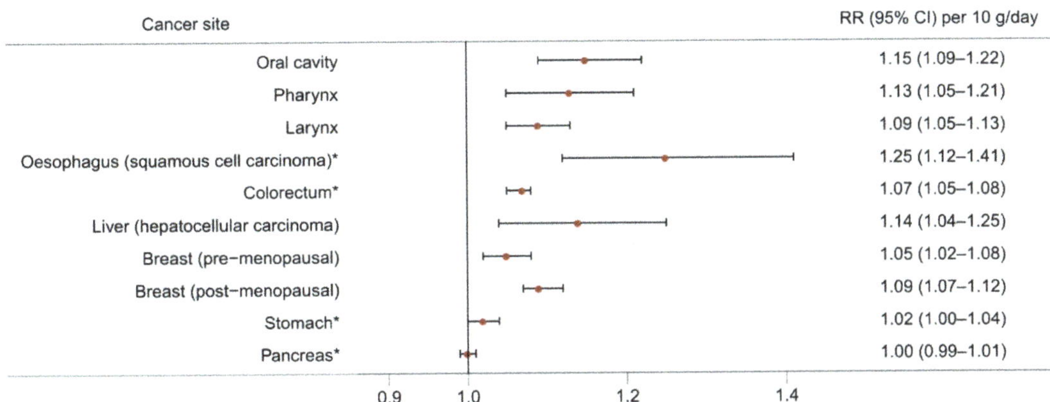

Figure 2. The dose-response relationship for the risk of cancer at different sites per 10 g/day increase in alcohol consumption. Source of relative risk estimates: WCRF Continuous Update Project [7]. RR = Relative risk; CI = Confidence interval. * Non-linear dose-response observed indicating threshold effect.

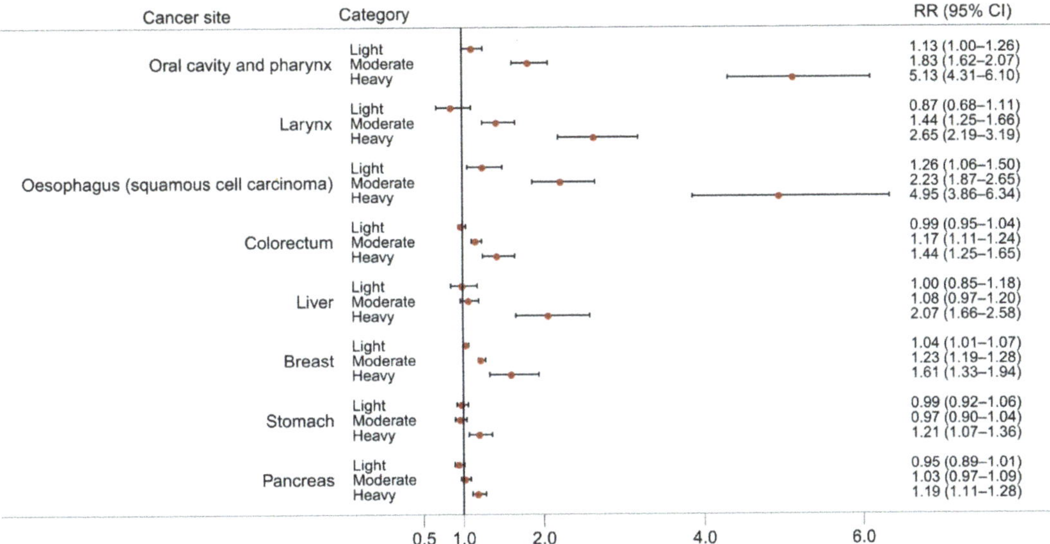

Figure 3. The dose-response relationship for the risk of cancer at different sites by three level of alcohol intake: light (up to 12.5 g/day), moderate (12.5 to 50 g/day), and heavy (more than 50 g/day). Source of relative risk estimates: Bagnardi and colleagues [8]. RR = Relative risk; CI = Confidence interval.

2.1. Oral Cavity Pharyngeal and Laryngeal Cancers

Drinking alcohol increases the risk of cancers of the upper aerodigestive tract. Consumption of 10 g alcohol per day was associated with a 15% increased risk of oral cavity cancer (RR 1.15 (95% CI 1.09–1.22)) in the most recent WCRF Continuous Update Project [7]. Pharyngeal cancer risk was also increased (RR 1.13 (95% CI 1.05–1.21) per 10 g alcohol per day) [7]. In Bagnardi and colleagues' meta-analysis the RR of cancer of the oral cavity and pharynx was increased from 1.13 (95% CI 1.00–1.26) for current light drinking (up to 12.5 g alcohol per day) to 5.13 (95% CI 4.31–6.10) for heavy drinking (more than 50 g per day) [8]. Cancers of the larynx were also observed to have an increased RR (1.09 (95% CI 1.05–1.13) per 10 g alcohol per day) in the WCRF meta-analysis [7]. Bagnardi and colleagues found significant increases in laryngeal cancer risk only in moderate and heavy drinking, with RRs of 1.44 (95% CI 1.25–1.66) and 2.65 (95% CI 2.19–3.19), respectively [8].

2.2. Oesophageal Cancer

Drinking alcohol increases the risk of squamous cell carcinoma of the oesophagus which is the most common histological subtype of oesophageal cancer globally, and contributed the most cases of cancer in 2020 attributable to alcohol (189,700 cases) [1,9]. An excess risk of oesophageal squamous cell carcinoma was found in the WCRF Continuous Update Project (RR 1.25 [95% CI 1.12–1.41] per 10 g alcohol per day) [7], and in Bagnardi and colleagues' meta-analysis, the pooled RR estimates for light and heavy drinking were 1.26 (95% CI 1.06–1.50) and 4.95 (95% CI 3.86–6.34), respectively [8]. There were differences in risk between geographic locations in both meta-analyses, with higher oesophageal squamous cell carcinoma risk among drinkers in studies conducted in Asia than those in North America or Europe. This observation possibly reflects the elevated risk of oesophageal squamous cell carcinoma among carriers of the *ALDH2*2* polymorphism of the gene that codes the enzyme aldehyde dehydrogenase 2 (ALDH2) [10]. The *ALDH2*2* variant allele is more common in Eastern Asian populations and confers nearly four times the risk of oesophageal cancer among drinkers compared with *ALDH2*1* carriers [10]. For oesophageal adenocarcinoma, the second most common histological subtype of oesophageal cancer, no increased risk was observed in the WCRF meta-analysis (RR 1.00 (95% CI 0.98–1.02) per 10 g per day) but an inverse association was found for oesophageal adenocarcinoma and gastric cardia cancer among light drinkers in the meta-analysis by Bagnardi and colleagues (RR 0.86 (95% CI 0.76–0.98)) [7,8].

Cancers of the upper aerodigestive tract can also be characterised as having a more than multiplicative increased risk when alcohol and tobacco are consumed together. This synergistic effect has been observed in several studies; for example a pooled analysis of 11,200 head and neck cancer cases and 16,200 controls found a 14 times risk of head and neck cancers among those who drank at least three alcoholic drinks per day and smoked more than 20 cigarettes per day, compared with never drinkers who had never smoked [11]. For oesophageal squamous cell carcinoma, a cohort study in the Netherlands observed an eight times risk among current smokers who drank 15 g alcohol or more per day, compared with never smokers who consumed less than 5 g alcohol per day [12].

2.3. Colorectal Cancer

The meta-analysis conducted by WCRF found a 7% increased risk of colorectal cancer (RR 1.07 (95% CI 1.05–1.08)) per 10 g alcohol per day [7]. WCRF also found some evidence of a threshold effect around 20 g per day with a weaker association at lower intake levels [7]. The meta-analysis by Bagnardi and colleagues did not find an effect of alcohol on colorectal cancer risk among light drinkers, but the RR increased to 1.17 (95% CI 1.11–1.24) for moderate drinking, and 1.44 (95% CI 1.25–1.65) for heavy drinking [8]. Differences between subsites were minimal, with the risk of colon cancer (RR 1.07 95% CI 1.05–1.09) similar to rectal cancer (RR 1.08 95% CI 1.07–1.10) [7]. Alcohol might also increase the risk of precancerous lesions in the colon, with a meta-analysis reporting a 27% increased risk of colorectal adenoma (RR 1.27 (95% CI 1.17–1.37)) per 25 g alcohol per day [13].

2.4. Liver Cancer

The most common histological subtype of liver cancer is hepatocellular carcinoma (HCC) and around 154,700 cases of HCC in 2020 were attributable to alcohol consumption [1]. When restricted to HCC only, meta-analysis of WCRF sources resulted in a 14% increased risk of HCC (RR 1.14 (95% CI 1.04–1.25)) per 10 g alcohol per day [7]. However, a possible threshold effect was observed in the non-linear dose-response analysis by WCRF, where less than 45 g alcohol per day did not significantly increase the risk of liver cancer. This was similar to the findings of Bagnardi and colleagues where light or moderate drinking did not significantly increase liver cancer risk but risk among heavy drinkers doubled (RR 2.07 (95% CI 1.66–2.58)) [8].

2.5. Breast Cancer

Female breast cancer is the most commonly diagnosed cancer globally and contributed the third largest number of alcohol-attributable cases in 2020 (98,300 cases) [1,14]. The WCRF found a 7% increased risk of breast cancer per 10 g alcohol per day (95% CI 1.05–1.09) [7]. Whether there is a difference in breast cancer risk by menopausal status is unclear, as risk of postmenopausal breast cancer overlapped with that of premenopausal breast cancer in the WCRF meta-analysis (RR 1.09 (95% CI 1.07–1.12) versus RR 1.05 (95% CI 1.02–1.08), respectively, per 10 g alcohol per day). It does, however, seem that risk of breast cancer among drinkers might be specific to hormone receptor status; the WCRF meta-analysis of postmenopausal women observed an excess risk of oestrogen-receptor-positive and progesterone receptor-positive (ER$^+$PR$^+$) tumours (RR 1.06 (95% CI 1.03–1.09)) and ER$^+$PR$^-$ tumours (RR 1.12 (95% CI 1.01–1.24)) per 10 g alcohol per day, and no significant association was observed for ER$^-$PR$^-$ tumours (RR 1.02 95% CI 0.98–1.06) [7]. In a meta-analysis by Sun and colleagues, current drinkers had an increased risk of all hormone receptor status breast tumours compared with never drinkers, but RRs were higher for ER$^+$PR$^+$ tumours (RR 1.40 95% CI 1.30–1.51) and ER$^+$PR$^-$ tumours (RR 1.39 95% CI 1.12–1.71) than ER$^-$PR$^-$ tumours (RR 1.21 95% CI 1.02–1.43) [15].

2.6. Stomach Cancer

Alcohol consumption might increase the risk of stomach cancer. The linear dose-response meta-analysis by WCRF resulted in a non-significant RR of 1.02 (95% CI 1.00–1.04) per 10 g alcohol per day, but the non-linear dose-response analysis found an increase in stomach cancer risk for intakes over 45 g alcohol per day [7]. The meta-analysis by Bagnardi and colleagues observed a 21% increased risk in heavy drinking (RR 1.21 95% CI 1.07–1.36), and no significant increase in light or moderate drinking categories [8].

2.7. Pancreatic Cancer

The meta-analysis by WCRF did not find an increased risk of pancreatic cancer per 10 g alcohol per day (RR 1.00 (95% CI 0.99–1.01)) but there was a possible threshold effect of increased risk for intakes of around 60 g per day (RR 1.17 (95% CI 1.05–1.29)) [7]. This was a similar finding to the meta-analysis by Bagnardi and colleagues which found no increased risk at light or moderate drinking but a significant RR of 1.19 (95% 1.11–1.28) for heavy drinking [8].

2.8. Other Cancer Types

The association between alcohol drinking and risk of other cancer types has been studied but without sufficient evidence to be classified in the IARC monographs or WCRF Continuous Update Project. Positive associations have been reported in some meta-analyses; for example, a 3% increase in lung cancer risk was observed per 10 g alcohol per day in the WCRF meta-analysis based on 28 studies (RR 1.03 (95% CI 1.01–1.04)) after excluding studies which did not control for smoking [7]. A positive association with lung cancer was only found for heavy drinkers in Bagnardi and colleagues' meta-analysis, but this was probably due to residual confounding from smoking because alcohol use did not increase the risk of lung cancer among non-smokers [8]. Little evidence of an association between alcohol consumption and gallbladder cancer was found in the WCRF Continuous Update Project, but Bagnardi and colleagues found an excess risk of gallbladder cancer among heavy drinkers (RR 2.64 (95% CI 1.62–4.30)). WCRF found an elevated risk of malignant melanoma per 10 g alcohol per day (RR 1.08 (95% CI 1.03–1.13)), but no effect on basal cell carcinoma (RR 1.04 (95% CI 0.99–1.10)) or squamous cell carcinoma (RR 1.03 (95% CI 0.97–1.09)) risk [7]. An increased risk of prostate cancer was observed for light and moderate drinking in Bagnardi and colleagues' meta-analysis but not in the dose-response analysis of one drink per day by WCRF [7,8].

WCRF found an inverse association between alcohol consumption and kidney cancer risk (RR 0.92 (95% CI 0.86–0.97) per 10 g per day) [7]. However, this association was restricted to light and moderate drinking in Bagnardi and colleagues' meta-analysis (RR 0.92 (95% CI 0.86–0.99) and 0.79 (95% CI 0.72–0.86), respectively) [8]. The same meta-analysis also found significant inverse associations for the risk of thyroid cancer, Hodgkin lymphoma and non-Hodgkin lymphoma [8].

2.9. Confirming the Causal Relation Reported in Observational Studies

Many observational studies have been conducted to identify and define the risks from drinking alcohol and cancer development. Some limitations in these studies have been identified, such as lack of sufficient adjustment of confounding factors, for example tobacco smoking and alcohol consumption are both common risk factors for oral cavity cancer. There are also concerns around reverse causality, with the reference categories of alcohol non-drinkers possibly including former drinkers who still have an elevated risk of cancer. There are other concerns over the accuracy of recording of alcohol exposure data where bias may be incorporated through non-participation of heavy drinkers in health studies, and under-reporting of alcohol consumption by the study subjects.

One method which might overcome some of the limitations in observational studies is Mendelian randomisation (MR), which uses genetic variants to explore the causal relationship between exposure and disease outcome. Assuming that analyses are conducted appropriately, due to the random distribution of these genetic variants at birth, MR studies should be less prone to conventional confounding and reverse causality.

For oral and oropharyngeal cancer, an MR study using genetic data on 6000 oral or oropharyngeal cancer cases and 6600 controls found a positive causal effect of alcohol consumption independent of smoking [16]. The authors concluded that previous estimates of the association between alcohol and oral and oropharyngeal cancer from observational studies may have been underestimated [16]. Another MR study on UK Biobank data found that drinking alcohol, especially above the UK's low-risk guideline of up to 14 units per week, was causally related with head and neck cancers, but not breast cancer [17]. A further updated MR study using UK Biobank data did not find an association between alcohol exposure and cancer of any site, though they noted limitations of a lack of precision in their analyses due to low variance explained by the single nucleotide polymorphisms [18]. An MR analysis by Ong and colleagues found no significant increase in breast cancer risk per genetically predicted drink per day (odds ratio 1.00 (95% CI 0.93–1.08)) [19].

The future potential of MR studies is yet to be discovered but disclosing potential sources of biases and confounding in observational studies is necessary to obtain robust estimates of the causal relationship between alcohol consumption and cancer risk.

3. Mechanisms of Alcohol-Driven Carcinogenesis

Following epidemiological evidence of the link between alcohol use and risk of cancer at multiple sites, several pathways have been investigated to explain the carcinogenic effects of alcohol. Here, we discuss the key mechanisms linking alcohol consumption to carcinogenesis, which are depicted in Figure 4.

3.1. Production of Acetaldehyde

Once consumed, alcohol is metabolised by enzymes including alcohol dehydrogenase (ADH), cytochrome P-450 2E1 (CYP2E1) and bacterial catalase, producing acetaldehyde [20]. Acetaldehyde is highly reactive towards DNA and has several carcinogenic and genotoxic properties.

As it is highly reactive towards DNA, acetaldehyde may bind to DNA to form DNA adducts which alter its physical shape and potentially block DNA synthesis and repair [21]. These DNA adducts are particularly genotoxic as they can induce DNA point mutations, double-strand breaks, sister chromatid exchanges, and structural changes to chromosomes [21,22]. The DNA adducts in question include N2-ethylidene-2′-deoxyguanosine, N2-ethyl-2′-deoxyguanosine, N2-propano-2′-deoxyguanosine (PdG), and N2-etheno-2′-deoxyguanosine [23]. The PdG adduct may form additional highly genotoxic structures such as DNA-protein cross-links and DNA interstrand cross-links which may confer carcinogenesis [24]. As well as DNA-protein cross-links, acetaldehyde may also bind to proteins directly causing structural and functional changes [21]; these proteins include glutathione, a protein involved in reducing oxidative stress caused by alcohol, and enzymes which contribute to DNA repair and methylation, among others.

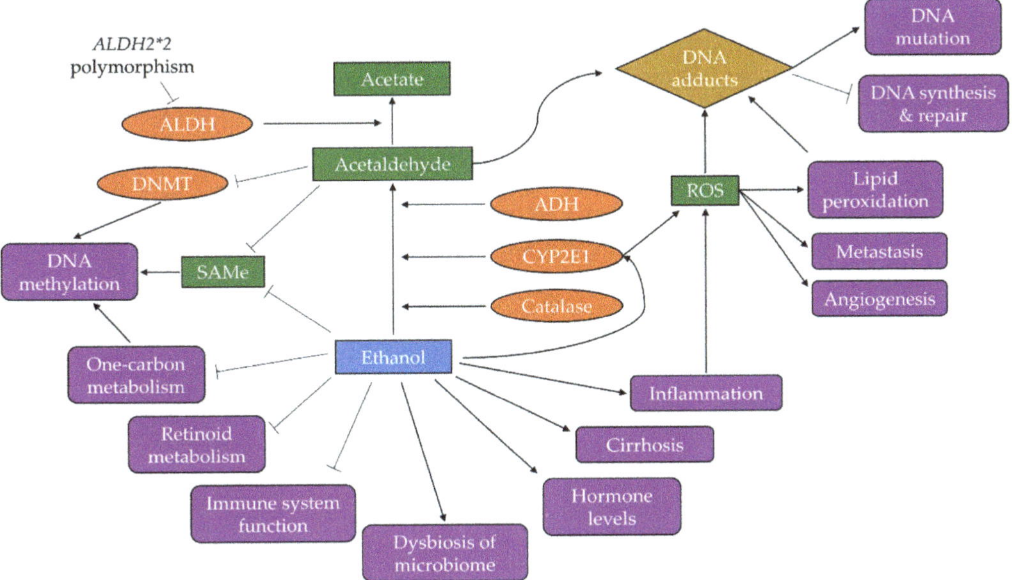

Figure 4. A simplification of the pathways by which alcohol, as ethanol, might drive carcinogenesis. The enzymes alcohol dehydrogenase (ADH), cytochrome P-450 2E1 (CYP2E1), and catalase metabolise ethanol to acetaldehyde; acetaldehyde dehydrogenase (ALDH) enzymes then metabolise acetaldehyde to acetate but common polymorphisms can reduce ALDH activity. Acetaldehyde forms DNA adducts causing mutations and blocking DNA synthesis and repair. Both ethanol and acetaldehyde can disrupt DNA methylation by inhibiting S-adenosyl-L-methionine (SAMe) synthesis and DNA methyltransferase (DNMT) activity, and ethanol can impair one-carbon metabolism. Cytochrome P-450 2E1 (CYP2E1) activity produces reactive oxygen species (ROS) leading to lipid peroxidation, metastasis, angiogenesis, and further formation of DNA adducts. Ethanol can also induce inflammation leading to production of ROS and their downstream effects. Retinoid metabolism and the normal function of the immune system are both impaired by ethanol, while ethanol may lead to increases in sex hormone levels, as well as dysbiosis of the microbiome and liver cirrhosis.

Both acetaldehyde and ethanol can impact DNA methylation which may lead to changes in the expression of oncogenes and tumour-suppressor genes [21]. Acetaldehyde can inhibit the activity of DNA methyltransferase (DNMT) which is essential for normal DNA methylation; acetaldehyde can also reduce DNMT mRNA levels leading to less production of DNMT [25]. Acetaldehyde and ethanol may also inhibit the synthesis of S-adenosyl-L-methionine (SAMe) which is essential to DNA methylation [21].

Acetaldehyde is not the end-product of ethanol metabolism, however, as under normal conditions, acetaldehyde dehydrogenase (ALDH) enzymes convert acetaldehyde to acetate. The group of ALDH enzymes contains ALDH1A1, ALDH2, and ALDH1B1, with ALDH2 being responsible for the majority of acetaldehyde oxidation in the liver. A common polymorphism of this enzyme is the *ALDH2*2* variant allele which dramatically reduces the activity of ALDH2 [10]. It is estimated that between 28% and 45% of East-Asian populations are carriers of the *ALDH2*2* allele [10], while the proportion is considerably lower among Caucasians. In carriers of this polymorphism, acetaldehyde is not metabolised quickly enough, leading to an accumulation of acetaldehyde and thus the prolonged possibility to exert its described genotoxic effects. Evidence shows that alcohol drinkers who carry the *ALDH2*2* variant allele have a substantially increased risk of cancers of the oesophagus and the upper aerodigestive tract [10], thus implicating the effects of acetaldehyde not only in the liver.

3.2. Induction of Oxidative Stress

Ethanol can also contribute to carcinogenesis through the induction of oxidative stress which is recognised as a key determinant of disease initiation [26]. Oxidative stress can be induced by activation of certain pathways which produce reactive oxygen species (ROS) such as superoxide anion and hydrogen peroxide. One pathway by which ethanol achieves this is through increased CYP2E1 activity which produces high quantities of ROS whilst oxidising ethanol to acetaldehyde [27]. Heavy alcohol use has been shown to increase *CYP2E1* expression in the oesophagus [27]. Other sources of ROS during ethanol metabolism include the mitochondrial respiratory chain and some cytosolic enzymes [28].

As ROS are highly reactive, their presence can lead to lipid peroxidation producing aldehydes which can bind to DNA forming etheno-DNA adducts [29,30]. These ethe-DNA adducts, namely 1,N6-ethenodeoxyadenosine and 3,N4-ethenodeoxycytidine, are highly mutagenic as they lead to mutations in several genes involved in key cell cycle regulation and tumour suppression [21]. Linhart and colleagues were able to demonstrate correlation between the amount of CYP2E1 and etheno-DNA adducts in cell, animal, and human tissue models, and highlighted their major importance in ethanol-mediated carcinogenesis in the liver, colorectum, and oesophagus, as well as other tissues [30].

Presence of ROS can also lead to changes in cell cycle behaviour. ROS can act as messengers in intracellular signalling pathways to activate the transcription factor nuclear factor κB (NF-κB). ROS can further promote cell proliferation and metastasis by interfering with mitogen-activated protein kinase signalling pathways and upregulating vascular endothelial growth factor (VEGF) and monocyte chemotactic protein-1 (MCP-1) which can stimulate angiogenesis [31]. In HCC tissue samples from alcohol drinkers, ROS accumulation and increased synthesis of VEGF, MCP-1 and NF-κB were observed, indicating alcohol-driven promotion and progression of HCC [32].

3.3. Increased Inflammation

Inflammation is a key pathway to cancer progression at several sites and is enhanced by alcohol use. Chronic alcohol consumption can recruit specific white blood cells (monocytes and macrophages) to the tumour microenvironment. These white blood cells produce pro-inflammatory cytokines, such as tumour necrosis factor α (TNF-α) and the interleukins IL-1, IL-6, and IL-8 [31,33], which activate oxidant-generating enzymes leading to downstream formation of ROS [30]. NF-κB is also activated by these cytokines, stimulating further ROS-producing enzymes.

In addition to its involvement in downstream ROS-producing pathways, it is hypothesised that IL-8 contributes to further accumulation of white blood cells (neutrophils, specifically) in the liver leading to acute inflammation. Elevated IL-8 levels have been found in patients with acute liver injury such as alcoholic hepatitis [34]. Additionally, the cytokine IL-6 stimulates production of the anti-apoptotic protein Mcl-1, thus avoiding cell death and exposing the cell to further DNA damage [35].

3.4. Disruption to One-Carbon Metabolism and Folate Absorption

There is mounting evidence that alcohol can negatively affect one-carbon metabolism which is essential for DNA methylation and DNA synthesis [25]. Ethanol and acetaldehyde can reduce the activity of enzymes involved in one-carbon metabolism that regulate DNA methylation, namely methionine synthase, methionine adenosyl transferase and DNMT, thus dysregulating epigenetic patterns and resulting in DNA hypomethylation [20].

Lipotropic nutrients such as folate are key sources of the methyl groups necessary for DNA methylation and influence the availability of SAMe, which is also essential to DNA methylation [25]. Alcohol intake may deplete folate levels, or indeed be a cause of folate and B vitamin deficiency if alcohol constitutes the majority of calories consumed, as observed in malnourished alcoholics [21,26]. Folate deficiency affects the availability of nucleotides needed for DNA synthesis leading to accumulation of deoxyuridine monophosphate which is incorporated into new DNA molecules causing double-strand breaks and chromosomal damage [25]. Interestingly, there is evidence that higher folate intake among alcohol drinkers may attenuate the increased risk of liver cancer mortality compared with those with low folate intake [36]. This attenuation was also observed for risk of postmenopausal breast cancer among women who drink alcohol and have higher folate levels [37]. The effect of alcohol on one-carbon metabolism and folate might also be important in colorectal cancer development [20].

3.5. Altered Retinoid Metabolism

Retinoids are important regulators against carcinogenesis as they can induce cell growth, cell differentiation, and apoptosis [31]. Alcohol can alter retinoid metabolism by inhibiting the oxidation of vitamin A to retinoic acid [21]. Alcohol increases CYP2E1 activity (Section 3.2) which also functions to metabolise retinoic acid resulting in the production of toxic metabolites [21]. This increased toxicity of retinoids may explain the observation of excess lung cancer risk in smokers who took β-carotene supplements and consumed 11 g or more of ethanol per day in the α-tocopherol, β-carotene cancer prevention study (ATBC trial) study [21].

Chronic alcohol consumption has been linked with decreased levels of retinoids in the liver [21], and low levels of retinol in the blood have been linked with higher risk of head and neck cancers [31]. Retinoids may also play a role in other signalling pathways implicated in cancer development, such as oestrogen and breast cancer [31].

3.6. Changes to Oestrogen Regulation

Alcohol might interfere with oestrogen pathways by increasing hormone levels and enhancing the activity of ERs, important in breast carcinogenesis [38]. Sex hormone levels may be increased by alcohol through oxidative stress and through inhibition of the steroid degradation enzymes sulfotransferase and 2-hydroxylase [39]. Heavy use of alcohol has also been linked with increased circulating levels of oestrone and oestradiol as well as dehydroepiandrosterone sulphate (DHEAS) [39]. DHEAS is metabolised to oestrogen by aromatase, the activity of which is also increased in chronic alcohol consumers [40]. A large cohort study found DHEAS levels 25% higher among women consuming at least 20 g alcohol per day compared with non-drinkers [41]. However, some of the associations among alcohol drinking premenopausal women were limited to those taking oral contraceptives [40]. Despite limited evidence of mediation of the association between alcohol and breast cancer by individual sex hormones, a case-control study nested within EPIC found that a hormonal signature reflecting lower levels of sex-hormone binding globulin and higher levels of sex hormones mediated 24% of the association, suggesting that an interplay of hormones may contribute to alcohol-mediated breast cancer development [42].

ERs are important transcription factors within cells and may provide the main pathway by which alcohol promotes breast tumour growth [40]. Elevated concentrations of oestrogen due to alcohol use may lead to increased transcriptional activity of ER (up to 15 times higher than normal activity), resulting in proliferation of ER$^+$ cells [39].

3.7. Reduced Function of the Immune System

Alcohol has multiple negative effects on the host immune system. Firstly, alcohol can disrupt the production of proteins such as perforin and granzymes A and B, which are necessary for natural killer (NK) cells to function in targeting and destroying potentially cancerous cells [33]. Alcohol can block NK cells from being released from the bone marrow [31]. Alcohol can also activate NKT cells which are associated with liver injury and hepatocyte apoptosis [33]. Additionally, alcohol may suppress T cell immune responses therefore decreasing the anti-tumour regulation of the immune system.

With the immune system being compromised, alcohol consumption can exacerbate damage from viral infections such as hepatitis C virus, which is common among chronic alcoholic liver disease patients [43]. In addition, heavy episodic alcohol use might reduce the immune system's defence against infection by disrupting the production of pro-inflammatory cytokines and increasing the expression of anti-inflammatory cytokines [33]. This is contrary to the increased expression of pro-inflammatory cytokines due to chronic alcohol exposure as discussed with other evidence on alcohol-induced inflammation (Section 3.3).

3.8. Dysbiosis of the Microbiome

Microbiota in the oral cavity metabolise ethanol to acetaldehyde by the enzyme catalase. However, these bacteria have limited capacity to break acetaldehyde down further into its non-harmful compound acetate, thus the oral epithelia are further exposed to acetaldehyde [21,44]. Acetaldehyde concentrations in the saliva of drinkers are between 10 and 100 times higher than in the blood; this is further doubled in smokers who drink alcohol as tobacco smoke contains high levels of acetaldehyde [21].

Increased ethanol consumption can induce microbial dysbiosis and bacterial overgrowth in the intestine [20]. This heightened bacterial presence may compromise the intestinal barrier resulting in "gut leakiness" where the permeability of the intestinal lumen is high enough such that bacterial products including lipopolysaccharides and peptidoglycan move from the intestine into the blood [20,45]. Once in the blood these bacterial products easily reach the liver where a variety of cells are activated (endothelial cells, liver macrophages, stellate cells and hepatocytes) producing a chronic inflammatory environment [33], which may confer an increased risk of liver cancer [46].

3.9. Liver Cirrhosis

Liver cirrhosis is a well-recognised pathway to hepatocellular carcinoma development in heavy alcohol users and manifests as pre-neoplastic lesions in the liver [47]. Chronic alcohol exposure is associated with reduced expression of the cytokine interferon-γ which is an inhibitor of liver fibrosis [33]. Furthermore, ROS (Section 3.2) may trigger the production of pro-fibrotic cytokines and collagen in liver cells [28].

3.10. Activation of Other Carcinogens

There is further hypothesis that alcohol consumption might activate the pathways of other carcinogenic agents; this could occur through the alcohol-induced activity of CYP2E1 which may metabolise pro-carcinogens in tobacco smoke and industrial chemicals [21]. It is also possible that ethanol might aid these carcinogens to penetrate cells, especially those of the mucosa of the upper aerodigestive tract [21,48], where tobacco and alcohol have a synergistic effect on the risk of cancer [11,12].

4. Conclusions

Alcohol and its metabolite acetaldehyde can drive cancer development through several pathways. Many of these pathways are interlinked and show the complexity and breadth of alcohol's harmful potential. For example, inflammation can result in oxidative stress, but inflammation is a reaction by the immune system which is itself compromised by alcohol use. Furthermore, DNA damage can occur through exposure to acetaldehyde and ROS which are both produced through CYP2E1 activity, with acetaldehyde also a product of ADH activity. Other potential pathways have been proposed including the dysregulation of carnitine metabolism [49]. We have only covered carcinogenesis in this review, but alcohol likely alters, through these pathways and others, other functions in the body which render it more susceptible to other diseases and injuries, as discussed in other articles in this Special Issue.

Alcohol consumption is a well-established risk factor for cancer and has been linked to cancers of the oral cavity and pharynx, oesophagus, liver, colorectum and breast. While studies have provided evidence on alcohol's carcinogenic potential, further understanding of alcohol's pathways to cancer development will inform the direction of future research. This information is useful to corroborate existing evidence, develop chemoprevention strategies, and could improve cancer therapy, but there is already a wealth of evidence to support the need for further alcohol control and cancer prevention efforts. We have discussed evidence on mechanistic and epidemiological research in the field, and this information must be used to decrease the burden of cancers, as well as other diseases and injuries, attributable to alcohol.

Author Contributions: Conceptualization, H.R. and I.S.; writing—original draft preparation, H.R.; review and editing, H.R., I.S., N.M. and P.F. All authors have read and agreed to the published version of the manuscript. Where authors are identified as personnel of the International Agency for Research on Cancer/World Health Organization, the authors alone are responsible for the views expressed in this article and they do not necessarily represent the decisions, policy, or views of the International Agency for Research on Cancer/World Health Organization.

Funding: The work by H.R. reported in this paper was undertaken during a PhD studentship at the International Agency for Research on Cancer.

Institutional Review Board Statement: Not applicable.

Informed Consent Statement: Not applicable.

Conflicts of Interest: The authors declare no conflict of interest.

References

1. Rumgay, H.; Shield, K.; Charvat, H.; Ferrari, P.; Sornpaisarn, B.; Obot, I.; Islami, F.; Lemmens, V.E.P.P.; Rehm, J.; Soerjomataram, I. Global burden of cancer in 2020 attributable to alcohol consumption: A population-based study. *Lancet Oncol.* **2021**, *22*, 1071–1080. [CrossRef]
2. Buykx, P.; Li, J.; Gavens, L.; Hooper, L.; Lovatt, M.; Gomes de Matos, E.; Meier, P.; Holmes, J. Public awareness of the link between alcohol and cancer in England in 2015: A population-based survey. *BMC Public Health* **2016**, *16*, 1194. [CrossRef]
3. Manthey, J.; Shield, K.D.; Rylett, M.; Hasan, O.S.M.; Probst, C.; Rehm, J. Global alcohol exposure between 1990 and 2017 and forecasts until 2030: A modelling study. *Lancet* **2019**, *393*, 2493–2502. [CrossRef]
4. International Agency for Research on Cancer. *IARC Monographs on the Evaluation of Carcinogenic Risks to Humans. Volume 44. Alcohol Drinking*; 1988; Volume 44. Available online: https://publications.iarc.fr/Book-And-Report-Series/Iarc-Monographs-On-The-Identification-Of-Carcinogenic-Hazards-To-Humans/Alcohol-Drinking-1988 (accessed on 20 June 2021).
5. International Agency for Research on Cancer. *IARC Monographs on the Evaluation of Carcinogenic Risks to Humans. Volume 96. Alcohol Consumption and Ethyl Carbamate*; 2010; Volume 96. Available online: https://publications.iarc.fr/Book-And-Report-Series/Iarc-Monographs-On-The-Identification-Of-Carcinogenic-Hazards-To-Humans/Alcohol-Consumption-And-Ethyl-Carbamate-2010 (accessed on 20 June 2021).
6. International Agency for Research on Cancer. *IARC Monographs on the Evaluation of Carcinogenic Risks to Humans. Volume 100E. Personal Habits and Indoor Combustions*; 2012. Available online: https://publications.iarc.fr/Book-And-Report-Series/Iarc-Monographs-On-The-Identification-Of-Carcinogenic-Hazards-To-Humans/Personal-Habits-And-Indoor-Combustions-2012 (accessed on 20 June 2021).
7. World Cancer Research Fund/American Institute for Cancer Research. *Diet, Nutrition, Physical Activity and Cancer: A Global Perspective*; Continuous Update Project Expert Report; 2018. Available online: https://www.wcrf.org/wp-content/uploads/2021/02/Summary-of-Third-Expert-Report-2018.pdf (accessed on 20 June 2021).
8. Bagnardi, V.; Rota, M.; Botteri, E.; Tramacere, I.; Islami, F.; Fedirko, V.; Scotti, L.; Jenab, M.; Turati, F.; Pasquali, E.; et al. Alcohol consumption and site-specific cancer risk: A comprehensive dose–response meta-analysis. *Br. J. Cancer* **2014**, *112*, 580. [CrossRef]
9. Arnold, M.; Ferlay, J.; van Berge Henegouwen, M.I.; Soerjomataram, I. Global burden of oesophageal and gastric cancer by histology and subsite in 2018. *Gut* **2020**, *69*, 1564–1571. [CrossRef] [PubMed]
10. Chang, J.S.; Hsiao, J.-R.; Chen, C.-H. ALDH2 polymorphism and alcohol-related cancers in Asians: A public health perspective. *J. Biomed. Sci.* **2017**, *24*, 19. [CrossRef] [PubMed]

11. Hashibe, M.; Brennan, P.; Chuang, S.-C.; Boccia, S.; Castellsague, X.; Chen, C.; Curado, M.P.; Dal Maso, L.; Daudt, A.W.; Fabianova, E.; et al. Interaction between tobacco and alcohol use and the risk of head and neck cancer: Pooled analysis in the International Head and Neck Cancer Epidemiology Consortium. *Cancer Epidemiol. Biomark. Prev.* **2009**, *18*, 541–550. [CrossRef]

12. Steevens, J.; Schouten, L.J.; Goldbohm, R.A.; van den Brandt, P.A. Alcohol consumption, cigarette smoking and risk of subtypes of oesophageal and gastric cancer: A prospective cohort study. *Gut* **2010**, *59*, 39–48. [CrossRef] [PubMed]

13. Ben, Q.; Wang, L.; Liu, J.; Qian, A.; Wang, Q.; Yuan, Y. Alcohol drinking and the risk of colorectal adenoma: A dose–response meta-analysis. *Eur. J. Cancer Prev.* **2015**, *24*, 286–295. [CrossRef]

14. Ferlay, J.; Ervik, M.; Lam, F.; Colombet, M.; Mery, L.; Piñeros, M.; Znaor, A.; Soerjomataram, I. Global Cancer Observatory: Cancer Today. Available online: https://gco.iarc.fr/today (accessed on 15 December 2020).

15. Sun, Q.; Xie, W.; Wang, Y.; Chong, F.; Song, M.; Li, T.; Xu, L.; Song, C. Alcohol Consumption by Beverage Type and Risk of Breast Cancer: A Dose-Response Meta-Analysis of Prospective Cohort Studies. *Alcohol Alcohol.* **2020**, *55*, 246–253. [CrossRef] [PubMed]

16. Gormley, M.; Dudding, T.; Sanderson, E.; Martin, R.M.; Thomas, S.; Tyrrell, J.; Ness, A.R.; Brennan, P.; Munafò, M.; Pring, M.; et al. A multivariable Mendelian randomization analysis investigating smoking and alcohol consumption in oral and oropharyngeal cancer. *Nat. Commun.* **2020**, *11*, 6071. [CrossRef]

17. Ingold, N.; Amin, H.A.; Drenos, F. Alcohol causes an increased risk of head and neck but not breast cancer in individuals from the UK Biobank study: A Mendelian randomisation analysis. *medRxiv* **2019**, 19002832. [CrossRef]

18. Larsson, S.C.; Carter, P.; Kar, S.; Vithayathil, M.; Mason, A.M.; Michaëlsson, K.; Burgess, S. Smoking, alcohol consumption, and cancer: A mendelian randomisation study in UK Biobank and international genetic consortia participants. *PLoS Med.* **2020**, *17*, e1003178. [CrossRef] [PubMed]

19. Ong, J.-S.; Derks, E.M.; Eriksson, M.; An, J.; Hwang, L.-D.; Easton, D.F.; Pharoah, P.P.; Berchuck, A.; Kelemen, L.E.; Matsuo, K.; et al. Evaluating the role of alcohol consumption in breast and ovarian cancer susceptibility using population-based cohort studies and two-sample Mendelian randomization analyses. *Int. J. Cancer* **2021**, *148*, 1338–1350. [CrossRef] [PubMed]

20. Rossi, M.; Jahanzaib Anwar, M.; Usman, A.; Keshavarzian, A.; Bishehsari, F. Colorectal Cancer and Alcohol Consumption-Populations to Molecules. *Cancers* **2018**, *10*, 38. [CrossRef] [PubMed]

21. Seitz, H.K.; Stickel, F. Molecular mechanisms of alcohol-mediated carcinogenesis. *Nat. Rev. Cancer* **2007**, *7*, 599–612. [CrossRef]

22. Garaycoechea, J.I.; Crossan, G.P.; Langevin, F.; Mulderrig, L.; Louzada, S.; Yang, F.; Guilbaud, G.; Park, N.; Roerink, S.; Nik-Zainal, S.; et al. Alcohol and endogenous aldehydes damage chromosomes and mutate stem cells. *Nature* **2018**, *553*, 171–177. [CrossRef] [PubMed]

23. Wang, M.; McIntee, E.J.; Cheng, G.; Shi, Y.; Villalta, P.W.; Hecht, S.S. Identification of DNA Adducts of Acetaldehyde. *Chem. Res. Toxicol.* **2000**, *13*, 1149–1157. [CrossRef]

24. Brooks, P.J.; Theruvathu, J.A. DNA adducts from acetaldehyde: Implications for alcohol-related carcinogenesis. *Alcohol* **2005**, *35*, 187–193. [CrossRef]

25. Varela-Rey, M.; Woodhoo, A.; Martinez-Chantar, M.-L.; Mato, J.M.; Lu, S.C. Alcohol, DNA methylation, and cancer. *Alcohol Res.* **2013**, *35*, 25–35.

26. Barve, S.; Chen, S.-Y.; Kirpich, I.; Watson, W.H.; McClain, C. Development, Prevention, and Treatment of Alcohol-Induced Organ Injury: The Role of Nutrition. *Alcohol Res.* **2017**, *38*, 289–302.

27. Millonig, G.; Wang, Y.; Homann, N.; Bernhardt, F.; Qin, H.; Mueller, S.; Bartsch, H.; Seitz, H.K. Ethanol-mediated carcinogenesis in the human esophagus implicates CYP2E1 induction and the generation of carcinogenic DNA-lesions. *Int. J. Cancer* **2011**, *128*, 533–540. [CrossRef]

28. Albano, E. Alcohol, oxidative stress and free radical damage. *Proc. Nutr. Soc.* **2006**, *65*, 278–290. [CrossRef]

29. Haorah, J.; Ramirez, S.H.; Floreani, N.; Gorantla, S.; Morsey, B.; Persidsky, Y. Mechanism of alcohol-induced oxidative stress and neuronal injury. *Free. Radic. Biol. Med.* **2008**, *45*, 1542–1550. [CrossRef] [PubMed]

30. Linhart, K.; Bartsch, H.; Seitz, H.K. The role of reactive oxygen species (ROS) and cytochrome P-450 2E1 in the generation of carcinogenic etheno-DNA adducts. *Redox Biol.* **2014**, *3*, 56–62. [CrossRef] [PubMed]

31. Ratna, A.; Mandrekar, P. Alcohol and Cancer: Mechanisms and Therapies. *Biomolecules* **2017**, *7*, 61. [CrossRef]

32. Wang, F.; Yang, J.-L.; Yu, K.-K.; Xu, M.; Xu, Y.-Z.; Chen, L.; Lu, Y.-M.; Fang, H.-S.; Wang, X.-Y.; Hu, Z.-Q.; et al. Activation of the NF-κB pathway as a mechanism of alcohol enhanced progression and metastasis of human hepatocellular carcinoma. *Mol. Cancer* **2015**, *14*, 10. [CrossRef]

33. Molina, P.E.; Happel, K.I.; Zhang, P.; Kolls, J.K.; Nelson, S. Focus on: Alcohol and the immune system. *Alcohol Res. Health* **2010**, *33*, 97–108.

34. Zimmermann, H.W.; Seidler, S.; Gassler, N.; Nattermann, J.; Luedde, T.; Trautwein, C.; Tacke, F. Interleukin-8 is activated in patients with chronic liver diseases and associated with hepatic macrophage accumulation in human liver fibrosis. *PLoS ONE* **2011**, *6*, e21381. [CrossRef] [PubMed]

35. Lin, M.-T.; Juan, C.-Y.; Chang, K.-J.; Chen, W.-J.; Kuo, M.-L. IL-6 inhibits apoptosis and retains oxidative DNA lesions in human gastric cancer AGS cells through up-regulation of anti-apoptotic gene mcl-1. *Carcinogenesis* **2001**, *22*, 1947–1953. [CrossRef] [PubMed]

36. Persson, E.C.; Schwartz, L.M.; Park, Y.; Trabert, B.; Hollenbeck, A.R.; Graubard, B.I.; Freedman, N.D.; McGlynn, K.A. Alcohol consumption, folate intake, hepatocellular carcinoma, and liver disease mortality. *Cancer Epidemio. Biomark. Prev.* **2013**, *22*, 415–421. [CrossRef]

37. Sellers, T.A.; Kushi, L.H.; Cerhan, J.R.; Vierkant, R.A.; Gapstur, S.M.; Vachon, C.M.; Olson, J.E.; Therneau, T.M.; Folsom, A.R. Dietary Folate Intake, Alcohol, and Risk of Breast Cancer in a Prospective Study of Postmenopausal Women. *Epidemiology* **2001**, *12*, 420–428. [CrossRef] [PubMed]

38. Dumitrescu, R.G.; Shields, P.G. The etiology of alcohol-induced breast cancer. *Alcohol* **2005**, *35*, 213–225. [CrossRef] [PubMed]

39. Liu, Y.; Nguyen, N.; Colditz, G.A. Links between alcohol consumption and breast cancer: A look at the evidence. *Womens Health* **2015**, *11*, 65–77. [CrossRef]

40. Singletary, K.W.; Gapstur, S.M. Alcohol and Breast CancerReview of Epidemiologic and Experimental Evidence and Potential Mechanisms. *JAMA* **2001**, *286*, 2143–2151. [CrossRef]

41. Endogenous Hormones Breast Cancer Collaborative Group; Key, T.J.; Appleby, P.N.; Reeves, G.K.; Roddam, A.W.; Helzlsouer, K.J.; Alberg, A.J.; Rollison, D.E.; Dorgan, J.F.; Brinton, L.A.; et al. Circulating sex hormones and breast cancer risk factors in postmenopausal women: Reanalysis of 13 studies. *Br. J. Cancer* **2011**, *105*, 709–722. [CrossRef] [PubMed]

42. Assi, N.; Rinaldi, S.; Viallon, V.; Dashti, S.G.; Dossus, L.; Fournier, A.; Cervenka, I.; Kvaskoff, M.; Turzanski-Fortner, R.; Bergmann, M.; et al. Mediation analysis of the alcohol-postmenopausal breast cancer relationship by sex hormones in the EPIC cohort. *Int. J. Cancer* **2020**, *146*, 759–768. [CrossRef]

43. Pessione, F.; Degos, F.; Marcellin, P.; Duchatelle, V.; Njapoum, C.; Martinot-Peignoux, M.; Degott, C.; Valla, D.; Erlinger, S.; Rueff, B. Effect of alcohol consumption on serum hepatitis C virus RNA and histological lesions in chronic hepatitis C. *Hepatology* **1998**, *27*, 1717–1722. [CrossRef]

44. Cao, Y.; Willett, W.C.; Rimm, E.B.; Stampfer, M.J.; Giovannucci, E.L. Light to moderate intake of alcohol, drinking patterns, and risk of cancer: Results from two prospective US cohort studies. *BMJ* **2015**, *351*, h4238. [CrossRef]

45. Shin, N.-R.; Whon, T.W.; Bae, J.-W. Proteobacteria: Microbial signature of dysbiosis in gut microbiota. *Trends Biotechnol.* **2015**, *33*, 496–503. [CrossRef] [PubMed]

46. Fedirko, V.; Tran, H.Q.; Gewirtz, A.T.; Stepien, M.; Trichopoulou, A.; Aleksandrova, K.; Olsen, A.; Tjønneland, A.; Overvad, K.; Carbonnel, F.; et al. Exposure to bacterial products lipopolysaccharide and flagellin and hepatocellular carcinoma: A nested case-control study. *BMC Med.* **2017**, *15*, 72. [CrossRef] [PubMed]

47. Yang, Y.M.; Kim, S.Y.; Seki, E. Inflammation and Liver Cancer: Molecular Mechanisms and Therapeutic Targets. *Semin. Liver Dis.* **2019**, *39*, 26–42. [CrossRef] [PubMed]

48. Howie, N.; Trigkas, T.; Cruchley, A.; Wertz, P.; Squier, C.; Williams, D. Short-term exposure to alcohol increases the permeability of human oral mucosa. *Oral Dis.* **2001**, *7*, 349–354. [CrossRef] [PubMed]

49. van Roekel, E.H.; Trijsburg, L.; Assi, N.; Carayol, M.; Achaintre, D.; Murphy, N.; Rinaldi, S.; Schmidt, J.A.; Stepien, M.; Kaaks, R.; et al. Circulating Metabolites Associated with Alcohol Intake in the European Prospective Investigation into Cancer and Nutrition Cohort. *Nutrients* **2018**, *10*, 654. [CrossRef]

Review

Margin of Exposure Analyses and Overall Toxic Effects of Alcohol with Special Consideration of Carcinogenicity

Alex O. Okaru [1] and **Dirk W. Lachenmeier** [2,*]

[1] Department of Pharmacy, University of Nairobi, Nairobi P.O. Box 19676-00202, Kenya; alex.okaru@gmail.com
[2] Chemisches und Veterinäruntersuchungsamt (CVUA) Karlsruhe, Weissenburger Straße 3, 76187 Karlsruhe, Germany
* Correspondence: lachenmeier@web.de; Tel.: +49-721-926-5434

Abstract: Quantitative assessments of the health risk of the constituents of alcoholic beverages including ethanol are reported in the literature, generally with hepatotoxic effects considered as the endpoint. Risk assessment studies on minor compounds such as mycotoxins, metals, and other contaminants are also available on carcinogenicity as the endpoint. This review seeks to highlight population cancer risks due to alcohol consumption using the margin of exposure methodology. The individual and cumulative health risk contribution of each component in alcoholic beverages is highlighted. Overall, the results obtained consistently show that the ethanol contributes the bulk of harmful effects of alcoholic beverages, while all other compounds only contribute in a minor fashion (less than 1% compared to ethanol). Our data provide compelling evidence that policy should be focused on reducing total alcohol intake (recorded and unrecorded), while measures on other compounds should be only secondary to this goal.

Keywords: alcohol; risk assessment; hepatotoxicity; dose–response relationship; margin of exposure; epidemiological methods

Citation: Okaru, A.O.; Lachenmeier, D.W. Margin of Exposure Analyses and Overall Toxic Effects of Alcohol with Special Consideration of Carcinogenicity. *Nutrients* **2021**, *13*, 3785. https://doi.org/10.3390/nu13113785

Academic Editor: Peter Anderson

Received: 29 July 2021
Accepted: 19 October 2021
Published: 25 October 2021

Publisher's Note: MDPI stays neutral with regard to jurisdictional claims in published maps and institutional affiliations.

1. Introduction

The epidemiological association of alcoholic beverages with cancer remains a topic that has continued to attract global attention for over a century with the first documented cases, cancer of the esophagus, being reported in 1910 [1,2]. Later in 1988, the World Health Organization (WHO)/International Agency for Research on Cancer (IARC) classified "alcohol drinking" as carcinogenic to humans (group 1) after establishing a causal link between alcohol use and malignancies of the oral pharynx, esophagus, and liver [1]. The promoters or causative factors in alcoholic beverages for developing carcinogenic lesions are a matter of continuing debate among scientists. However, alcohol being a multicomponent mixture, the potential contribution of each or all the compounds to carcinogenesis should not be overlooked. These substances occur as residues, contaminants, or even adulterants, in addition to being naturally occurring in either raw materials or fermentation by-products.

Ethanol, the principal component of alcoholic beverages, is classified as a human carcinogen (group 1) by IARC. Other than ethanol, other IARC-classified carcinogenic compounds such as acetaldehyde, formaldehyde, acrylamide, aflatoxins, ochratoxin A, arsenic, lead, cadmium, ethyl carbamate, furan, safrole, 4-methylimidazole, N-nitrosodimethylamine (NDMA), 3-Monochloropropane-1,2-diol (3-MCPD), and benzene have occurred in alcoholic beverages. The contribution of these compounds to cancer is either synergistic or independent of each other. Understanding the contribution of each component is important in disentangling the mechanisms of carcinogenicity due to alcohol and ultimately aids in alcohol control policies. Nevertheless, epidemiological research has reported that only ethanol achieves the requisite threshold to explain the carcinogenic risk of alcoholic beverages. This review seeks to highlight population cancer risks due to alcohol consumption using the margin of exposure (MOE) approach with emphasis on the cancer-risk contribution of individual components of alcoholic beverages. This review identifies ethanol as the main oncogenic component in alcoholic beverages and lays emphasis on the need for policy geared towards the reduction in drinking per se and not target on other minor carcinogens that may require strict implementation of industry best practices, i.e., as low as reasonably achievable (ALARA) guidelines and good manufacturing practices.

2. The Margin of Exposure Method and Its Application to Alcoholic Beverages

Despite there being other methods for evaluating the health risks associated with alcohol intake, the margin of exposure (MOE) method is recommended for comparing the risks of different alcoholic beverage components [1]. MOE compares exposure levels to a toxicological threshold. The toxicological thresholds are derived from the dose–response evaluations for both carcinogens and non-carcinogens.

The ratio between the benchmark dose's lower one-sided confidence limit (BMDL) and predicted human consumption/exposure of the same substance is known as the margin of exposure. MOE is typically used to compare the health risks of various chemicals and, as a result, to prioritize risk management efforts. The lower the MOE, the greater the risk to people; typically, a value of less than 10,000 is used to indicate health risk.

The benchmark dose (BMD) is the dose of a chemical that, based on the dose–response modeling, causes a specified change in the response rate (benchmark response) of an undesirable impact compared to the background. The benchmark response is typically suggested to be set near the lower limit of what can be measured (e.g., for animal experiment in the 1–10% range). BMD–response modeling results can then be used with exposure data to create a MOE for quantitative risk assessment. No observed effect level (NOEL) or no observed adverse effect level (NOAEL) values can be used as surrogate thresholds where BMDL values are unavailable in the literature. Consequently, the MOEs can be determined by dividing the NO(A)EL by estimated human intake [1].

The human intakes for each beverage group (i.e., beer, wine, spirits, and unrecorded alcohol) for various drinking scenarios (e.g., low risk drinking and heavy drinking) can be based on drinking guidelines such as the Canadian ones, which consider 13.6 g of pure alcohol a standard drink [1]. MOEs for average and maximum contamination with the various substances can also be determined for both drinking scenarios to give a range for average and worst-case contamination scenarios [1].

The most recent detailed IARC reviews were suggested to be used to select compounds and their levels in alcoholic beverages. For the established and probable human carcinogens, toxicological endpoints and BMD are primarily based on literature data [1]. Suitable risk assessment studies, including endpoints and dose–response modeling results, were typically identified in monographs published by national and international risk assessment bodies such as the United States Environmental Protection Agency (US EPA), the World Health Organization International Programme on Chemical Safety (WHO-IPCS), the Joint FAO/WHO Expert Committee on Food Additives (JECFA), and the European Food Safety Authority (EFSA). Data from peer-reviewed scientific research can be used for compounds without accessible monographs or those with missing data on dose–response modeling findings [1].

3. Occurrence of Carcinogenic Compounds in Alcoholic Beverages

Ethanol and acetaldehyde (ethanal), both categorized by IARC as group 1 carcinogens, are the primary carcinogens occurring in alcoholic beverages accounting for approximately 5.5% of all cancer cases worldwide [3]. Although the inherent cancer risk of alcoholic beverages parallels consumption volumes, even light alcohol drinking has been associated with cancer with ethanol and acetaldehyde being central to the pathogenicity. At the molecular level, ethanol and acetaldehyde are postulated to cause cancer in similar mechanistic fashion, since acetaldehyde, a genotoxic compound, is a metabolite of ethanol resulting from the alcohol dehydrogenase or CYP 450 E1 pathways. Since ethanol and acetaldehyde have similar carcinogenesis mechanisms, the computation of cancer risk can be be undertaken cumulatively. Additionally, ethanol plays a promoting role in oncogenesis by solvating other carcinogens [1].

Besides metabolism, acetaldehyde occurs naturally, albeit in small amounts in alcoholic beverages with the highest contents reported to be in fortified wines (118 mg/L) and some spirit drinks (66 mg/L) [4]. Additionally, acetaldehyde occurs at high levels in certain unrecorded alcohols [5]. The average daily acetaldehyde exposure from alcoholic beverages has been calculated to be 0.112 mg/kg body weight, with a MOE of 498 [5].

IARC has classified formaldehyde (methanal), a naturally occurring substance found in various plants, mainly fruits and vegetables, and animal products such as meat, dairy products, and fish [1], as a group 1 carcinogen [6]. Formaldehyde is a carcinogen linked to the development of leukemia and naso-pharyngeal cancer in humans. Alcoholic beverages contain a substantial quantity of formaldehyde [7]. In a sampling of 500 beverages including wine, beer, spirits, and unrecorded alcohol, lower formaldehyde contamination (1.8 percent) was found, which was however more than the WHO IPCS permissible concentrations [7]. To surpass the daily US EPA reference dose (RfD) of 0.2 mg/kg bodyweight [8], a person weighing 60 kg would need to partake daily 800 mL of alcohol containing 14.37 mg/L formaldehyde. Even in the worst-case scenario, this level of exposure is exceedingly unlikely.

Acrylamide, considered by IARC as probably carcinogenic, may produce cancer through its metabolite, glycidamide, that forms DNA adducts [9]. Nevertheless, there are only a few reports on the occurrence of acrylamide in alcoholic beverages with one study reporting acrylamide levels of 22 µg/kg [10]. The group 2B carcinogen, 3-monochloropropane-1,2-diol, is a heat-induced contaminant resulting from the thermal processing of malt [11]. In experimental animals, 3-MCPD causes renal tubule adenocarcinomas. Although 3-MCPD is detected in some dark specialty malts used for beer production [11–13], it only occurs in low levels in most beers. It typically ranges from <10 µg/L to 14 µg/L [14,15].

IARC has classified the mycotoxins, aflatoxin B_1 and ochratoxin A, found in some alcoholic beverages as carcinogenic to humans (group 1) and possibly carcinogenic to humans (group 2B), respectively [16]. Aflatoxin B_1, as well as other aflatoxins (B_2, G_1, and G_2), is a naturally occurring toxin in barley, corn, and sorghum malts that enters beer due to the use of contaminated cereals [17–19]. The occurrence of aflatoxins is climate-related with aflatoxins thriving in warm climates, especially in the tropics. Indeed, higher contamination of beer is reported to be in warm climatic countries such as South Africa, India, Mexico, and Kenya, among others [19,20]. Aflatoxin B_1 has been found in the greatest concentrations (up to 6.8 µg/L) in artisanal beers from Kenya [20,21]. Similarly, ochratoxin A (OTA) occurs as a contaminant in grapes and in raw materials for beer, such as barley, malt, or cereal derivatives. Unlike aflatoxin B_1, OTA is partially detoxified during fermentation [22], and its concentration remains unchanged in wine for one year [23].

Among heavy metals, arsenic, cadmium, and lead are possibly the ones of carcinogenic concern. The IARC classifies metalloid arsenic and inorganic arsenic compounds as group 1 carcinogens [24]. Lung, skin, liver, kidney, prostate, and urinary bladder malignancies have all been linked to inorganic arsenic compounds [24]. The reported levels of arsenic in beer are 0–102.4 µg/L [24], while those in spirits and wines are 0–27 and 0–14.6 µg/L, respectively [25]. The IARC designated cadmium as a group 1 carcinogenic agent because it causes cancers of the lungs, kidneys, and prostate [26]. According to an EFSA report [25], the amount of arsenic in various beverages varies. Fortified and liqueur wines had a Cd concentration of 0.5 µg/L, whereas liqueur had a level of 6.0 µg/L. The average concentration of Cd in wines and beers is 1.2 and 1.8 µg/L, respectively [25]. Organic lead compounds are "not classifiable as to their carcinogenicity to humans" (group 3) [27], whereas inorganic lead and lead compounds are "probably carcinogenic" (group 2A) [28]. The concentrations of lead vary across alcohol types. The average content of Pb in wines is 29 µg/kg with no significant differences in the amounts between the red and white varieties. Beer and beer-like beverages contain 12 µg/kg Pb on average [29].

Benzene, a heat-induced contaminant, is classified as a group I carcinogen, and it arises in alcoholic beverages. Benzene is a genotoxic compound that targets pluripotent hematopoietic stem cells leading to a raft of chromosomal aberrations [30]. The compound can occur in soft beverages that contain benzoic acid (a preservative) [31–33] or in beers manufactured with benzene-contaminated industrial carbon dioxide [34,35].

Furan, a group 2B carcinogen [36], is touted to intercalate with DNA via its cytochrome P-450-mediated metabolite, *cis*-2-butene-1,4-dial [37,38] leading to carcinogenesis. Furan has been found in beer samples at amounts as high as 28 µg/kg. Lower furan concentrations have been reported in wines and liqueurs, 6.5 and 28 µg/kg, respectively [39].

In 2015, IARC classified the controversial herbicide glyphosate as "probably carcinogenic to humans" based on some evidence in humans due to a correlation with non-Hodgkin lymphoma and significant evidence for glyphosate's carcinogenicity in experimental animals [40]. In 2013, Nagatomi et al. observed that glyphosate content in 15 commercial canned beers from Japan was below the limit of quantitation (10 µg/L) [41]. From a risk assessment standpoint, these observed amounts are unlikely to cause harm.

Ethyl carbamate, a probable human carcinogen (group 2A) [42], has been found in small concentrations in wines and beers (in µg/L) [43] and in larger proportions in stone-fruit spirits (in mg/L) [43]. Another group 2A carcinogenic compound, *N*-nitrosodimethylamine (NDMA), is hepatotoxic [27]. Ethanol through its solvation effect or via alteration of cellular metabolism and suppression of DNA repair, enhances the carcinogenicity of NDMA [44]. NDMA in alcoholic beverages can arise from the manufacturing processes or from storage. During the production process, *N*-nitroso compounds can emerge by activities such as when malt is directly heated or when polluted water is used, or when foods and beverages are stored [45,46]. In a follow-up screening of German beers conducted between 1992 and 2006, NDMA was found in 29 malt samples (43%) and 81 beer samples (7%), with only 4% of the beer samples (n = 1242) having concentrations above the technical threshold value [47].

Pulegone, a component of essential oil-containing plants of the mint family, is found in mint-flavored alcoholic beverages [48]. Pulegone has been linked to liver and bladder cancer in animal models, prompting the IARC to classify it as probably carcinogenic to humans (group 2B) [48]. Despite being recognized as a potential carcinogen, occurrence data on pulegone are still scanty with only the National Toxicology Programme (NTP) reporting a mean value of 10.5 µg/L [49].

Safrole, a substituted benzodioxole, is a genotoxic agent that naturally occurs in several spices such as sweet basil, black pepper, cinnamon nutmeg, mace, cinnamon, and aniseed. Moreover, safrole occurs in food and beverages that are flavored with it. The IARC categorizes safrole as "possibly carcinogenic to humans" (group 2B) [27]. Since safrole occurs in cola drinks [50], it has the potential to occur in alcoholic beverages [51] especially admixtures of cola and alcohol. On average, humans consume 0.3 mg of safrole per day, with the 97.5th percentile consuming 0.5 mg. The presence of possibly carcinogenic compounds in alcoholic beverages is summarized in Table 1.

Table 1. Summary of the occurrence of potentially carcinogenic compounds in alcoholic beverages (reprinted with modifications with permission from Springer Nature, Archives of Toxicology, Pflaum et al. [1], copyright 2016).

Agent (IARC Group [a])	Beverage Type	Concentration		Reference
		Average	Maximum	
Acetaldehyde in alcoholic beverages (1)	Beer	9 mg/L	63 mg/L	[4]
	Spirit	66 mg/L	1159 mg/L	
	Wine	34 mg/L	211 mg/L	

Table 1. *Cont.*

Agent (IARC Group [a])	Beverage Type	Concentration		Reference
		Average	Maximum	
Acrylamide [b] (2A)	Beer	0–72 µg/kg	363 µg/kg	[15,52]
Aflatoxins (1)	Commercial beer	0.002 µg/L	0.230 µg/L	[17]
	Artisanal beer	3.5 µg/L	6.8 µg/L	[20]
Arsenic (1)	Beer	0 µg/L	102.4 µg/L	[1]
	Spirit	13 µg/L	27 µg/L	
	Wine	13 µg/L	27 µg/L	
Benzene (1)	Beer	10 µg/L	20 µg/L	[1]
Cadmium (1)	Beer	0.9 µg/L	14.3 µg/L	[1]
	Spirits	6 µg/L	40 µg/L	
	Wine	1.0 µg/L	30 µg/L	
Ethanol (1)	Varies	2% vol.	80% vol.	[1]
Ethyl carbamate (2A)	Beer	0 µg/kg	33 µg/kg	[53]
	Spirits	93 µg/kg	6730	
	Stone spirits	744 µg/kg	22,000 µg/kg	
	Wine	5 µg/kg	180 µg/kg	
Formaldehyde (1)	Beer	0 mg/L	0 mg/L	[8]
	Spirits	0.50 mg/L	14.37 mg/L	
	Wine	0.13 mg/L	1.15 mg/L	
Furan (2B)	Beer	3.3 µg/kg	28 µg/kg	[39]
Glyphosate [c] (2A)	Beer	0–30 µg/L		[1]
Lead compounds, inorganic (2A)	Beer	2 µg/L	15 µg/L	[54]
	Spirits	31 µg/L	600 µg/L	[1]
	Wine	57 µg/L	236 µg/L	[55]
MCPD [d] (2B)	Beer	0–14 µg/kg		[12]
4-Methylimidazole [e] (2B)	Beer[e]	9 µg/L	28 µg/L	[56]
	Spirit	0 µg/L	0.014 µg/L	[57]
NMDA (2A)	Beer	0.1 µg/kg	1.3 µg/kg	[1]
Ochratoxin A (2B)	Beer	0.05 µg/L	1.5 µg/L	[1]
	Wine	0.23 µg/L	7.0 µg/L	
Pulegone [f] (2B)		10.5 mg/kg	100 mg/kg	[49,58]
Safrole (2B)	Liqueurs, aperitifs, and bitters	ND	6.6 mg/L	[51]

Abbreviations: MCPD—3-Monochloropropane-1,2-diol, NMDA—N-Nitrosodimethylamine, ND—below the limit of quantitation. [a] Only compounds present in alcoholic beverages that fall under IARC groups 1 (carcinogenic to humans), 2A (probably carcinogenic to humans), and 2B (possibly carcinogenic to humans) were included in this list. [b] There are few studies on acrylamide in alcoholic beverages. Most samples examined had levels that were below the detection limit. A single sample of wheat beer had a level of 72 µg/kg, while craft beers found in Poland and the Czech Republic had 363 µg/kg [52]. [c] Except for the "worst-case" scenario, upper level of 30 µg/L [59] was used, since there is a dearth of systematic data on the occurrence glyphosate in beer. [d] There was limited research on the presence of 3-MCPD in alcoholic beverages. As a result, the upper limit was set at less than 10 µg/L from a study on beers [12]. [e] Caramelized alcoholic beverages. [f] Studies on the occurrence of pulegone in alcoholic beverages are scanty. Thus, 10.5 mg/kg [49] and 100 mg/kg [58] were utilized as the minimum and maximum amounts of pulegone, respectively, in alcohol products.

4. Comparative Risk Assessment of Compounds in Alcoholic Beverages

The presence of a carcinogenic compound in an alcoholic beverage does not directly impute an inherent risk of consumers of the drink. However, the quantitative risk assessment serves to ascribe harm due to a compound if it exceeds the toxicological threshold. The margin of exposure (MOE) methodology as described in the literature is applicable to conduct a comparative risk assessment for compounds in alcoholic beverages [1,5,60–62]. Where human data were unavailable, animal data were used instead for risk assessment. Moreover, non-cancer endpoints were chosen for substances such as Pb where there was no dose–response modeling data for cancer effects available. However, non-cancer endpoints may be more sensitive than cancer endpoints. Additionally, the most sensitive endpoint was chosen if dose–response data for several organ sites were available. Table 2 lists the toxicological endpoints and points of departure used in dose–response modeling and risk assessment.

Table 2. Dose–response modeling for potential human carcinogens occurring in alcoholic beverages (reprinted with permission from Springer Nature, Archives of Toxicology, Pflaum et al. [1] copyright 2016).

Carcinogenic Agent	Modeling Toxicological Endpoint	Animal Model	Route/Mode of Exposure	BMDL [a] (mg/kg bw/Day)	Reference
Acetaldehyde	Animal tumors [60]	Male rats	Oral	56	[60]
Acrylamide	Harderian gland tumors [63]	Mice	Oral	0.18	[64]
Aflatoxin B_1	Cancer of the lungs in humans [65]	NA	Food	0.00087	[66]
Arsenic	Cancer of the lungs in humans [67]	NA	Water	$BMDL_{0.5}$: 0.003	[68]
Benzene	Human lymphocyte count [69]	NA	Inhalation extrapolated to oral	1.2 [b]	[70]
Cadmium	Human studies [70]	NA	Food	NOAEL: 0.01 [c]	[70]
Ethanol	Hepatocellular adenoma or carcinoma [71]	Rats	Oral	700	[72,73]
Ethyl carbamate	Bronchiolar alveolar carcinoma [74]	Mice	Oral	0.3	[73]
Formaldehyde	The aerodigestive tract, comprising the oral and gastrointestinal mucosa, undergoes histological alterations [75]	Rats	Oral	NOEL: 15 [c]	[76]
Furan	Adenomas and carcinomas of the liver [77]	Female mice	Oral	0.96	[78]
Glyphosate [b]	There are no dose–response data for the cancer outcome			NOAEL: 50	[79]
Lead	Human cardiovascular effects [29]	NA	Diet	BMDL10: 0015 [d]	[80]
3-MCPD	Hyperplasia of the tubules of the kidneys [e] [81]	Rats	Oral	0.27	[82]
4-Methylimidazole	Lung cancer [83]	Mice	Oral	NOAEL: 80 [c]	[84]
N-Nitrosodimethylamine	Hepatocellular carcinoma [85]		Oral	0.029	[86,87]
Ochratoxin A	Renal adeno-carcinoma [88]	Male rats	Oral	0.025	[89]
Pulegone	Urinary bladder tumors [90]	Rats	Oral	LOAEL: 20 [c]	[49]
Safrole	Hepatic tumors [91]	Mice	Oral	3 [f]	[92,93]

NA—not applicable. [a] For an x % occurrence of health effect, BMDLx is the lower one-sided confidence limit of the benchmark dose (BMD). [b] Inhalation exposure was used as the original endpoint. Route-to-route extrapolation was used to calculate the BMDL for oral exposure [69]. [c] The no effect level (NOEL), no observed adverse effect level (NOAEL), or lowest observed adverse effect level (LOAEL) were utilized because no appropriate BMD modeling for exposure through the mouth has been documented. [d] Overall exposure to lead is determined in blood, and the figures are based on that. The BMDL that was employed was determined based on dietary exposure [29]. [e] Renal tubular hyperplasia, rather than renal tubule adenoma or cancer, was a more sensitive endpoint. [f] This was a conservative minimal concentration based on the literature's BMDL10 range of "about 3–29 mg/kg bw/day" for safrole [91].

For daily consumption of four standard alcoholic drinks, MOEs were calculated for the average and worst-case scenarios. For ethanol, the lowest MOE was achieved (0.8). Inorganic lead and arsenic showed MOEs ranging from 10 to 300, while acetaldehyde, cadmium, ethyl carbamate, and pulegone had MOEs ranging from 1000 to 10,000. Safrole, ochratoxin A, NDMA, 4-methylimidazole, 3-MCPD, glyphosate, furan, formaldehyde, and acrylamide had average MOEs exceeding 10,000 even in these extreme contexts such as binge drinking (Figure 1). However, the MOE for aflatoxin B_1 from Kenyan artisanal beer that was significantly tainted ranged from 15 to 58 with a mean of 36. As a result, ethanol is the most significant carcinogen found in alcoholic beverages, with a clear dose–response relationship. Other contaminants (lead, arsenic, ethyl carbamate, acetaldehyde) may pose risks below those tolerated for food contaminants, but from a cost-effectiveness standpoint, the focus should be on reducing alcohol consumption in general rather than on mitigative actions for some contaminants that contribute only a small (if any) portion of the total health risk. This review again highlights the fact that ethanol remains the compound with the highest carcinogenic potential that is present in alcoholic beverages. This finding is consistent with other studies reported in the literature [1,21,60,62]. Aflatoxin B$_1$ also emerged as a compound of interest in unrecorded artisanal beers that clearly requires attention in the warm climatic countries where the consumption of such beers is prevalent [20,21]. Figure 1 shows the comparative MOEs for carcinogens.

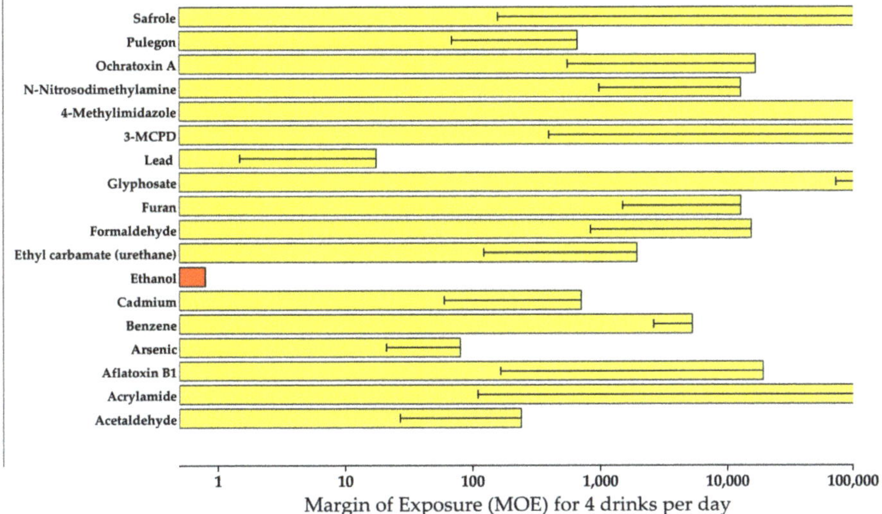

Figure 1. Comparative MOEs for IARC-classified carcinogens in alcoholic beverages (reprinted with permission from Springer Nature, Archives of Toxicology, Pflaum et al. [1], copyright 2016).

5. Overall Toxic Effects of Alcoholic Beverages

According to studies, no amount of alcohol use promotes health [94]. Alcohol consumption significantly contributes to death, disability, and ill health worldwide [94–96]. Alcohol is the sixth most common cause of mortality and disability-adjusted life years (DALYs) in both men and women, accounting for 22% of female fatalities and 68% of male deaths [94]. There is a link between harmful alcohol consumption and various mental and behavioral illnesses, as well as other non-communicable diseases such as tuberculosis and HIV/AIDS and injuries. Injuries constitute the greatest negative consequence of alcohol consumption after cancer. Cardiovascular disease accounts for 15% of alcohol-attributable morbidity, while liver cirrhosis accounts for 13% of all alcohol-attributable deaths [97]. Besides the health risks, irresponsible alcohol use results in social and economic losses for consumers and the community as a whole [98,99].

6. Conclusions

Despite there being other methods for evaluating the health risks associated with alcohol intake, the margin of exposure method is recommended for comparing the risks of different alcoholic beverage components. From this review, ethanol remains the most prominent carcinogen in alcoholic beverages, according to quantitative comparative risk assessment. Therefore, the reduction in alcohol intake ought to be prioritized in combating harm due to alcoholic beverages [100]. Since the dose–response relationship holds for alcohol harm, reduction in alcoholic strength would be beneficial in minimizing the harmful effects of alcohol [101]. For illustration, drinking four bottles of 5.5 percent vol. ethanol beer generates a MOE of 0.5, whereas drinking the same volume of light beer (1.5 percent vol. ethanol) yields a substantially greater MOE of 1.9 [1]. Moreover, consumers may not be typically able to discriminate different alcohol strengths in beer and, thus, may not ingest more volumes to compensate for the lower alcoholic strength beer [102,103].

Other carcinogens besides ethanol require mitigative steps as well, which may require strict adoption of industry best practices such as keeping contaminants/components as low as can reasonably be achieved (ALARA). We urge the relevant regulatory authorities to implement the available mitigative measures to protect consumers from potentially carcinogenic substances.

Author Contributions: D.W.L. was in charge of conceptualization; D.W.L. and A.O.O. were in charge of methodology; A.O.O. was in charge of writing—original draft preparation; D.W.L. was in charge of writing—review and editing. All authors have read and agreed to the published version of the manuscript.

Funding: This research received no external funding.

Institutional Review Board Statement: Not applicable.

Informed Consent Statement: Not applicable.

Data Availability Statement: Extracted data are presented in the main tables.

Conflicts of Interest: The authors declare no conflict of interest.

References

1. Pflaum, T.; Hausler, T.; Baumung, C.; Ackermann, S.; Kuballa, T.; Rehm, J.; Lachenmeier, D.W. Carcinogenic compounds in alcoholic beverages: An update. *Arch. Toxicol.* **2016**, *90*, 2349–2367. [CrossRef] [PubMed]
2. Lamy, L. Etude de statistique clinique de 134 cas de cancer de l' oesophage en du cardia. *Arch. Mal. Appar. Dig.* **1910**, *4*, 451–475.
3. Praud, D.; Rota, M.; Rehm, J.; Shield, K.; Zatoński, W.; Hashibe, M.; La Vecchia, C.; Boffetta, P. Cancer incidence and mortality attributable to alcohol consumption. *Int. J. Cancer* **2016**, *138*, 1380–1387. [CrossRef] [PubMed]
4. Lachenmeier, D.W.; Sohnius, E.-M. The role of acetaldehyde outside ethanol metabolism in the carcinogenicity of alcoholic beverages: Evidence from a large chemical survey. *Food Chem. Toxicol.* **2008**, *46*, 2903–2911. [CrossRef]
5. Lachenmeier, D.W.; Kanteres, F.; Rehm, J. Carcinogenicity of acetaldehyde in alcoholic beverages: Risk assessment outside ethanol metabolism. *Addiction* **2009**, *104*, 533–550. [CrossRef]
6. IARC Working Group on the Evaluation of Carcinogenic Risks to Humans. Formaldehyde. *IARC Monogr. Eval. Carcinog. Risks Hum.* **2006**, *88*, 39–325.
7. Monakhova, Y.B.; Jendral, J.A.; Lachenmeier, D.W. The margin of exposure to formaldehyde in alcoholic beverages. *Arch. Ind. Hyg. Toxicol.* **2012**, *63*, 227–237. [CrossRef]
8. Jendral, J.A.; Monakhova, Y.B.; Lachenmeier, D.W. Formaldehyde in Alcoholic Beverages: Large Chemical Survey Using Purpald Screening Followed by Chromotropic Acid Spectrophotometry with Multivariate Curve Resolution. *Int. J. Anal. Chem.* **2011**, *2011*, 797604. [CrossRef]
9. Boettcher, M.I.; Schettgen, T.; Kütting, B.; Pischetsrieder, M.; Angerer, J. Mercapturic acids of acrylamide and glycidamide as biomarkers of the internal exposure to acrylamide in the general population. *Mutat. Res.-Genet. Toxicol. Environ. Mutagen.* **2005**, *580*, 167–176. [CrossRef]
10. Mo, W.; He, H.; Xu, X.; Huang, B.; Ren, Y. Simultaneous determination of ethyl carbamate, chloropropanols and acrylamide in fermented products, flavoring and related foods by gas chromatography–triple quadrupole mass spectrometry. *Food Control* **2014**, *43*, 251–257. [CrossRef]
11. Wenzl, T.; Lachenmeier, D.W.; Gökmen, V. Analysis of heat-induced contaminants (acrylamide, chloropropanols and furan) in carbohydrate-rich food. *Anal. Bioanal. Chem.* **2007**, *389*, 119–137. [CrossRef]
12. Breitling-Utzmann, C.M.; Köbler, H.; Harbolzheimer, D.; Maier, A. 3-MCPD - Occurrence in bread crust and various food groups as well as formation in toast. *Dtsch. Leb.* **2003**, *99*, 280–285.
13. Svejkovská, B.; Novotný, O.; Divinová, V.; Réblová, Z.; Doležal, M.; Velíšek, J. Esters of 3-chloropropane-1,2-diol in foodstuffs. *Czech J. Food Sci.* **2018**, *22*, 190–196. [CrossRef]
14. Baxter, E.D.; Booer, C.D.; Muller, R.E.; O'Shaugnessy, C.; Slaiding, I.R. Minimizing acrylamide and 3-MCPD in crystal malts; effects on flavour. *Proc. Congr. Eur. Brew Conv.* **2005**, *30*, 163-1–163-6.

15. Sadowska-Rociek, A.; Surma, M. A survey on thermal processing contaminants occurrence in dark craft beers. *J. Food Compos. Anal.* **2021**, *99*, 103888. [CrossRef]
16. IARC Working Group on the Evaluation of Carcinogenic Risks to Humans. Aflatoxins. *IARC Monogr. Eval. Carcinog. Risks Hum.* **2012**, *100F*, 225–248.
17. Mably, M.; Mankotia, M.; Cavlovic, P.; Tam, J.; Wong, L.; Pantazopoulos, P.; Calway, P.; Scott, P.M. Survey of aflatoxins in beer sold in Canada. *Food Addit. Contam.* **2005**, *22*, 1252–1257. [CrossRef]
18. Gilbert, J.; Michelangelo, P. Analytical methods for mycotoxins in the wheat chain. In *Mycotoxin Reduction in Grain Chain*; Leslie, J.F., Logrieco, A.F., Eds.; Wiley Blackwell: Oxford, UK, 2014; pp. 169–188.
19. Odhav, B.; Naicker, V. Mycotoxins in South African traditionally brewed beers. *Food Addit. Contam.* **2002**, *19*, 55–61. [CrossRef]
20. Okaru, A.O.; Abuga, K.O.; Kibwage, I.O.; Hausler, T.; Luy, B.; Kuballa, T.; Rehm, J.; Lachenmeier, D.W. Aflatoxin contamination in unrecorded beers from Kenya – A health risk beyond ethanol. *Food Control* **2017**, *79*, 344–348. [CrossRef]
21. Okaru, A.O.; Rehm, J.; Sommerfeld, K.; Kuballa, T.; Walch, S.G.; Lachenmeier, D.W. The Threat to Quality of Alcoholic Beverages by Unrecorded Consumption. In *Alcoholic Beverages*; Woodhead Publishing: Cambridge, UK, 2019; pp. 1–34.
22. Esti, M.; Benucci, I.; Liburdi, K.; Acciaro, G. Monitoring of ochratoxin A fate during alcoholic fermentation of wine-must. *Food Control* **2012**, *27*, 53–56. [CrossRef]
23. Lopez de Cerain, A.; González-Peñas, E.; Jiménez, A.M.; Bello, J. Contribution to the study of ochratoxin A in Spanish wines. *Food Addit. Contam.* **2002**, *19*, 1058–1064. [CrossRef]
24. IARC Working Group on the Evaluation of Carcinogenic Risks to Humans. Arsenic and arsenic compounds. *IARC Monogr. Eval. Carcinog. Risks Hum.* **2012**, *100C*, 41–93.
25. Barbaste, M.; Medina, B.; Perez-Trujillo, J.P. Analysis of arsenic, lead and cadmium in wines from the Canary Islands, Spain, by ICP/MS. *Food Addit. Contam.* **2021**, *4*, 141–148. [CrossRef]
26. IARC Working Group on the Evaluation of Carcinogenic Risks to Humans. Cadmium and cadmium compounds. *IARC Monogr. Eval. Carcinog. Risks Hum.* **2012**, *100C*, 121–145.
27. *IARC Monographs on the Evaluation of Carcinogenic Risks to Humans*; Suppl. S7; International Agency for Research on Cancer: Lyon, France, 1987.
28. IARC Working Group on the Evaluation of Carcinogenic Risks to Humans. Inorganic and organic lead compounds. *IARC Monogr. Eval. Carcinog. Risks Hum.* **2006**, *87*, 39–468.
29. EFSA. Scientifc opinion on lead in food. *EFSA J.* **2010**, *8*, 1570.
30. IARC Working Group on the Evaluation of Carcinogenic Risks to Humans. Benzene. *IARC Monogr. Eval. Carcinog. Risks Hum.* **2012**, *100F*, 249–294.
31. Lachenmeier, D.W.; Kuballa, T.; Reusch, H.; Sproll, C.; Kersting, M.; Alexy, U. Benzene in infant carrot juice: Further insight into formation mechanism and risk assessment including consumption data from the DONALD study. *Food Chem. Toxicol.* **2010**, *48*, 291–297. [CrossRef]
32. Loch, C.; Reusch, H.; Ruge, I.; Godelmann, R.; Pflaum, T.; Kuballa, T.; Schumacher, S.; Lachenmeier, D.W. Benzaldehyde in cherry flavour as a precursor of benzene formation in beverages. *Food Chem.* **2016**, *206*, 74–77. [CrossRef]
33. Steinbrenner, N.; Löbell-Behrends, S.; Reusch, H.; Kuballa, T.; Lachenmeier, D.W. Benzol in Lebensmitteln – ein Überblick. *J. Verbrauchersch. Lebensm.* **2010**, *5*, 443–452. [CrossRef]
34. Long, D.G. From cobalt to chloropropanol: De tribulationibus aptis cerevisiis imbibendis. *J. Inst. Brew.* **1999**, *105*, 79–84. [CrossRef]
35. Wu, Q.-J.; Lin, H.; Fan, W.; Dong, J.-J.; Chen, H.-L. Investigation into Benzene, Trihalomethanes and Formaldehyde in Chinese Lager Beers. *J. Inst. Brew.* **2006**, *112*, 291–294. [CrossRef]
36. IARC Working Group on the Evaluation of Carcinogenic Risks to Humans. Furan. *IARC Monogr. Eval. Carcinog. Risks Hum.* **1995**, *63*, 393–407.
37. Chen, L.J.; Hecht, S.S.; Peterson, L.A. Identification of cis-2-butene-1,4-dial as a microsomal metabolite of furan. *Chem. Res. Toxicol.* **1995**, *8*, 903–906. [CrossRef]
38. Peterson, L.A.; Cummings, M.E.; Vu, C.C.; Matter, B.A. Glutathione trapping to measure microsomal oxidation of furan toto cis-2-butene-1,4-dial. *Drug Metab. Dispos.* **2005**, *33*, 1453–1458. [CrossRef]
39. EFSA. Update on furan levels in food from monitoring years 2004–2010 and exposure assessment. *EFSA J.* **2011**, *9*, 2347.
40. IARC Working Group on the Evaluation of Carcinogenic Risks to Humans. Glyphosate. *IARC Monogr. Eval. Carcinog. Risks Hum.* **2015**, *112*, 321–399.
41. Nagatomi, Y.; Yoshioka, T.; Yanagisawa, M.; Uyama, A.; Mochizuki, N. Simultaneous LC-MS/MS Analysis of Glyphosate, Glufosinate, and Their Metabolic Products in Beer, Barley Tea, and Their Ingredients. *Biosci. Biotechnol. Biochem.* **2013**, *77*, 2218–2221. [CrossRef]
42. IARC Working Group on the Evaluation of Carcinogenic Risks to Humans. Alcohol consumption and ethyl carbamate. *IARC Monogr. Eval. Carcinog. Risks Hum.* **2010**, *96*, 1–1428.
43. Uthurry, C.A.; Varela, F.; Colomo, B.; Suárez Lepe, J.A.; Lombardero, J.; García del Hierro, J.R. Ethyl carbamate concentrations of typical Spanish red wines. *Food Chem.* **2004**, *88*, 329–336. [CrossRef]
44. Anderson, L.M.; Souliotis, V.L.; Chhabra, S.K.; Moskal, T.J.; Harbaugh, S.D.; Kyrtopoulos, S.A. N-nitrosodimethylamine-derived O(6)-methylguanine in DNA of monkey gastrointestinal and urogenital organs and enhancement by ethanol. *Int. J. Cancer* **1996**, *66*, 130–134. [CrossRef]

45. Lijinsky, W. N-Nitroso compounds in the diet. *Mutat. Res. Toxicol. Environ. Mutagen.* **1999**, *443*, 129–138. [CrossRef]

46. Tricker, A.R.; Kubacki, S.J. Review of the occurrence and formation of non-volatile N-nitroso compounds in foods. *Food Addit. Contam.* **1992**, *9*, 39–69. [CrossRef]

47. Lachenmeier, D.W.; Fügel, D. Reduction of nitrosamines in beer—Review of a success story. *Brew Sci.* **2007**, *60*, 84–89.

48. IARC Working Group on the Evaluation of Carcinogenic Risks to Humans. Pulegone. *IARC Monogr. Eval. Carcinog. Risks Hum.* **2015**, *108*, 141–154.

49. National Toxicology Program. Toxicology and carcinogenesis studies of pulegone (CAS No. 89-82-7) in F344/N rats and B6C3F1 mice (gavage studies). *Natl. Toxicol. Program Tech. Rep. Ser.* **2011**, *563*, 1–201.

50. SCF. *Opinion of the Scientifc Committee on Food on the Safety of the Presence of Safrole (1-allyl-3,4-methylene Dioxy Benzene) in Flavouring and Other Food Ingredients with Flavouring Properties*; European Commission: Brussels, Belgium, 2002.

51. Curro, P.; Micali, G.; Lanuzza, F. Determination of beta-asarone, safrole, isosafrole and anethole in alcoholic drinks by high-performance liquid chromatography. *J Chromatogr.* **1987**, *404*, 273–278. [CrossRef]

52. Gutsche, B.; Weißhaar, R.; Buhlert, J. Acrylamide in food - Screening results from food control in Baden-Württemberg. *Deut. Lebensm. Rundsch.* **2002**, *98*, 437–443.

53. EFSA. Opinion of the Scientific Panel on Contaminants in the Food Chain on a request from the European Commission on Ethyl Carbamate and Hydrocyanic Acid in Food and Beverages (Question No EFSA-Q-2006-076). *EFSA J.* **2007**, 1–44.

54. Donhauser, S.; Wagner, D.J.F. Critical trace-elements in brewing technology. 2. Occurrence of arsenic, lead, cadmium, chromium, mercury and selenium in beer. *Monatsschr. Brauwiss.* **1987**, *40*, 328–333.

55. Andrey, D. A simple gas chromatography method for the determination of ethylcarbamate in spirits. *Z. Lebensm. Unters. Forsch.* **1987**, *185*, 21–23. [CrossRef]

56. Klejdus, B.; Moravcová, J.; Lojková, L.; Vacek, J.; Kubán, V. Solid-phase extraction of 4(5)-methylimidazole (4MeI) and 2-acetyl-4(5)-(1,2,3,4-tetrahydroxybutyl)-imidazole (THI) from foods and beverages with subsequent liquid chromatographic-electrospray mass spectrometric quantification. *J. Sep. Sci.* **2006**, *29*, 378–384. [CrossRef]

57. Yoshikawa, S.; Fujiwara, M. Determination of 4 (5)-Methylimidazole in Food by Thin Layer Chromatography. *Food Hyg. Saf. Sci. (Shokuhin Eiseigaku Zasshi)* **1981**, *22*, 189–196. [CrossRef]

58. European Commission. European Commission Regulation (EC) No 1334/2008 of the European Parliament and of the Council of 16 December 2008 on flavourings and certain food ingredients with flavouring properties for use in and on foods and amending Council Regulation (EEC) No 1601/91, Regulations (EC). *Off. J. Eur. Union* **2008**, *L 354/34*, 34–50.

59. *BfR Provisional Assessment of Glyphosate in Beer*; BfR Communication No. 005/2016; Bundesinstitut für Risikobewertung (BfR): Berlin, Germany, 25 February 2016.

60. Lachenmeier, D.W.; Przybylski, M.C.; Rehm, J. Comparative risk assessment of carcinogens in alcoholic beverages using the margin of exposure approach. *Int. J. Cancer* **2012**, *131*, 995–1003. [CrossRef]

61. Lachenmeier, D.W.; Gill, J.S.; Chick, J.; Rehm, J. The total margin of exposure of ethanol and acetaldehyde for heavy drinkers consuming cider or vodka. *Food Chem. Toxicol.* **2015**, *83*, 210–214. [CrossRef]

62. Lachenmeier, D.W.; Rehm, J. Comparative risk assessment of alcohol, tobacco, cannabis and other illicit drugs using the margin of exposure approach. *Sci. Rep.* **2015**, *5*, 8126. [CrossRef]

63. Mueller, U.; Agudo, A.; Carrington, C.; Doerge, D.; Hellenäs, K.E.; Leb-lanc, J.C.; Rao, M.; Renwick, A.; Slob, W.; Wu, Y. Acrylamide (Addendum). In *Safety Evaluation of Certain Contaminants in Food. Prepared by the Seventysecond Meeting of the Joint FAO/WHO Expert Committee on Food Additives (JECFA)*; WHO Food Additives Series 63; WHO and FAO: Geneva, Switzerland, 2011; pp. 1–51.

64. National Toxology Program NTP. Technical Report on the Toxicology and Carcinogenesis Studies of Acrylamide in F344/N Rats and B6C3F1 Mice (Feed and Drinking Water Studies). *Natl. Toxicol. Progr. Tech. Rep. Ser.* **2012**, *575*, 1–236.

65. EFSA. Opinion of the scientific panel on contaminants in the food chain [CONTAM] related to the potential increase of consumer health risk by a possible increase of the existing maximum levels for aflatoxins in almonds, hazelnuts and pistachios and derived prod. *EFSA J.* **2007**, *446*, 1–127.

66. Yeh, F.S.; Yu, M.C.; Mo, C.C.; Luo, S.; Tong, M.J.; Henderson, B.E. Hepatitis B virus, aflatoxins, and hepatocellular carcinoma in southern Guangxi, China. *Cancer Res.* **1989**, *49*, 2506–2509. [CrossRef]

67. Benford, D.J.; Alexander, J.; Baines, J.; Bellinger, D.C.; Carrington, C.; Peréz, V.A.; Uxbury, J.; Fawell, J.; Hailemariam, K.; Montoro, R.; et al. Arsenic (Addendum). In *Safety Evaluation of Certain Contaminants in Food. Prepared by the Seventy-Second Meeting of the Joint FAO/WHO Expert Committee on Food Additives (JECFA)*; WHO Food Additives Series 63; WHO and FAO: Geneva, Switzerland, 2011; pp. 153–316.

68. Chen, C.-L.; Chiou, H.-Y.; Hsu, L.-I.; Hsueh, Y.-M.; Wu, M.-M.; Chen, C.-J. Ingested arsenic, characteristics of well water consumption and risk of different histological types of lung cancer in northeastern Taiwan. *Environ. Res.* **2010**, *110*, 455–462. [CrossRef] [PubMed]

69. US EPA. *Benzene (CASRN 71-43-2). Integrated Risk Information System. Document 0276*; US Environmental Protection Agency: Washington, DC, USA, 2003.

70. Rothman, N.; Li, G.L.; Dosemeci, M.; Bechtold, W.E.; Marti, G.E.; Wang, Y.Z.; Linet, M.; Xi, L.Q.; Lu, W.; Smith, M.T.; et al. Hematotoxicity among Chinese workers heavily exposed to benzene. *Am. J. Ind. Med.* **1996**, *29*, 236–246. [CrossRef]

71. Lachenmeier, D.W.; Kanteres, F.; Rehm, J. Epidemiology-based risk assessment using the benchmark dose/margin of exposure approach: The example of ethanol and liver cirrhosis. *Int. J. Epidemiol.* **2011**, *40*, 210–218. [CrossRef] [PubMed]
72. Beland, F.A.; Benson, R.W.; Mellick, P.W.; Kovatch, R.M.; Roberts, D.W.; Fang, J.-L.; Doerge, D.R. Effect of ethanol on the tumorigenicity of urethane (ethyl carbamate) in B6C3F1 mice. *Food Chem. Toxicol.* **2005**, *43*, 1–19. [CrossRef]
73. National Toxicology Program. NTP technical report on the toxicology and carcino- gensis. Studies of urethane, ethanol, and urethane/ethanol (urethane, CAS No. 51-79-6; ethanol, CAS No. 64-17-5) in B6C3F1 mice (drinking water studies). *Natl. Toxicol. Progr. Tech. Rep. Ser.* **2004**, *510*, 1–346.
74. Vavasour, E.; Renwick, A.G.; Engeli, B.; Barlow, S.; Castle, L.; DiNovi, M.; Slob, W.; Schlatter, J.; Bolger, M. Ethyl carbamate. In *Safety Evaluation of Certain Contaminants in Food. Prepared by the Sixty-Fourth Meeting of the Joint FAO/WHO Expert Committee on Food Additives (JECFA)*; WHO Food Additives Series 55; WHO and FAO: Geneva, Switzerland, 2006.
75. IPCS. *Environmental Health Criteria 239: Principles for Modelling Dose-Response for the Risk Assessment of Chemicals*; World Health Organization: Geneva, Switzerland, 2009.
76. Til, H.P.; Woutersen, R.A.; Feron, V.J.; Hollanders, V.H.; Falke, H.E.; Clary, J.J. Two-year drinking-water study of formaldehyde in rats. *Food Chem. Toxicol.* **1989**, *27*, 77–87. [CrossRef]
77. Williams, G.M.; Arisseto, A.P.; Baines, J.; DiNovi, M.; Feeley, M.; Schlatter, J.; Slob, W.; Toledo, M.C.F.; Vavasour, E. Furan. In *Safety Evaluation of Certain Contaminants in food. Prepared by the Seventy-Second Meeting of the Joint FAO/WHO Expert Committee on Food Additives (JECFA)*; WHO Food Additives Series 63; WHO and FAO: Geneva, Switzerland, 2011.
78. Moser, G.J.; Foley, J.; Burnett, M.; Goldsworthy, T.L.; Maronpot, R. Furan-induced dose–response relationships for liver cytotoxicity, cell proliferation, and tumorigenicity (furan-induced liver tumorigenicity). *Exp. Toxicol. Pathol.* **2009**, *61*, 101–111. [CrossRef]
79. EFSA. Conclusion on the peer review of the pesticide risk assessment of the active substance glyphosate. *EFSA J.* **2015**, *13*, 4302.
80. Navas-Acien, A.; Tellez-Plaza, M.; Guallar, E.; Muntner, P.; Silbergeld, E.; Jaar, B.; Weaver, V. Blood Cadmium and Lead and Chronic Kidney Disease in US Adults: A Joint Analysis. *Am. J. Epidemiol.* **2009**, *170*, 1156–1164. [CrossRef]
81. Abraham, K.; Mielke, H.; Lampen, A. Hazard characterization of 3-MCPD using benchmark dose modeling: Factors influencing the outcome. *Eur. J. Lipid Sci. Technol.* **2012**, *114*, 1225–1226. [CrossRef]
82. Cho, W.S.; Han, B.S.; Nam, K.T.; Park, K.; Choi, M.; Kim, S.H.; Jeong, J.; Jang, D.D. Carcinogenicity study of 3-monochloropropane-1,2-diol in Sprague-Dawley rats. *Food Chem. Toxicol.* **2008**, *46*, 3172–3177. [CrossRef]
83. EFSA. Scientific Opinion on the re-evaluation of caramel colours (E 150 a,b,c,d) as food additives. *EFSA J.* **2011**, *9*, 2004.
84. National Toxicology Program. Toxicology and carcinogenesis studies of 4-methylimidazole (Cas No. 822-36-6) in F344/N rats and B6C3F1 mice (feed studies). *Natl. Toxicol. Program Tech. Rep. Ser.* **2007**, *535*, 1–274.
85. Zeilmaker, M.J.; Bakker, M.I.; Schothorst, R.; Slob, W. Risk Assessment of N-nitrosodimethylamine Formed Endogenously after Fish-with-Vegetable Meals. *Toxicol. Sci.* **2010**, *116*, 323–335. [CrossRef]
86. Peto, R.; Gray, R.; Brantom, P.; Grasso, P. Dose and time relationships for tumor induction in the liver and esophagus of 4080 inbred rats by chronic ingestion of N-nitrosodiethylamine or N-nitrosodimethylamine. *Cancer Res.* **1991**, *51*, 6452–6469.
87. Peto, R.; Gray, R.; Brantom, P.; Grasso, P. Effects on 4080 rats of chronic ingestion of N-nitrosodiethylamine or N-nitrosodimethylamine: A detailed dose-response study. *Cancer Res.* **1991**, *51*, 6415–6451.
88. Barlow, S.; Bolger, M.; Pitt, J.I.; Verger, P. Ochratoxin A (addendum). In *Safety Evaluation of Certain Contaminants in Food. Prepared by the Sixty-Eighth Meeting of the Joint FAO/WHO Expert Committee on Food Additives (JECFA)*; WHO Food Additives Series 59; WHO and FAO: Geneva, Switzerland, 1989.
89. National Toxicology Program. Toxicology and Carcinogenesis Studies of Ochratoxin A (CAS No. 303-47-9) in F344/N Rats (Gavage Studies). *Natl. Toxicol. Program Tech. Rep. Ser.* **1989**, *358*, 1–142.
90. E.M.A. *Public Statement on the Use of Herbal Medicinal Products Containing Pulegone and Menthofuran. EMEA/HMPC/138386/2005*; European Medicines Agency: London, UK, 2016.
91. Martati, E.; Boersma, M.G.; Spenkelink, A.; Khadka, D.B.; Punt, A.; Vervoort, J.; Bladeren, P.J.; Rietjens, I.M. Physiologically based biokinetic (PBBK) model for safrole bioactivation and detoxification in rats. *Chem. Res. Toxicol.* **2011**, *24*, 818–834. [CrossRef]
92. Boberg, E.W.; Miller, E.C.; Miller, J.A.; Poland, A.; Liem, A. Strong evidence from studies with brachymorphic mice and pentachlorophenol that 1′-sulfoöxysafrole is the major ultimate electrophilic and carcinogenic metabolite of 1′-hydroxysafrole in mouse liver. *Cancer Res.* **1983**, *43*, 5163–5173.
93. Miller, E.C.; Swanson, A.B.; Phillips, D.H.; Fletcher, T.L.; Liem, A.; Miller, J.A. Structure-activity studies of the carcinogenicities in the mouse and rat of some naturally occurring and synthetic alkenylbenzene derivatives related to safrole and estragole. *Cancer Res.* **1983**, *43*, 1124–1134.
94. Burton, R.; Sheron, N. No level of alcohol consumption improves health. *Lancet* **2018**, *392*, 987–988. [CrossRef]
95. Astrup, A.; Estruch, R. Alcohol and the global burden of disease. *Lancet* **2019**, *393*, 2390. [CrossRef]
96. Room, R.; Babor, T.; Rehm, J. Alcohol and public health. *Lancet* **2005**, *365*, 519–530. [CrossRef]
97. Rehm, J.; Room, R.; Monteiro, M.; Gmel, G.; Graham, K.; Rehn, N.; Sempos, C.T.; Jernigan, D. Alcohol as a risk factor for global burden of disease. *Eur. Addict. Res.* **2003**, *9*, 157–164. [CrossRef]
98. Rehm, J.; Mathers, C.; Popova, S.; Thavorncharoensap, M.; Teerawattananon, Y.; Patra, J. Global burden of disease and injury and economic cost attributable to alcohol use and alcohol-use disorders. *Lancet* **2009**, *373*, 2223–2233. [CrossRef]

99. Bardach, A.E.; Alcaraz, A.O.; Ciapponi, A.; Garay, O.U.; Riviere, A.P.; Palacios, A.; Cremonte, M.; Augustovski, F. Alcohol consumption's attributable disease burden and cost-effectiveness of targeted public health interventions: A systematic review of mathematical models. *BMC Public Health* **2019**, *19*, 1378. [CrossRef]

100. Babor, T.; Holder, H.; Caetano, R.; Homel, R.; Casswell, S.; Livingston, M.; Edwards, G.; Österberg, E.; Giesbrecht, N.; Rehm, J.; et al. Strategies and interventions to reduce alcohol-related harm. In *Alcohol: No Ordinary Commodity*; Oxford University Press Inc.: New York, NY, USA, 2010; Volume 58, pp. 103–108.

101. Rehm, J.; Lachenmeier, D.W.; Llopis, E.J.; Imtiaz, S.; Anderson, P. Evidence of reducing ethanol content in beverages to reduce harmful use of alcohol. *Lancet Gastroenterol. Hepatol.* **2016**, *1*, 78–83. [CrossRef]

102. Segal, D.S.; Stockwell, T. Low alcohol alternatives: A promising strategy for reducing alcohol related harm. *Int. J. Drug Policy* **2009**, *20*, 183–187. [CrossRef] [PubMed]

103. Geller, E.S.; Kalsher, M.J.; Clarke, S.W. Beer versus mixed-drink consumption at fraternity parties: A time and place for low-alcohol alternatives. *J. Stud. Alcohol* **1991**, *52*, 197–204. [CrossRef] [PubMed]

Review
Alcohol and the Brain

David Nutt *, Alexandra Hayes, Leon Fonville , Rayyan Zafar, Emily O.C. Palmer , Louise Paterson
and Anne Lingford-Hughes

Neuropsychopharmacology Unit, Division of Psychiatry, Department of Brain Sciences, Hammersmith Hospital,
Imperial College London, London W12 ONN, UK; alexandra.hayes17@imperial.ac.uk (A.H.);
l.fonville@imperial.ac.uk (L.F.); r.zafar19@imperial.ac.uk (R.Z.); e.palmer@imperial.ac.uk (E.O.C.P.);
l.paterson@imperial.ac.uk (L.P.); anne.lingford-hughes@imperial.ac.uk (A.L.-H.)
* Correspondence: d.nutt@imperial.ac.uk; Tel.: +44-2075946628

Abstract: Alcohol works on the brain to produce its desired effects, e.g., sociability and intoxication, and hence the brain is an important organ for exploring subsequent harms. These come in many different forms such as the consequences of damage during intoxication, e.g., from falls and fights, damage from withdrawal, damage from the toxicity of alcohol and its metabolites and altered brain structure and function with implications for behavioral processes such as craving and addiction. On top of that are peripheral factors that compound brain damage such as poor diet, vitamin deficiencies leading to Wernicke-Korsakoff syndrome. Prenatal alcohol exposure can also have a profound impact on brain development and lead to irremediable changes of fetal alcohol syndrome. This chapter briefly reviews aspects of these with a particular focus on recent brain imaging results. Cardiovascular effects of alcohol that lead to brain pathology are not covered as they are dealt with elsewhere in the volume.

Keywords: alcohol; brain; addiction; alcoholism

Citation: Nutt, D.; Hayes, A.; Fonville, L.; Zafar, R.; Palmer, E.O.C.; Paterson, L.; Lingford-Hughes, A. Alcohol and the Brain. *Nutrients* **2021**, *13*, 3938. https://doi.org/10.3390/nu13113938

Academic Editor: Peter Anderson

Received: 22 September 2021
Accepted: 2 November 2021
Published: 4 November 2021

Publisher's Note: MDPI stays neutral with regard to jurisdictional claims in published maps and institutional affiliations.

Terminology describing alcohol consumption has varied over the years but generally drinking at a level causing mental and/or physical ill health is referred to as harmful (ICD-10/11) or abuse (DSM-IV). Alcohol dependence or addiction or alcoholism is a complex behavioral syndrome that has at its core the inability to control consumption despite adverse social, occupational or health consequences (ICD-10/11; DSM-IV). More recently DSM-5 criteria use the term alcohol use disorder (AUD) as a continuum with mild equivalent to abuse and moderate-severe equivalent to dependence [1]. Whilst most social drinkers remain in control of their engagement with alcohol, a small but significant minority transition to dependence. Alcoholism is a global public health emergency with the World Health Organization estimating prevalence of 4% and is associated with c. 3 million deaths annually [2]. There is a high rate of relapse; around 80% of dependent individuals relapse within a year with current available therapies [3]. Current adjunctive pharmacotherapies have only mild-moderate effects on alcohol consumption and relapse prevention [4,5] and no there are no rescue medications available to counteract the adverse effects of intoxication [6]. Neuroimaging studies have revolutionized our understanding of brain neurochemistry and function in response to chronic alcohol consumption and dependence, but so far, this has not led to huge improvements in treatment availability or efficacy. In order to improve treatment outcomes, a detailed understanding of the neurobiological mechanisms responsible for vulnerability, relapse and successful recovery and the identification of novel biomarkers to develop more efficacious therapeutic targets are warranted.

1. Structural and Volumetric Changes

1.1. Structural Alterations in Adults

The link between alcohol use and cerebral atrophy goes back decades, with early findings coming from post-mortem investigations [7] and subsequent in vivo examinations of gross morphology using computerized tomography (CT) [8–10]. The emergence of magnetic resonance imaging (MRI) brought vast improvements to image resolution and allowed for differentiation of brain tissue. Whilst the early CT scans pointed towards ventricular enlargement and widening of cortical sulci, MRI studies have provided a wealth of evidence towards distinct differences in grey and white matter associated with chronic alcohol use [11]. Grey matter volume loss is commonly observed in alcohol dependence and effects are widespread across cortical and subcortical regions [12,13], though meta-analytic efforts have pointed specifically towards volume loss in the prefrontal cortex, cingulate cortex, insula, and striatum in particular [14–16]. A meta-regression analysis further showed that the impact on grey matter was linked to lifetime alcohol consumption and duration of alcohol dependence [16]. Macrostructural differences can also be observed in white matter volume [17] but more importantly there are noticeable differences in microstructure [18], most notably in the corpus callosum, highlighting potential disruption to myelination and axonal integrity. It should be noted that the impact of alcohol on the cerebellar structure has been relatively understudied and most MRI research has focused on cortical and subcortical structures. The cerebellum is known to be affected in alcohol dependence, even more so in those with additional neurological complications such as Wernicke-Korsakoff syndrome, and studies have found a loss of cerebellar volume that further increases with age [19].

Reductions in brain volume are not necessarily irreversible and early CT studies had already shown that brain volume appears to partially recover with abstinence from alcohol [20,21]. Longitudinal MRI studies further showed that changes to volume follow a non-linear pattern with greater increases occurring in the early stages of abstinence [22–24]. Furthermore, reducing alcohol consumption to an average of 20 drinks per month rather than abstaining completely was sufficient to produce increases in brain volume compared with those who returned to patterns of heavy drinking that matched pre-treatment levels (consuming an average of 155 drinks per month) [25]. Though evidence in white matter is limited, it does suggest a similar pattern of recovery with abstinence exists [26,27]. An interesting finding from longitudinal MRI studies has been that people prone to future relapses are distinguishable from those able to abstain [28–31], suggesting there might be biological differences that play a role in treatment progression.

There is a longstanding notion that alcohol has an interactive effect on the biological aging processes, whereby the brains of alcohol dependent individuals resemble those of chronologically older individuals who do not have alcohol dependence [32]. Imaging studies have long found that the loss of grey matter volume as well as the disturbances to white matter microstructure typically seen in alcohol dependence are exacerbated with age [10,27,33–38]. This phenomenon has also been investigated using the brain age paradigm, an approach that investigates healthy brain aging by estimating chronological age from neuroimaging data and examines the difference between an individual's predicted and actual age [39]. One study found that individuals with alcohol dependence showed a difference of up to 11.7 years between their chronological and predicted biological age based on their grey matter volume [33]. Crucially, the difference showed a linear increase with age and was at its greatest in old age which further offers support to the notion of a greater vulnerability to the effects of alcohol in later life.

Low levels of alcohol consumption have historically been viewed as harmless or even beneficial due to its potentially favorable effects on cardiovascular health [40], as described in more detail elsewhere in this special issue. It has been suggested a similar J- or U-shaped relationship exists with brain structure, but only a few MRI studies have found support for this assumption [41–43]. More recent large-scale studies of the general population have in fact shown negative linear associations between alcohol consumption and brain structure, showing widespread reductions in grey and white matter volume as well as white matter microstructure and cortical thickness [44–47]. Taken together, these studies highlight that there is no protective effect of light drinking (e.g., 1–<7 drinks per week) over abstinence and in fact show, similar to alcohol dependence though to a lesser degree, that any level of alcohol consumption can affect brain volume and white matter microstructure. It should be noted that these are cross-sectional association studies, and it is not possible to infer causality or estimate the impact of alcohol on the brain over time. Furthermore, only a small proportion of the variance in brain structure is explained by alcohol consumption [44,46]. A study on brain age further found that only those who drank daily or almost daily showed a difference between chronological and predicted brain age, and there were no differences in those who drank less frequently or abstained from alcohol [48]. This suggests that whilst low levels of alcohol consumption may affect brain structure, this may not negatively impact healthy brain aging. An important caveat to the generalizability is the relatively small amount of large-scale imaging datasets and the findings discussed here primarily relate to the UK BioBank dataset [44,46–48].

1.2. Structural Alterations in Adolescents

Alcohol use is typically initiated during adolescence, and studies have found that alcohol can impact neurodevelopmental trajectories during this period. Typical brain maturation can be characterized as a loss in grey matter density due to synaptic pruning alongside ongoing growth of white matter volume that reflects increased myelination to strengthen surviving connections [49]. In adolescents who engage in heavy and binge drinking, as defined by the National Institute on Alcohol Abuse and Alcoholism [50], there is a greater decrease in grey matter volume along with an attenuated increase of white matter volume as well as disturbances in white matter microstructure in comparison to non-drinkers [51]. These effects are found in prefrontal, cingulate, and temporal regions as well as the corpus callosum and may reflect an acceleration of typical age-related developmental processes similar to what we have described in adults with alcohol dependence. Less is known about the dose-response mechanism, though it has been suggested moderate drinking lies somewhere intermediate [52,53]. This would again imply that the impact of alcohol consumption on brain structure is not limited to heavy alcohol consumption. However, it has been noted there are differences in brain structure that predate alcohol initiation and may predispose individuals to heavy alcohol use. Structural precursors have mostly been found in the prefrontal cortex and fronto-limbic white matter and show considerable overlap with structural differences found in individuals with a family history of alcohol dependence [54]. Nevertheless, there are studies that have suggested differences are not solely attributable to familial risk [55,56], and more research is needed to better understand these risk factors.

1.3. Pre-Natal Alcohol Exposure

A far more severe disruption to neurodevelopment comes from prenatal exposure to alcohol—collectively known as fetal alcohol spectrum disorders (FASD)—which can encompass cognitive deficiencies, neurobehavioral disorders, growth retardation, craniofacial dysmorphism, and deficits to the central nervous system including brain malformations [57]. Early case studies highlighted striking morphological anomalies, most notably thinning of the corpus callosum and enlargement of ventricles, but subsequent radiological investigations have highlighted there is considerable variability in the impact of FASD on brain development [58]. Quantitative analyses of brain macrostructure in FASD have repeatedly found lower grey and white matter volume along with increased thickness and density of cortical grey matter [59]. Crucially, findings have found no morphological differences in the occipital lobe, suggesting that not all brain structures are affected equally. Brain phenotypes of FASD have consistently been recapitulated in animal models and highlight the modulating role of timing and alcohol exposure [60]. Taken together, it is clear that the teratogenic effects of alcohol on brain structure are widespread and can be seen across the spectrum of FASD. However, understanding the link between these structural alterations and other parameters of FASD remains an ongoing challenge.

1.4. Conclusion

In summary, MRI studies have offered invaluable insight into the effects of alcohol and have typically found a loss of volume and reduced myelination throughout the brain. The findings described here fit the notion that alcohol affects healthy brain aging and this effect becomes more pronounced with higher levels of consumption. It also suggests that there may be a greater vulnerability to the effects of alcohol on brain health with old age. The impact of alcohol can be observed early on, moderate to heavy drinking during adolescence leads to observable differences to non-drinkers, but this is further confounded by risk factors to unhealthy drinking patterns and alcohol dependence. However, though MRI research will be important in advancing our understanding of the impact of alcohol on the brain we cannot infer harm solely from alterations to brain structure.

2. Neurotoxic Properties of Alcohol

2.1. Thiamine Deficiency

Thiamine, also known as Vitamin B_1, is an essential cofactor that is critical for nerve function. It is required for the functioning of several enzymes responsible for carbohydrate and lipid metabolism and the generation of several essential molecules including nucleic acids and neurotransmitters [61,62]. The human body is incapable of producing thiamine itself and therefore it must be derived solely from diet. Thiamine deficiency predominantly occurs as a result of malnutrition, and in most western countries is most commonly (90%) associated with alcoholism [63]. Over time, thiamine deficiency can result in nerve damage, leading to alcoholic neuropathy.

Much of the calorific intake of individuals with alcoholism comes from their consumption of alcoholic beverages, which are generally low in nutrient content. Vitamin supplementation in alcoholic beverages can play a role in mitigation of these deficiencies, e.g., some Danish beer brands contain vitamins (B6 in addition to thiamine) to 'normalize' blood thiamine levels in those with alcoholic neuropathy [64]. However, individuals with alcoholism require levels of thiamine supplementation much higher than that required from the average diet, so control of dietary intake alone is not the whole story [61]. In addition to thiamine deficiency AUD is associated with other malnutrition and other micronutrient deficiencies which are comprehensively reviewed by McLean et al. [65].

Alcohol use can also cause thiamine deficiency by disrupting absorption in the gastrointestinal tract. Alcohol damages the mucosa of the gut and reduces intestinal thiamine transport. Studies in both humans and rodents have demonstrated that thiamine is transported via an active sodium independent transporter and therefore requires both energy and a normal pH level [66–68], both of which are reduced in alcoholism. Additionally, thiamine absorption can further be depleted by diarrhoea or vomiting which are common occurrences in alcoholism. It is also important to note that thiamine absorption in the gut can be altered by several genetic variants that affect thiamine transport and metabolism [69].

Alcohol also hinders the ability of cells to utilize the thiamine. Thiamine requires phosphorylation by thiamine pyrophosphokinase to be converted to its active co-enzyme form. Thiamine pyrophosphokinase is inhibited by alcohol, which also increases the rate of thiamine metabolism [63]. This phosphorylation step requires magnesium as a cofactor, which is also depleted in alcoholism [70]. Cumulatively, alcoholism leads to thiamine deficiency via the reduction of intake, uptake, and utilization.

Chronic thiamine deficiency ultimately leads to neurotoxicity. As previously mentioned, thiamine is an essential cofactor required for the synthesis and function of several essential enzymes. One of these enzymes is transketolase which is required for glucose breakdown via the pentose phosphate pathway. This pathway produces several essential products. The first is Ribose-5-Phosphate which is required for the synthesis of nucleic acids and other complex sugars. The second is nicotinamide adenine dinucleotide phosphate (NADPH) which is required in the assembly of coenzymes, steroids, fatty acids, amino acids, neurotransmitters, and glutathione [61]. The reduction in production of these factors in addition to thiamine deficiency interrupts the cells' defense mechanisms, notably the ability to reduce reactive oxygen species (ROS), leading to cellular damage. This damage then triggers apoptosis (cell death). Another mechanism by which thiamine deficiency leads to cytotoxicity is by affecting carbohydrate metabolism leading to the reduction of the enzyme α-Ketoglutarate Dehydrogenase, leading to mitochondrial damage, which in turn induces necrosis [61].

The brain is the most energy-utilizing organ in the body, necessitating a constant supply of energy to function. Without the breakdown of glucose and the subsequent production of essential molecules, thiamine deficiency leads to brain dysfunction and degeneration. The most extreme outcome of alcohol-induced thiamine deficiency-related neurotoxicity is the development of Wernicke–Korsakoff Syndrome (WKS). WKS refers to the closely associated conditions of Wernicke's encephalopathy (WE) and Korsakoff's Psychosis (KP). WE is an acute, but reversable condition characterized by confusion, oculomotor disturbances and ataxia [61]. However, these are all rarely seen together so a high index of clinical suspicion should be maintained. WE is caused by thiamine deficiency induced energy deficits and glutamate increases leading to cytotoxicity [63]. WE can develop into non-reversable brain damage (KP) relating to behavior abnormalities and memory impairments.

Both WE and KP are well characterized disorders associated with a distinct clinical and neuropathological presentation. More broadly neurological structural and functional consequences of alcoholism are called alcohol-related brain damage (ARBD). Reduced cognitive functioning is prevalent in between 50% and 80% of individuals with AD [71]. These deficits are often transient [72] and are not normally as severe as those observed in patients with WE or KP [73], but instead it has been hypothesized that ARBD may evolve to KP via progressive brain damage and associated cognitive impairment [74]. Despite individual variations in severity, it is well established that thiamine deficiency leads to neurotoxicity with negative consequences for cognitive functioning.

2.2. Neurotoxicity of Acetaldehyde

Alcohol is metabolized to acetaldehyde, via the action of alcohol dehydrogenase (ADH), CYP2E1 and catalase. Acetaldehyde is known to be toxic active metabolite, it is implicated in; the induction of alcoholic cardiomyopathy [75], the development of cancers [76] and to have some neurobehavioral effects [77]. During intoxication the production of acetaldehyde can cause flushing, increased heart rate, dry mouth, nausea and headache [78]. Notably, Acetaldehyde contributes to toxic effects of chronic alcohol on the brain leading to neuronal degeneration [79]. Acetaldehyde induces cell damage and cytotoxicity by inducing DNA malfunction and protein adducts [78]. Additionally, this protein adduct formation can also induce an immune response which can further damage tissues.

2.3. Alcohol and Neuroinflammation

In addition to thiamine-deficiency and acetaldehyde related toxicity, alcohol can also cause damage via peripheral and neuro-inflammatory mechanisms. Studies in rodents have demonstrated that alcohol stimulates intestinal inflammation by irritating the stomach and gut, causing the release of the nuclear protein high-mobility group box 1 (HMGB1), which subsequently activate Toll-like receptor 4 (TLR4) and makes the gut "leaky" [80]. This makes alcohol and endotoxins more likely to cross the lining of the gut and travel via the circulation to the liver. Further alcohol metabolism and increases in bacteria cause the liver to produce inflammatory factors such as pro-inflammatory cytokines [81]. This cumulatively increases levels of circulating pro-inflammatory cytokines which can cross the blood brain barrier (BBB) and cause inflammation in the brain [82].

Alcohol also induces neuroinflammation via alterations in neurotransmitter levels. Alcohol is known to increase glutamate levels via the inhibition of the N-methyl-D-aspartate (NMDA) receptor [83] and its cellular action on glutamatergic neurons [84]. Elevated levels of glutamate in a rodent binge drinking model are associated with increased microglial activation in the hippocampus [85]. In addition, alcohol also activates the body's main stress response system, the hypothalamic–pituitary–adrenal (HPA) axis. When activated, the HPA axis results in the release of corticotropin-releasing hormone (CRH), which acts to suppress peripheral inflammation but increases neuroinflammation via a complex regulation of NK-cells, [81] and by potentiating NF-κB activation in the rodent prefrontal cortex [86]. This combination of increased glutamate and CRH levels enhance the ability of alcohol to induce neuroinflammation and cause subsequent tissue damage.

Alcohol can also directly induce neuroinflammation via the activation of resident neuro-immune cells (microglia and astrocytes). Microglia respond to pathogens, tissue damage, cell death and degeneration. They can respond in a pro- or anti- inflammatory way and depending on the type of activation will produce neurotoxic or neuroprotective mediators [81]. Once activated, alterations in microglial gene expression can potentially lead to increases in inflammatory mediators such as cytokines, glutamate and reactive oxygen species (ROS). These mediators are associated with microglial-dependent neuronal loss [87]. Microglial activation is characterized not only by changes in gene expression, but also by characteristic changes in morphology. Microglia normally survey the brain tissue in a ramified shape with several projecting processes, however once activated the processes shrink and thicken, and the cells gradually become ameboid in shape [88]. Chronic alcohol consumption can therefore cause the de-regulation of microglial activation. This in turn can lead to degeneration of brain tissue and is likely associated with brain volume loss [46], as covered in Section 1.

Alcohol is thought to activate microglia partially via TLR4 receptors, indeed TLR4 deficiency protected against alcohol induced glial activation and neurotoxicity in a rodent model of chronic alcohol consumption [89]. Several studies have investigated the effect of alcohol administration on microglia. Analysis of post-mortem brains of patients with Alcohol Use Disorder showed in increase in microglial markers (Iba1 and GluT5) compared with controls [82]. Binge alcohol administration in adolescent rats established microglial proliferation and morphological changes [90]. However, the activation was described as only partial due to the lack of alteration alcohol had on levels of MHC-II or TNF-α expression. Conversely, microglial activation and neurodegeneration were clearly shown in rats exposed to intermittent alcohol treatment [91]. Indeed two-photon microscopy has been used to demonstrate the rapid response of microglia to even single acute alcohol exposure [92]. Microglial activation has also been investigated in response to heavy session intermittent drinking in rodents [93]. It has been suggested that peripheral inflammation could be caused by stimulation of systemic monocytes and macrophages or by causing gastrointestinal mucosal injury [93]. This innate response was linked to the perpetuation of the immune cascade via microglial activation which produces neuroinflammation [94] this, in turn has been shown to affect cognitive function [93]. Initial transcriptome studies indicated that alcohol increased levels of TSPO (18 kDa translocator protein, that is upregulated in activated microglia). These findings were supported by PET studies performed in baboons [95]. However, when TSPO binding was analyzed using PET in alcohol dependent individuals and individuals undergoing detoxification these findings were not replicated [96,97]. Cumulatively, this evidence suggests that alcohol is clearly an activator of microglia, and as previously described upregulation of microglial activation can result in neurotoxicity. However, the extent of alcohol induced microglial activation may well be dependent on the extent and pattern of alcohol exposure.

2.4. Conclusion

In summary, alcohol can contribute to neurotoxicity via thiamine deficiency, metabolite toxicity and neuroinflammation. Alcohol reduces the uptake and metabolism of thiamine, the essential co-factor without which glucose breakdown and the production of essential molecules cannot occur. This leads to neurotoxicity and can lead to the development of conditions of WE and KP. The metabolism of alcohol itself can also lead directly to neurotoxicity as the metabolite acetaldehyde is toxic and can lead to neurodegeneration.

Finally, alcohol can lead to neurotoxicity via the induction of both the central and peripheral immune system, causing damaging levels of inflammation.

3. Functional Brain Changes

In addition to structural alterations, evidence suggests that chronic exposure to alcohol can lead to functional dysregulation of key brain systems that control behaviour such as reward processing, impulse control and emotional regulation. This likely contributes to the pathophysiology of alcohol misuse and addiction. In recent years, functional magnetic resonance imaging (fMRI) has been used to probe these pathways via blood oxygen level dependent (BOLD) signal in the brain both at rest and during the performance of neurocognitive tasks in an MRI scanner.

3.1. Reward Processing

The reward system is in part controlled by the dopaminergic mesolimbic pathway. Originating in the ventral tegmental area. (VTA), dopaminergic projections extend through the striatum and prefrontal regions of the brain. The reward system is responsible for goal-directed behavior by means of reinforcement and responds to conventional rewards such as food and money, as well as all known drugs of abuse. Drugs of abuse, including alcohol, interact with and influence this system and several fMRI paradigms have been developed to probe such effects. One of the most commonly used to probe non-drug related reward sensitivity is the monetary incentive delay (MID) task [98], whereas to measure drug-related reward, cue-reactivity tasks are usually employed [99]. Most commonly these tasks consist of presenting the individual with static or video imagery of a 'cue', typically drug or related paraphernalia, however, smell and taste can also be used.

In response to alcohol-related cues, abstinent alcohol-dependent individuals demonstrate an increased BOLD signal in reward-related fronto-striatal brain regions compared with healthy controls, notably prefrontal cortex, ventral striatum (VS), orbitofrontal cortex (OFC) and anterior cingulate cortex (ACC), which is often associated with increased craving [99]. In contrast, a blunted BOLD response to anticipation of non-drug rewards has been observed in the VS and dorsal striatum (DS) [100]. Together, these findings suggest that in alcohol dependence the reward system attributes excess salience to alcohol-related stimuli while simultaneously responding less to conventional rewards.

Interestingly, evidence suggests that dysregulation of the reward system in abstinent alcohol-dependent individuals can be ameliorated by pharmacological intervention. For example, naltrexone, a µ-opioid receptor antagonist, can attenuate the increased BOLD response to alcohol-related cues in the putamen and reduce risk of relapse [101].

Alcohol-related functional differences in the brain are not exclusively observed in dependent individuals. When comparing the neural response of light (consuming ~0.4 drinks per day) and heavy (consuming ~5 drinks per day) drinkers to alcohol cues, light drinkers have been found to have a higher BOLD signal in VS, while heavy drinkers show an increased BOLD signal in DS [102]. The DS response in the heavy drinkers suggests the initiation of a shift from experimental to compulsive alcohol use during which a shift in neural processing is thought to occur from VS to DS control [103]. However, such cross-sectional studies are unable to establish whether such differences are prodromal or consequential of alcohol exposure. A recent longitudinal study in adolescents showed that blunted BOLD response to non-drug reward was predictive of subsequent problematic alcohol use [104]. Similarly, college students who transitioned from moderate (consuming less than 30 drinks per month) to heavy (consuming more than 30 drinks per month) alcohol consumption exhibited hyperactivity in the striatum, OFC, ACC and insula in response to alcohol-related cues compared with those whose alcohol consumption did not alter over time [105]. These results suggests that certain functional differences in reward processing may predate problematic alcohol consumption.

3.2. Impulsivity

Impulsivity, a term used to describe a lack of inhibitory control characterized by reckless behavior in the absence of premeditation, has multiple domains including choice, trait, and response inhibition [106]. Increased impulsivity is thought to be a determinant and a consequence of alcohol use [107]. At the behavioral level, alcohol intoxication has been shown to increase risky behaviors such as risky driving, criminal behavior, and sexual promiscuity [108], whilst trait impulsivity has often been found to be increased in alcohol dependent individuals [109].

To probe impulsiveness through fMRI, response inhibition tasks are commonly used, such as the Go/no-go (GNG) task and Stop Signal Task (SST). Several longitudinal studies have probed response inhibition in adolescent drinkers. Such studies have found that adolescents who later transitioned into heavy drinking had lower BOLD activation at baseline and increased activation in frontal regions when subsequently drinking heavily compared with continuous non-drinkers [110,111]. This supports the role of impaired response inhibition as a risk factor rather than a consequence of alcohol consumption.

Choice impulsivity, the tendency to make choices that lead to suboptimal, immediate or risky outcomes is often measured using a delay discounting task to assess an individual's preference for a smaller, immediate reward compared with a larger, delayed reward [112]. The literature regarding the effects of alcohol on choice impulsivity is varied with findings that alcohol (0.7 g of alcohol/kg body weight) consumption decreased choice impulsivity in non-dependent drinkers [113], whereas another found alcohol (0.2–0.8 g of alcohol/kg body weight) intoxication increased choice impulsivity [114]. Individuals who scored higher in trait impulsivity measures exhibited greater choice impulsivity than their lower trait impulsive counterparts [115].

3.3. Emotional Regulation

Altered emotional processing has been found both during alcohol intoxication and dependence and appears to worsen as consumption increases. At the behavioral level, binge drinkers, as defined by scoring in the top third of the Alcohol Use Questionnaire (AUQ), report reduced positive mood and alcohol dependent individuals are more likely to interpret disgusted faces as angry faces and demonstrate a bias for fear recognition in facial expressions when fearful faces are morphed with happy, surprised, sad, disgusted or angry faces [116].

The brain mechanisms of emotional regulation can be measured using imagery tasks where participants are shown either faces expressing emotions or evocative/aversive images designed to evoke emotional responses. In heavy drinkers, who consume more than 15 drinks a week, a blunted BOLD response to fearful faces in the amygdala, a region of the brain involved in fear conditioning and stress responses, has been found to be associated with drinking level in the previous 3 months and higher scores on an obsessive-compulsive drinking scale [117]. These findings suggest that acute intoxication diminishes one's ability to process emotional information accurately and that this may be perpetuated with heavy alcohol consumption.

3.4. Resting State Functional Connectivity

Resting state functional connectivity (RSFC) is a technique that quantifies connections between brain regions based on temporal correlation of BOLD signal change. In a recent UK BioBank study of 25,378 individuals, increased within-network connectivity was identified within the default mode network (DMN) in those with higher alcohol consumption [46]. The DMN is believed to be involved in the processing of self-awareness, negative emotions, and rumination, so increased connectivity within this network may infer a decreased responsiveness to external incentives and increased rumination towards alcohol-related cues [118].

Interestingly, in abstinent alcohol-dependent individuals, RSFC was increased between the amygdala and the substantia nigra/VTA and associated with increased lifetime exposure to alcohol [119]. Such differences in abstinent individuals can suggest a pathological change in brain function after chronic exposure to alcohol or a mechanism for successful abstinence. The latter proposal is corroborated by Beck at al., 2012 [120] who found that hyperconnectivity between these regions during a cue reactivity task was associated with successful maintenance of abstinence.

3.5. Conclusion

fMRI studies have allowed us to identify the effects of alcohol use and dependence on brain function as well as vulnerability to heavy use. Typically, exposure to alcohol sensitizes the reward system to alcohol related cues, interferes with the processing of non-drug reward, increases impulsivity, and disrupts emotional regulation. However, the findings discussed here also highlight the variability of individual differences in the presence and magnitude of such neurocognitive deficits which may be driven by exposure, trait factors or abstinence. Finally, an important caveat to much of the present evidence is the generalizability of small cohort cross-sectional studies. To better characterize brain function and behavior following exposure to alcohol both acute and chronic, as well as improve treatment outcome and reduce risk of relapse, it is imperative that large-scale studies with longitudinal designs are conducted. This information is critical for development of alcohol regulation and abuse prevention.

4. Neurochemical Dysfunction in Alcoholism

Neuroimaging studies have also dramatically advanced our understanding of the brain's response to alcohol and the neurochemical basis of alcohol dependence. Positron emission tomography (PET) and single photon emission computed tomography (SPECT) use radiotracers that bind specifically to key receptors of interest, to quantify receptor location and availability. Neurotransmitter release can also be indirectly quantified using PET, through measurement of the amount of tracer that is 'displaced' from the receptor when endogenous neurotransmitter is released in response to a pharmacological (or other) challenge. Such techniques have been instrumental in the investigation of key neurotransmitter systems and identification of molecular dysfunction in the human brain. The use of PET to study the effects of chronic alcohol consumption has advanced our understanding of reward mechanisms, neuroadaptations resulting from chronic use that led to tolerance and withdrawal and has identified key regions and circuits implicated in loss of control and motivation to drink. This section summarizes PET studies that investigate the key neurotransmitter systems and review the evidence in case-control studies (summarized in Table 1).

Table 1. Strength of evidence to show direction of effects on receptor radioligand binding in human PET imaging studies in alcohol dependence.

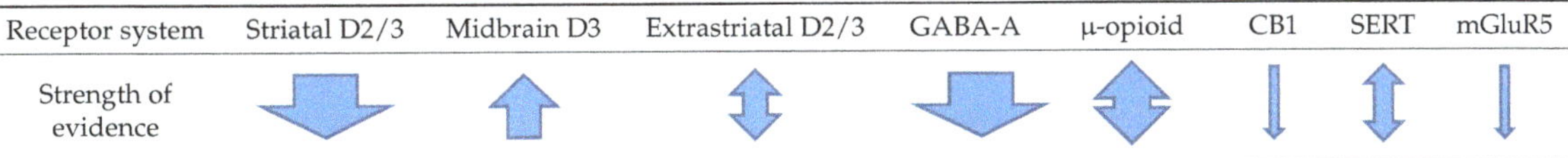

Receptor system	Striatal D2/3	Midbrain D3	Extrastriatal D2/3	GABA-A	μ-opioid	CB1	SERT	mGluR5
Strength of evidence	⬇	⬆	⬍	⬇	⬍	↓	↕	↓

Thickness of arrow indicates the relative strength of evidence of research in the receptor system as assessed by the author based on studies reported in the chapter.

4.1. Dopamine

The dopamine system has been the most extensively studied neurotransmitter system in addiction and several targets in both pre- and post-synaptic locations have been evaluated for their respective roles in alcoholism. Dopamine receptor number (availability) and dopamine release can both be measured, in receptor availability and neurotransmitter challenge PET studies, respectively. In studies investigating receptor availability, the key target to date has been the dopamine D2 receptor in the striatum, primarily using the antagonist radiotracer [^{11}C]Raclopride which binds selectively to both D2 and D3 dopamine receptors (hereafter referred to as DRD2/3). Data from 7 studies with 105 alcohol dependent individuals and 113 healthy controls were compared in a meta-analysis, revealing an overall reduction in DRD2/3 availability in the alcohol dependent group with an effect size of −0.78 (95% CI, −1.21, −0.35, $p < 0.001$) [121]. Lower DRD2/3 receptors in alcoholism in turn has been associated with decreased metabolic activity in prefrontal brain regions necessary for cognitive control and executive functioning, as assessed with the radiotracer [18F]Fluorodeoxyglucose (FDG) [122]. This could explain the vulnerability of such individuals to both compulsive and impulsive drinking, due to disrupted self-regulation [123]. Combined PET/fMRI studies have indicated that reduced striatal DRD2/3 availability was associated with greater frontal BOLD reactivity to alcohol-induced cues [124] indicating a relationship with reward processing. Moreover, the severity of clinical impairment has been shown to correlate with cortical hypometabolism as measured with FDG PET in alcoholism [125], providing several potential functional implications for D2 and/or D3 receptor loss.

Using other dopaminergic tracers, reduced levels of DRD2/3 availability and dopamine synthesis capacity, as measured by [18F]DMFP and [18F]DOPA, respectively, have been showed to be associated with increased craving and relapse [126], suggesting these receptors have prognostic value and may represent a target for drug development through upregulation of dopamine receptor function or dopaminergic transmission. To support such hypotheses, Rominger et al. identified that DRD2/3 receptor numbers, as assessed with [18F]Fallypride, recovered by 30% in individuals who successfully abstained from alcohol at one year, to a level comparable with healthy controls [127], whereas in those that relapsed the DRD2/3 receptor levels did not change. [18F]Fallypride has additional utility as it can quantify extra-striatal DRD2/3 receptors due to its very high affinity. Accordingly, studies have found lower DRD2/3 availability amongst alcohol dependent individuals in brain regions outside the striatum, such as the thalamus, insular cortex, hippocampus, and temporal cortex in comparison to matched healthy controls [123], although two other studies found no such difference in temporal [128] or frontal binding [129] using this tracer.

The development of novel radiotracers with greater specificity for the dopamine D3 receptor allowed characterization of this subtype which has been shown in preclinical models to regulate alcohol consumption. Notably, no difference in binding in the ventral striatum or caudate or putamen was found, however, there was a significantly higher D3 receptor availability in the hypothalamus that was linked to higher lifetime use of alcohol [130]. Preclinical imaging has identified D3 receptor antagonism as a plausible therapeutic target to ameliorate alcoholism and its potential efficacy as an intervention is currently under investigation using fMRI [131] and combined PET/MR techniques [132].

PET studies using dopamine-sensitive tracers such as [11C]Raclopride have successfully been employed to detect changes in dopamine release, demonstrating that dopaminergic deficits exist in alcoholism. For example, amphetamine-induced striatal dopamine release was found to be blunted in alcohol dependence relative to controls [133], highlighting a lower release potential which may explain the reward-deficiency phenomena associated with addiction described earlier.

Dopaminergic function following chronic alcohol consumption has been extensively investigated with several targets for potential therapeutics being discovered. Whilst promising early clinical work has identified some novel pharmacological targets that could be used to treat alcohol dependent individuals, further large-scale studies are required to validate their use and further exploration of dopaminergic dysregulation is warranted to better characterize the extent of pathology induced by alcoholism.

4.2. GABA

GABA is the brain's main inhibitory neurotransmitter and alcohol acutely enhances GABAergic neurotransmission [134]. A host of in-vivo PET imaging studies have observed an association between alcoholism and lower GABA-A receptors in the cortex (medial prefrontal, OFC, parietal, temporal, and ACC) and the cerebellum [135–138]. These studies have found lower radiotracer binding of between 6–20% using non-subtype selective GABA-A receptor tracers [11C]Flumazenil PET and [123I]Iomazenil SPECT imaging in alcohol dependence relative to controls. More recently, the alpha-5 subunit selective PET tracer [^{11}C]Ro15-4513 identified lower availability of this subunit in the nucleus accumbens (NAc) and hippocampus in abstinent alcohol dependent individuals when compared with matched controls [138]. A functional [^{18}F]FDG PET study investigating the differential effects of a benzodiazepine challenge on cerebellar metabolism indicating that dysregulation of GABA-A receptor may serve as a predisposing trait to alcoholism rather than as a result of chronic alcohol consumption: cerebellar hypo-metabolism was evident in those with a positive family history of alcohol dependence compared with family negative history individuals [139].

4.3. Opioids

The opioid system is acutely involved in the reinforcing effects of alcohol. The μ-opioid receptor (MOR) binds β-endorphins and enkephalins which, in turn, increase dopamine release in the NAc [140]. [^{11}C]Carfentanil is a PET tracer that can be used to define MOR receptor availability and is also sensitive to endogenous endorphin release. Endorphin release in the NAc and OFC was measured in light versus heavy drinkers through displacement of [^{11}C]Carfentanil following acute alcohol consumption of an alcoholic drink. Changes in OFC binding correlated significantly with problematic drinking and subjective high in heavy drinkers but not in controls [141]. In abstinent alcohol dependent individuals a greater MOR availability in the ventral striatum, as measured by [^{11}C]Carfentanil, compared with healthy controls was correlated with a greater craving for alcohol [142]. Increased MOR binding could be due to higher receptor levels or reduced release of endogenous endorphins. It was later postulated that greater [^{11}C]Carfentanil binding could be related to reduced β-endorphins in alcoholism. Post-mortem studies have noted a 23–51% reduction in MOR binding [143] in alcohol dependent individuals when compared with controls. Reduced MOR binding in post-mortem tissue could be interpreted as a neuroadaptive response to alcohol-induced release of endogenous β-endorphins in patients with severe alcohol dependence and could explain why naltrexone remains relatively ineffective in this subpopulation [140]. Preclinical data suggests that nalmefene counters alcohol-induced dysregulations of the MOR/endorphin and the KOR/dynorphin system [141]. Drugs that antagonize these receptors, including the licensed drug naltrexone have been found to attenuate alcohol seeking in rats and have been shown to clinically reduce alcohol consumption [144].

4.4. Other Neurochemical Systems

The endocannabinoid system is implicated in modulating alcohol rewards [145]. Although limited in scope, one small PET study using [^{18}F]FMPEP-d2 reported increased cannabinoid CB1 receptor in alcohol dependence in early withdrawal [146]. A more long-term PET study found that alcohol dependence is associated with widespread reduction of cannabinoid CB1 receptor binding in the human brain and this reduction persists at least 2–4 weeks into abstinence. The correlation of reduced binding with years of alcohol abuse suggests an involvement of CB1 receptors in alcohol dependence in humans [147].

PET studies investigating the serotonin system in alcohol dependence are very limited in number, and so a consensus opinion on their importance has not been reached. Studies have focused on the serotonin transporter (SERT) using [^{11}C] DASB, revealing mixed results with some [148,149] reporting increased levels of SERT whereas others have found no difference or reduced levels of SERT [150].

Only recently have radiotracers specific for characterizing excitatory glutamate receptors been developed. Early findings indicate impaired mGluR5 signaling to be involved in compulsive alcohol consumption [151]. These effects are found to be reversible following 28 days of abstinence and so can be viewed as a target to aid withdrawal [152].

4.5. Conclusion

The dopamine, GABA and opioid systems are by far the most researched using PET and SPECT imaging techniques to measure neurochemical dysfunction in alcohol dependence, due to the availability of selective radiolabeled tracers for the targets of DRD2/3, GABA-A and MOR receptors, respectively. Well validated tracers for other targets such as those in the serotonergic system do exist, but their use in alcohol dependent individuals is not well characterized. Studies using novel radioligands to assess other receptor targets and neurochemical systems including the endocannabinoid and glutamatergic systems is less advanced, but a few selective tracers do exist. It must be acknowledged that PET/SPECT is somewhat limited as a technique because of its radioactivity meaning that young people and repeat scanning cannot be carried out. Nevertheless, PET/SPECT imaging is still the only way to directly image neurotransmitter receptors and neurotransmitter release (when sensitive tracers are available) in the living human brain. Further studies are required to elucidate receptor changes in response to alcohol consumption and dependence across all known neurotransmitter systems.

5. Summary

Chronic alcohol consumption is thought to contribute directly to neurotoxicity via thiamine deficiency, metabolite toxicity and neuroinflammation, leading to the development of serious conditions of WE and KP, and the acceleration of neurodegeneration more generally. In addition, neuroimaging of the brain in response to alcohol dependence has found important structural, functional, and neurochemical differences compared with healthy brains which have shone light on possible chronic effects of alcohol consumption, revealed potential vulnerability markers which may be of clinical relevance, identified prognostic biomarkers associated with relapse and recovery and identified possible biomarkers for drug development. The picture is complex with modulation of brain systems by alcohol differing according to the time course of the disorder, the severity and quantity of alcohol used and with an important role for family history in which genetics also plays a role. The sometimes-contradictory findings could also be related to differences in duration of alcohol abstinence and different characteristics of patients being assessed. Another consideration for PET and fMRI imaging studies in alcohol dependence in general is its association with significant cortical grey matter loss, such that in theory, certain reductions observed in receptor availability or changes in BOLD response may in part be explained by changes in brain volume. Further effort is required to determine the interaction between cortical atrophy and the observed functional and/or neurochemical changes.

There is evidence of gender- and sex-related differences in consumption of alcohol as well as its effects on the brain [153]. However, neuroimaging studies on the effects of alcohol use and dependence have either excluded women or shown low female enrolment [154]. Consideration of gender- and sex-related effects has also been limited, in part due to a lack of power [154]. Rates of alcohol dependence have increased drastically in women and many of the harmful health effects are more severe and occur more rapidly in women [155]. This underscores the need to examine sex- and gender-related alterations on brain function and structure in alcohol use; improving our understanding of these effects may enable tailoring of pharmacotherapeutic treatments to improve outcomes.

Another important area for further research is to determine whether alterations in brain structure, brain function and receptor availability in alcohol dependent individuals occur as a direct result of alcohol toxicity or whether they represent vulnerability factors for the onset, development or persistence of problematic or dependent drinking, or whether both contribute to the observed changes. Larger prospective studies and those with a longitudinal design are needed to better understand trait markers that may exist prior to the development of addiction and how they may change across the whole trajectory of the disorder to assess causality, and to stratify and target patients most at risk. Continued efforts to identify suitable biological targets to reduce craving, withdrawal, increase cognition and maintain abstinence for those affected are warranted because despite 30 years of neuroimaging and the huge advances in technologies to understand the brain basis of alcohol use disorder, it remains the case that, to date, only three pharmacotherapies are licensed for alcohol dependence and only 9% of such individuals receive such treatment [156].

Funding: Some of the research reviewed here from our group was funded by the UK MRC.

Conflicts of Interest: The authors declare no conflict of interest.

References

1. *Diagnostic and Statistical Manual of Mental Disorders: DSM-5TM*, 5th ed.; American Psychiatric Publishing: Arlington, VA, USA, 2013; p. xliv.
2. World Health Organization. *Global Status Report on Alcohol and Health 2018*; World Health Organization: Geneva, Switzerland, 2018.
3. Weisner, C.; Matzger, H.; Kaskutas, L.A. How important is treatment? One-year outcomes of treated and untreated alcohol-dependent individuals. *Addiction* **2003**, *98*, 901–911.
4. Kranzler, H.R.; Van Kirk, J. Efficacy of naltrexone and acamprosate for alcoholism treatment: A meta-analysis. *Alcohol. Clin. Exp. Res.* **2001**, *25*, 1335–1341. [CrossRef]
5. Pettinati, H.M.; Rabinowitz, A.R. New pharmacotherapies for treating the neurobiology of alcohol and drug addiction. *Psychiatry* **2006**, *3*, 14.
6. Akbar, M.; Egli, M.; Cho, Y.-E.; Song, B.-J.; Noronha, A. Medications for alcohol use disorders: An overview. *Pharmacol. Ther.* **2018**, *185*, 64–85. [CrossRef]
7. Courville, C.B. *Effects of Alcohol on the Nervous System of Man*; San Lucas Press: Oxford, UK, 1955.
8. Brewer, C.; Perrett, L. Brain damage due to alcohol consumption: An air-encephalographic, psychometric and electroencephalographic study. *Br. J. Addict.* **1971**, *66*, 170–182. [CrossRef]
9. Jernigan, T.L.; Zatz, L.M.; Ahumada, A.J., Jr.; Pfefferbaum, A.; Tinklenberg, J.R.; Moses, J.A., Jr. CT measures of cerebrospinal fluid volume in alcoholics and normal volunteers. *Psychiatry Res.* **1982**, *7*, 9–17. [CrossRef]
10. Pfefferbaum, A.; Rosenbloom, M.; Crusan, K.; Jernigan, T.L. Brain CT changes in alcoholics: Effects of age and alcohol consumption. *Alcohol. Clin. Exp. Res.* **1988**, *12*, 81–87. [CrossRef]
11. Bühler, M.; Mann, K. Alcohol and the human brain: A systematic review of different neuroimaging methods. *Alcohol. Clin. Exp. Res.* **2011**, *35*, 1771–1793. [CrossRef]
12. Chye, Y.; Mackey, S.; Gutman, B.A.; Ching, C.R.; Batalla, A.; Blaine, S.; Brooks, S.; Caparelli, E.C.; Cousijn, J.; Dagher, A. Subcortical surface morphometry in substance dependence: An ENIGMA addiction working group study. *Addict. Biol.* **2020**, *25*, e12830. [CrossRef]
13. Mackey, S.; Allgaier, N.; Chaarani, B.; Spechler, P.; Orr, C.; Bunn, J.; Allen, N.B.; Alia-Klein, N.; Batalla, A.; Blaine, S. Mega-analysis of gray matter volume in substance dependence: General and substance-specific regional effects. *Am. J. Psychiatry* **2019**, *176*, 119–128. [CrossRef]
14. Li, L.; Yu, H.; Liu, Y.; Meng, Y.-j.; Li, X.-j.; Zhang, C.; Liang, S.; Li, M.-l.; Guo, W.; Deng, W. Lower regional grey matter in alcohol use disorders: Evidence from a voxel-based meta-analysis. *BMC Psychiatry* **2021**, *21*, 1–11. [CrossRef] [PubMed]
15. Xiao, P.; Dai, Z.; Zhong, J.; Zhu, Y.; Shi, H.; Pan, P. Regional gray matter deficits in alcohol dependence: A meta-analysis of voxel-based morphometry studies. *Drug Alcohol Depend.* **2015**, *153*, 22–28. [CrossRef]
16. Yang, X.; Tian, F.; Zhang, H.; Zeng, J.; Chen, T.; Wang, S.; Jia, Z.; Gong, Q. Cortical and subcortical gray matter shrinkage in alcohol-use disorders: A voxel-based meta-analysis. *Neurosci. Biobehav. Rev.* **2016**, *66*, 92–103. [CrossRef] [PubMed]
17. Monnig, M.A.; Tonigan, J.S.; Yeo, R.A.; Thoma, R.J.; McCrady, B.S. White matter volume in alcohol use disorders: A meta-analysis. *Addict. Biol.* **2013**, *18*, 581–592. [CrossRef]
18. Hampton, W.H.; Hanik, I.M.; Olson, I.R. Substance abuse and white matter: Findings, limitations, and future of diffusion tensor imaging research. *Drug Alcohol Depend.* **2019**, *197*, 288–298. [CrossRef]
19. Zahr, N.M.; Pitel, A.-L.; Chanraud, S.; Sullivan, E.V. Contributions of studies on alcohol use disorders to understanding cerebellar function. *Neuropsychol. Rev.* **2010**, *20*, 280–289. [CrossRef] [PubMed]
20. Carlen, P.L.; Wortzman, G.; Holgate, R.; Wilkinson, D.; Rankin, J. Reversible cerebral atrophy in recently abstinent chronic alcoholics measured by computed tomography scans. *Science* **1978**, *200*, 1076–1078. [CrossRef]
21. Ishikawa, Y.; Meyer, J.S.; Tanahashi, N.; Hata, T.; Velez, M.; Fann, W.E.; Kandula, P.; Motel, K.F.; Rogers, R.L. Abstinence improves cerebral perfusion and brain volume in alcoholic neurotoxicity without Wernicke-Korsakoff syndrome. *J. Cereb. Blood Flow Metab.* **1986**, *6*, 86–94. [CrossRef]
22. Durazzo, T.C.; Mon, A.; Gazdzinski, S.; Yeh, P.-H.; Meyerhoff, D.J. Serial longitudinal MRI data indicates non-linear regional gray matter volume recovery in abstinent alcohol dependent individuals. *Addict. Biol.* **2015**, *20*, 956. [CrossRef]
23. Gazdzinski, S.; Durazzo, T.C.; Meyerhoff, D.J. Temporal dynamics and determinants of whole brain tissue volume changes during recovery from alcohol dependence. *Drug Alcohol Depend.* **2005**, *78*, 263–273. [CrossRef]
24. Zou, X.; Durazzo, T.C.; Meyerhoff, D.J. Regional brain volume changes in alcohol-dependent individuals during short-term and long-term abstinence. *Alcohol. Clin. Exp. Res.* **2018**, *42*, 1062–1072. [CrossRef]
25. Meyerhoff, D.J.; Durazzo, T.C. Not all is lost for relapsers: Relapsers with low WHO risk drinking levels and complete abstainers have comparable regional gray matter volumes. *Alcohol. Clin. Exp. Res.* **2020**, *44*, 1479–1487. [CrossRef]

26. Gazdzinski, S.; Durazzo, T.C.; Mon, A.; Yeh, P.-H.; Meyerhoff, D.J. Cerebral white matter recovery in abstinent alcoholics—A multimodality magnetic resonance study. *Brain* **2010**, *133*, 1043–1053. [CrossRef]

27. Pfefferbaum, A.; Rosenbloom, M.J.; Chu, W.; Sassoon, S.A.; Rohlfing, T.; Pohl, K.M.; Zahr, N.M.; Sullivan, E.V. White matter microstructural recovery with abstinence and decline with relapse in alcohol dependence interacts with normal ageing: A controlled longitudinal DTI study. *Lancet Psychiatry* **2014**, *1*, 202–212. [CrossRef]

28. Cardenas, V.A.; Durazzo, T.C.; Gazdzinski, S.; Mon, A.; Studholme, C.; Meyerhoff, D.J. Brain morphology at entry into treatment for alcohol dependence is related to relapse propensity. *Biol. Psychiatry* **2011**, *70*, 561–567. [CrossRef]

29. Cardenas, V.A.; Studholme, C.; Gazdzinski, S.; Durazzo, T.C.; Meyerhoff, D.J. Deformation-based morphometry of brain changes in alcohol dependence and abstinence. *Neuroimage* **2007**, *34*, 879–887. [CrossRef]

30. Durazzo, T.C.; Mon, A.; Gazdzinski, S.; Meyerhoff, D.J. Regional brain volume changes in alcohol-dependent individuals during early abstinence: Associations with relapse following treatment. *Addict. Biol.* **2017**, *22*, 1416–1425. [CrossRef]

31. Zou, Y.; Murray, D.E.; Durazzo, T.C.; Schmidt, T.P.; Murray, T.A.; Meyerhoff, D.J. White matter microstructural correlates of relapse in alcohol dependence. *Psychiatry Res. Neuroimaging* **2018**, *281*, 92–100. [CrossRef]

32. Oscar-Bemian, M.; Schendan, H.E. Asymmetries of brain function in alcoholism: Relationship to aging. In *Neurobehavior of Language and Cognition*; Springer: Berlin/Heidelberg, Germany, 2002; pp. 213–240.

33. Guggenmos, M.; Schmack, K.; Sekutowicz, M.; Garbusow, M.; Sebold, M.; Sommer, C.; Smolka, M.N.; Wittchen, H.-U.; Zimmermann, U.S.; Heinz, A. Quantitative neurobiological evidence for accelerated brain aging in alcohol dependence. *Transl. Psychiatry* **2017**, *7*, 1–7. [CrossRef]

34. Pfefferbaum, A.; Adalsteinsson, E.; Sullivan, E.V. Dysmorphology and microstructural degradation of the corpus callosum: Interaction of age and alcoholism. *Neurobiol. Aging* **2006**, *27*, 994–1009. [CrossRef]

35. Pfefferbaum, A.; Lim, K.O.; Zipursky, R.B.; Mathalon, D.H.; Rosenbloom, M.J.; Lane, B.; Ha, C.N.; Sullivan, E.V. Brain gray and white matter volume loss accelerates with aging in chronic alcoholics: A quantitative MRI study. *Alcohol. Clin. Exp. Res.* **1992**, *16*, 1078–1089. [CrossRef]

36. Sorg, S.F.; Squeglia, L.M.; Taylor, M.J.; Alhassoon, O.M.; Delano-Wood, L.M.; Grant, I. Effects of aging on frontal white matter microstructure in alcohol use disorder and associations with processing speed. *J. Stud. Alcohol Drugs* **2015**, *76*, 296–306. [CrossRef]

37. Sullivan, E.V.; Zahr, N.M.; Sassoon, S.A.; Thompson, W.K.; Kwon, D.; Pohl, K.M.; Pfefferbaum, A. The role of aging, drug dependence, and hepatitis C comorbidity in alcoholism cortical compromise. *JAMA Psychiatry* **2018**, *75*, 474–483. [CrossRef]

38. Thayer, R.E.; Hagerty, S.L.; Sabbineni, A.; Claus, E.D.; Hutchison, K.E.; Weiland, B.J. Negative and interactive effects of sex, aging, and alcohol abuse on gray matter morphometry. *Hum. Brain Mapp.* **2016**, *37*, 2276–2292. [CrossRef]

39. Cole, J.H.; Franke, K. Predicting age using neuroimaging: Innovative brain ageing biomarkers. *Trends Neurosci.* **2017**, *40*, 681–690. [CrossRef]

40. Kawano, Y. Physio-pathological effects of alcohol on the cardiovascular system: Its role in hypertension and cardiovascular disease. *Hypertens. Res.* **2010**, *33*, 181–191. [CrossRef]

41. Kubota, M.; Nakazaki, S.; Hirai, S.; Saeki, N.; Yamaura, A.; Kusaka, T. Alcohol consumption and frontal lobe shrinkage: Study of 1432 non-alcoholic subjects. *J. Neurol. Neurosurg. Psychiatry* **2001**, *71*, 104–106. [CrossRef]

42. McEvoy, L.K.; Fennema-Notestine, C.; Elman, J.A.; Eyler, L.T.; Franz, C.E.; Hagler, D.J., Jr.; Hatton, S.N.; Lyons, M.J.; Panizzon, M.S.; Dale, A.M. Alcohol intake and brain white matter in middle aged men: Microscopic and macroscopic differences. *NeuroImage Clin.* **2018**, *18*, 390–398. [CrossRef]

43. Sachdev, P.S.; Chen, X.; Wen, W.; Anstry, K.J. Light to moderate alcohol use is associated with increased cortical gray matter in middle-aged men: A voxel-based morphometric study. *Psychiatry Res. Neuroimaging* **2008**, *163*, 61–69. [CrossRef]

44. Daviet, R.; Aydogan, G.; Jagannathan, K.; Spilka, N.; Koellinger, P.D.; Kranzler, H.R.; Nave, G.; Wetherill, R.R. Multimodal brain imaging study of 36,678 participants reveals adverse effects of moderate drinking. *bioRxiv* **2021**. [CrossRef]

45. Morris, V.L.; Owens, M.M.; Syan, S.K.; Petker, T.D.; Sweet, L.H.; Oshri, A.; MacKillop, J.; Amlung, M. Associations between drinking and cortical thickness in younger adult drinkers: Findings from the Human Connectome Project. *Alcohol. Clin. Exp. Res.* **2019**, *43*, 1918–1927. [CrossRef] [PubMed]

46. Topiwala, A.; Ebmeier, K.P.; Maullin-Sapey, T.; Nichols, T.E. No safe level of alcohol consumption for brain health: Observational cohort study of 25,378 UK Biobank participants. *medRxiv* **2021**. [CrossRef]

47. Topiwala, A.; Allan, C.L.; Valkanova, V.; Zsoldos, E.; Filippini, N.; Sexton, C.; Mahmood, A.; Fooks, P.; Singh-Manoux, A.; Mackay, C.E. Moderate alcohol consumption as risk factor for adverse brain outcomes and cognitive decline: Longitudinal cohort study. *BMJ* **2017**, *357*, j2353. [CrossRef]

48. Ning, K.; Zhao, L.; Matloff, W.; Sun, F.; Toga, A.W. Association of relative brain age with tobacco smoking, alcohol consumption, and genetic variants. *Sci. Rep.* **2020**, *10*, 1–10. [CrossRef] [PubMed]

49. Sowell, E.R.; Thompson, P.M.; Toga, A.W. Mapping changes in the human cortex throughout the span of life. *Neurosci.* **2004**, *10*, 372–392. [CrossRef]

50. National Institute on Alcohol Abuse and Alcoholism. Drinking Levels Defined. Available online: https://www.niaaa.nih.gov/alcohol-health/overview-alcohol-consumption/moderate-binge-drinking (accessed on 1 November 2021).

51. Lees, B.; Meredith, L.R.; Kirkland, A.E.; Bryant, B.E.; Squeglia, L.M. Effect of alcohol use on the adolescent brain and behavior. *Pharmacol. Biochem. Behav.* **2020**, *192*, 172906. [CrossRef]

52. Pfefferbaum, A.; Kwon, D.; Brumback, T.; Thompson, W.K.; Cummins, K.; Tapert, S.F.; Brown, S.A.; Colrain, I.M.; Baker, F.C.; Prouty, D. Altered brain developmental trajectories in adolescents after initiating drinking. *Am. J. Psychiatry* **2018**, *175*, 370–380. [CrossRef] [PubMed]

53. Sullivan, E.V.; Brumback, T.; Tapert, S.F.; Brown, S.A.; Baker, F.C.; Colrain, I.M.; Prouty, D.; De Bellis, M.D.; Clark, D.B.; Nagel, B.J. Disturbed cerebellar growth trajectories in adolescents who initiate alcohol drinking. *Biol. Psychiatry* **2020**, *87*, 632–644. [CrossRef]

54. Squeglia, L.M.; Cservenka, A. Adolescence and drug use vulnerability: Findings from neuroimaging. *Curr. Opin. Behav. Sci.* **2017**, *13*, 164–170. [CrossRef]

55. Jones, S.A.; Nagel, B.J. Altered frontostriatal white matter microstructure is associated with familial alcoholism and future binge drinking in adolescence. *Neuropsychopharmacology* **2019**, *44*, 1076–1083. [CrossRef]

56. Weiland, B.J.; Korycinski, S.T.; Soules, M.; Zubieta, J.-K.; Zucker, R.A.; Heitzeg, M.M. Substance abuse risk in emerging adults associated with smaller frontal gray matter volumes and higher externalizing behaviors. *Drug Alcohol Depend.* **2014**, *137*, 68–75. [CrossRef] [PubMed]

57. Nguyen, V.T.; Chong, S.; Tieng, Q.M.; Mardon, K.; Galloway, G.J.; Kurniawan, N.D. Radiological studies of fetal alcohol spectrum disorders in humans and animal models: An updated comprehensive review. *Magn. Reson. Imaging* **2017**, *43*, 10–26. [CrossRef] [PubMed]

58. Treit, S.; Jeffery, D.; Beaulieu, C.; Emery, D. Radiological findings on structural magnetic resonance imaging in fetal alcohol spectrum disorders and healthy controls. *Alcohol. Clin. Exp. Res.* **2020**, *44*, 455–462. [CrossRef]

59. Lebel, C.; Roussotte, F.; Sowell, E.R. Imaging the impact of prenatal alcohol exposure on the structure of the developing human brain. *Neuropsychol. Rev.* **2011**, *21*, 102–118. [CrossRef] [PubMed]

60. O'Leary-Moore, S.K.; Parnell, S.E.; Lipinski, R.J.; Sulik, K.K. Magnetic resonance-based imaging in animal models of fetal alcohol spectrum disorder. *Neuropsychol. Rev.* **2011**, *21*, 167–185. [CrossRef]

61. Martin, P.R.; Singleton, C.K.; Hiller-Sturmhöfel, S. The role of thiamine deficiency in alcoholic brain disease. *Alcohol Res. Health* **2003**, *27*, 134.

62. Thomson, A.D. Mechanisms of vitamin deficiency in chronic alcohol misusers and the development of the Wernicke-Korsakoff syndrome. *Alcohol Alcohol.* **2000**, *35*, 2. [CrossRef]

63. Thomson, A.D.; Marshall, E.J. The natural history and pathophysiology of Wernicke's encephalopathy and Korsakoff's psychosis. *Alcohol Alcohol.* **2006**, *41*, 151–158. [CrossRef]

64. Behse, F.; Buchthal, F. Alcoholic neuropathy: Clinical, electrophysiological, and biopsy findings. *Ann. Neurol. Off. J. Am. Neurol. Assoc. Child Neurol. Soc.* **1977**, *2*, 95–110. [CrossRef]

65. McLean, C.; Tapsell, L.; Grafenauer, S.; McMahon, A.T. Systematic review of nutritional interventions for people admitted to hospital for alcohol withdrawal. *Nutr. Diet.* **2020**, *77*, 76–89. [CrossRef] [PubMed]

66. Thomson, A.D.; Baker, H.; Leevy, C.M. Patterns of 35S-thiamine hydrochloride absorption in the malnourished alcoholic patient. *J. Lab. Clin. Med.* **1970**, *76*, 34–45.

67. Gastaldi, G.; Casirola, D.; Ferrari, G.; Rindi, G. Effect of chronic ethanol administration on thiamine transport in microvillous vesicles of rat small intestine. *Alcohol Alcohol.* **1989**, *24*, 83–89. [CrossRef] [PubMed]

68. Reidling, J.C.; Said, H.M. Adaptive regulation of intestinal thiamin uptake: Molecular mechanism using wild-type and transgenic mice carrying hTHTR-1 and-2 promoters. *Am. J. Physiol.-Gastrointest. Liver Physiol.* **2005**, *288*, G1127–G1134. [CrossRef] [PubMed]

69. Marcé-Grau, A.; Martí-Sánchez, L.; Baide-Mairena, H.; Ortigoza-Escobar, J.D.; Pérez-Dueñas, B. Genetic defects of thiamine transport and metabolism: A review of clinical phenotypes, genetics, and functional studies. *J. Inherit. Metab. Dis.* **2019**, *42*, 581–597. [CrossRef]

70. Traviesa, D. Magnesium deficiency: A possible cause of thiamine refractoriness in Wernicke-Korsakoff encephalopathy. *J. Neurol. Neurosurg. Psychiatry* **1974**, *37*, 959–962. [CrossRef]

71. Bates, M.E.; Bowden, S.C.; Barry, D. Neurocognitive impairment associated with alcohol use disorders: Implications for treatment. *Exp. Clin. Psychopharmacol.* **2002**, *10*, 193. [CrossRef] [PubMed]

72. Pitel, A.L.; Rivier, J.; Beaunieux, H.; Vabret, F.; Desgranges, B.; Eustache, F. Changes in the episodic memory and executive functions of abstinent and relapsed alcoholics over a 6-month period. *Alcohol. Clin. Exp. Res.* **2009**, *33*, 490–498. [CrossRef]

73. Oscar-Berman, M.; Marinković, K. Alcohol: Effects on neurobehavioral functions and the brain. *Neuropsychol. Rev.* **2007**, *17*, 239–257. [CrossRef]

74. Brion, M.; Pitel, A.-L.; Beaunieux, H.; Maurage, P. Revisiting the continuum hypothesis: Toward an in-depth exploration of executive functions in korsakoff syndrome. *Front. Hum. Neurosci.* **2014**, *8*, 498. [CrossRef] [PubMed]

75. Guo, R.; Ren, J. Alcohol and Acetaldehyde in Public Health: From Marvel to Menace. *Int. J. Environ. Res. Public Health* **2010**, *7*, 1285–1301. [CrossRef]

76. Niemelä, O. Distribution of ethanol-induced protein adducts in vivo: Relationship to tissue injury. *Free Radic. Biol. Med.* **2001**, *31*, 1533–1538. [CrossRef]

77. Quertemont, E.; Eriksson, C.J.P.; Zimatkin, S.M.; Pronko, P.S.; Diana, M.; Pisano, M.; Rodd, Z.A.; Bell, R.R.; Ward, R.J. Is Ethanol a Pro-Drug? Acetaldehyde Contribution to Brain Ethanol Effects. *Alcohol. Clin. Exp. Res.* **2005**, *29*, 1514–1521. [CrossRef]

78. Eriksson, C.J.P. The Role of Acetaldehyde in the Actions of Alcohol (Update 2000). *Alcohol. Clin. Exp. Res.* **2001**, *25*, 15S–32S. [CrossRef]

79. Takeuchi, M.; Saito, T. Cytotoxicity of Acetaldehyde-Derived Advanced Glycation End-Products (AA-AGE) in Alcoholic-Induced Neuronal Degeneration. *Alcohol. Clin. Exp. Res.* **2005**, *29*, 220S–224S. [CrossRef] [PubMed]

80. Ferrier, L.; Bérard, F.; Debrauwer, L.; Chabo, C.; Langella, P.; Buéno, L.; Fioramonti, J. Impairment of the intestinal barrier by ethanol involves enteric microflora and mast cell activation in rodents. *Am. J. Pathol.* **2006**, *168*, 1148–1154. [CrossRef] [PubMed]

81. Crews, F.T.; Sarkar, D.K.; Qin, L.; Zou, J.; Boyadjieva, N.; Vetreno, R.P. Neuroimmune function and the consequences of alcohol exposure. *Alcohol Res. Curr. Rev.* **2015**, *37*, 331.

82. He, J.; Crews, F.T. Increased MCP-1 and microglia in various regions of the human alcoholic brain. *Exp. Neurol.* **2008**, *210*, 349–358. [CrossRef] [PubMed]

83. Lovinger, D.M.; White, G.; Weight, F.F. Ethanol inhibits NMDA-activated ion current in hippocampal neurons. *Science* **1989**, *243*, 1721–1724. [CrossRef]

84. Fliegel, S.; Brand, I.; Spanagel, R.; Noori, H.R. Ethanol-induced alterations of amino acids measured by in vivo microdialysis in rats: A meta-analysis. *Silico Pharmacol.* **2013**, *1*, 7. [CrossRef] [PubMed]

85. Ward, R.J.; Lallemand, F.; de Witte, P. Biochemical and neurotransmitter changes implicated in alcohol-induced brain damage in chronic or 'binge drinking' alcohol abuse. *Alcohol Alcohol* **2009**, *44*, 128–135. [CrossRef]

86. Munhoz, C.D.; Sorrells, S.F.; Caso, J.R.; Scavone, C.; Sapolsky, R.M. Glucocorticoids Exacerbate Lipopolysaccharide-Induced Signaling in the Frontal Cortex and Hippocampus in a Dose-Dependent Manner. *J. Neurosci.* **2010**, *30*, 13690. [CrossRef]

87. Socodato, R.; Portugal, C.C.; Canedo, T.; Domith, I.; Oliveira, N.A.; Paes-de-Carvalho, R.; Relvas, J.B.; Cossenza, M. c-Src deactivation by the polyphenol 3-O-caffeoylquinic acid abrogates reactive oxygen species-mediated glutamate release from microglia and neuronal excitotoxicity. *Free Radic. Biol. Med.* **2015**, *79*, 45–55. [CrossRef] [PubMed]

88. Henriques, J.F.; Portugal, C.C.; Canedo, T.; Relvas, J.B.; Summavielle, T.; Socodato, R. Microglia and alcohol meet at the crossroads: Microglia as critical modulators of alcohol neurotoxicity. *Toxicol. Lett.* **2018**, *283*, 21–31. [CrossRef]

89. Alfonso-Loeches, S.; Pascual-Lucas, M.; Blanco, A.M.; Sanchez-Vera, I.; Guerri, C. Pivotal Role of TLR4 Receptors in Alcohol-Induced Neuroinflammation and Brain Damage. *J. Neurosci.* **2010**, *30*, 8285–8295. [CrossRef] [PubMed]

90. McClain, J.A.; Morris, S.A.; Deeny, M.A.; Marshall, S.A.; Hayes, D.M.; Kiser, Z.M.; Nixon, K. Adolescent binge alcohol exposure induces long-lasting partial activation of microglia. *Brain Behav. Immun.* **2011**, *25* (Suppl. 1), S120–S128. [CrossRef] [PubMed]

91. Zhao, Y.N.; Wang, F.; Fan, Y.X.; Ping, G.F.; Yang, J.Y.; Wu, C.F. Activated microglia are implicated in cognitive deficits, neuronal death, and successful recovery following intermittent ethanol exposure. *Behav. Brain Res.* **2013**, *236*, 270–282. [CrossRef] [PubMed]

92. Stowell, R.D.; Majewska, A.K. Acute ethanol exposure rapidly alters cerebellar and cortical microglial physiology. *Eur. J. Neurosci.* **2020**, *54*, 5834–5843. [CrossRef]

93. Ward, R.J.; Lallemand, F.; de Witte, P. Influence of adolescent heavy session drinking on the systemic and brain innate immune system. *Alcohol Alcohol* **2014**, *49*, 193–197. [CrossRef]

94. Qin, L.; He, J.; Hanes, R.N.; Pluzarev, O.; Hong, J.S.; Crews, F.T. Increased systemic and brain cytokine production and neuroinflammation by endotoxin following ethanol treatment. *J. Neuroinflamm.* **2008**, *5*, 10. [CrossRef]

95. Saba, W.; Goutal, S.; Auvity, S.; Kuhnast, B.; Coulon, C.; Kouyoumdjian, V.; Buvat, I.; Leroy, C.; Tournier, N. Imaging the neuroimmune response to alcohol exposure in adolescent baboons: A TSPO PET study using 18F-DPA-714. *Addict. Biol.* **2018**, *23*, 1000–1009. [CrossRef]

96. Hillmer, A.T.; Sandiego, C.M.; Hannestad, J.; Angarita, G.A.; Kumar, A.; McGovern, E.M.; Huang, Y.; O'Connor, K.C.; Carson, R.E.; O'Malley, S.S.; et al. In vivo imaging of translocator protein, a marker of activated microglia, in alcohol dependence. *Mol. Psychiatry* **2017**, *22*, 1759–1766. [CrossRef]

97. Kalk, N.; Guo, Q.; Owen, D.; Cherian, R.; Erritzoe, D.; Gilmour, A.; Ribeiro, A.; McGonigle, J.; Waldman, A.; Matthews, P. Decreased hippocampal translocator protein (18 kDa) expression in alcohol dependence: A [^{11}C] PBR28 PET study. *Transl. Psychiatry* **2017**, *7*, e996. [CrossRef]

98. Knutson, B.; Adams, C.M.; Fong, G.W.; Hommer, D. Anticipation of increasing monetary reward selectively recruits nucleus accumbens. *J. Neurosci.* **2001**, *21*, RC159. [CrossRef]

99. Courtney, K.E.; Schacht, J.P.; Hutchison, K.; Roche, D.J.; Ray, L.A. Neural substrates of cue reactivity: Association with treatment outcomes and relapse. *Addict. Biol.* **2016**, *21*, 3–22. [CrossRef] [PubMed]

100. Luijten, M.; Schellekens, A.F.; Kühn, S.; Machielse, M.W.; Sescousse, G. Disruption of reward processing in addiction: An image-based meta-analysis of functional magnetic resonance imaging studies. *JAMA Psychiatry* **2017**, *74*, 387–398. [CrossRef]

101. Bach, P.; Weil, G.; Pompili, E.; Hoffmann, S.; Hermann, D.; Vollstädt-Klein, S.; Mann, K.; Perez-Ramirez, U.; Moratal, D.; Canals, S. Incubation of neural alcohol cue reactivity after withdrawal and its blockade by naltrexone. *Addict. Biol.* **2020**, *25*, e12717. [CrossRef] [PubMed]

102. Vollstädt-Klein, S.; Wichert, S.; Rabinstein, J.; Bühler, M.; Klein, O.; Ende, G.; Hermann, D.; Mann, K. Initial, habitual and compulsive alcohol use is characterized by a shift of cue processing from ventral to dorsal striatum. *Addiction* **2010**, *105*, 1741–1749. [CrossRef] [PubMed]

103. Everitt, B.J.; Robbins, T.W. Neural systems of reinforcement for drug addiction: From actions to habits to compulsion. *Nat. Neurosci.* **2005**, *8*, 1481–1489. [CrossRef]

104. Büchel, C.; Peters, J.; Banaschewski, T.; Bokde, A.L.; Bromberg, U.; Conrod, P.J.; Flor, H.; Papadopoulos, D.; Garavan, H.; Gowland, P. Blunted ventral striatal responses to anticipated rewards foreshadow problematic drug use in novelty-seeking adolescents. *Nat. Commun.* **2017**, *8*, 1–11. [CrossRef]

105. Dager, A.D.; Anderson, B.M.; Rosen, R.; Khadka, S.; Sawyer, B.; Jiantonio-Kelly, R.E.; Austad, C.S.; Raskin, S.A.; Tennen, H.; Wood, R.M. Functional magnetic resonance imaging (fMRI) response to alcohol pictures predicts subsequent transition to heavy drinking in college students. *Addiction* **2014**, *109*, 585–595. [CrossRef] [PubMed]
106. Mitchell, M.R.; Potenza, M.N. Recent insights into the neurobiology of impulsivity. *Curr. Addict. Rep.* **2014**, *1*, 309–319. [CrossRef]
107. De Wit, H. Impulsivity as a determinant and consequence of drug use: A review of underlying processes. *Addict. Biol.* **2009**, *14*, 22–31. [CrossRef]
108. Leigh, B.C. Peril, chance, adventure: Concepts of risk, alcohol use and risky behavior in young adults. *Addiction* **1999**, *94*, 371–383. [CrossRef]
109. Taylor, E.M.; Murphy, A.; Boyapati, V.; Ersche, K.D.; Flechais, R.; Kuchibatla, S.; McGonigle, J.; Metastasio, A.; Nestor, L.; Orban, C. Impulsivity in abstinent alcohol and polydrug dependence: A multidimensional approach. *Psychopharmacology* **2016**, *233*, 1487–1499. [CrossRef]
110. Norman, A.L.; Pulido, C.; Squeglia, L.M.; Spadoni, A.D.; Paulus, M.P.; Tapert, S.F. Neural activation during inhibition predicts initiation of substance use in adolescence. *Drug Alcohol Depend.* **2011**, *119*, 216–223. [CrossRef]
111. Wetherill, R.R.; Squeglia, L.M.; Yang, T.T.; Tapert, S.F. A longitudinal examination of adolescent response inhibition: Neural differences before and after the initiation of heavy drinking. *Psychopharmacology* **2013**, *230*, 663–671. [CrossRef] [PubMed]
112. Kirby, K.N.; Maraković, N.N. Delay-discounting probabilistic rewards: Rates decrease as amounts increase. *Psychon. Bull. Rev.* **1996**, *3*, 100–104. [CrossRef] [PubMed]
113. Ortner, C.N.; MacDonald, T.K.; Olmstead, M.C. Alcohol intoxication reduces impulsivity in the delay-discounting paradigm. *Alcohol Alcohol.* **2003**, *38*, 151–156. [CrossRef]
114. Dougherty, D.M.; Marsh-Richard, D.M.; Hatzis, E.S.; Nouvion, S.O.; Mathias, C.W. A test of alcohol dose effects on multiple behavioral measures of impulsivity. *Drug Alcohol Depend.* **2008**, *96*, 111–120. [CrossRef]
115. McCarthy, D.M.; Niculete, M.E.; Treloar, H.R.; Morris, D.H.; Bartholow, B.D. Acute alcohol effects on impulsivity: Associations with drinking and driving behavior. *Addiction* **2012**, *107*, 2109–2114. [CrossRef] [PubMed]
116. Herman, A.M.; Duka, T. Facets of impulsivity and alcohol use: What role do emotions play? *Neurosci. Biobehav. Rev.* **2019**, *106*, 202–216. [CrossRef] [PubMed]
117. Gowin, J.L.; Vatsalya, V.; Westman, J.G.; Schwandt, M.L.; Bartlett, S.; Heilig, M.; Momenan, R.; Ramchandani, V.A. The effect of varenicline on the neural processing of fearful faces and the subjective effects of alcohol in heavy drinkers. *Alcohol. Clin. Exp. Res.* **2016**, *40*, 979–987. [CrossRef] [PubMed]
118. Zhang, R.; Volkow, N.D. Brain default-mode network dysfunction in addiction. *Neuroimage* **2019**, *200*, 313–331. [CrossRef] [PubMed]
119. Orban, C.; McGonigle, J.; Flechais, R.S.; Paterson, L.M.; Elliott, R.; Erritzoe, D.; Ersche, K.D.; Murphy, A.; Nestor, L.J.; Passetti, F. Chronic alcohol exposure differentially modulates structural and functional properties of amygdala: A cross-sectional study. *Addict. Biol.* **2021**, *26*, e12980. [CrossRef]
120. Beck, A.; Wüstenberg, T.; Genauck, A.; Wrase, J.; Schlagenhauf, F.; Smolka, M.N.; Mann, K.; Heinz, A. Effect of brain structure, brain function, and brain connectivity on relapse in alcohol-dependent patients. *Arch. Gen. Psychiatry* **2012**, *69*, 842–852. [CrossRef]
121. Erritzoe, D.; Zafar, R.; Nutt, D. Meta-analysis and systematic review of the dopamine DRD2/3 receptors in Addiction. Manuscript in prep.
122. Volkow, N.D.; Wang, G.-J.; Telang, F.; Fowler, J.S.; Logan, J.; Jayne, M.; Ma, Y.; Pradhan, K.; Wong, C. Profound decreases in dopamine release in striatum in detoxified alcoholics: Possible orbitofrontal involvement. *J. Neurosci.* **2007**, *27*, 12700–12706. [CrossRef] [PubMed]
123. Volkow, N.D.; Wiers, C.E.; Shokri-Kojori, E.; Tomasi, D.; Wang, G.-J.; Baler, R. Neurochemical and metabolic effects of acute and chronic alcohol in the human brain: Studies with positron emission tomography. *Neuropharmacology* **2017**, *122*, 175–188. [CrossRef]
124. Heinz, A.; Siessmeier, T.; Wrase, J.; Hermann, D.; Klein, S.; Grüsser-Sinopoli, S.M.; Flor, H.; Braus, D.F.; Buchholz, H.G.; Gründer, G. Correlation between dopamine D2 receptors in the ventral striatum and central processing of alcohol cues and craving. *Am. J. Psychiatry* **2004**, *161*, 1783–1789. [CrossRef]
125. Gilman, S.; Adams, K.M.; Johnson-Greene, D.; Koeppe, R.A.; Junck, L.; Kluin, K.J.; Martorello, S.; Heumann, M.; Hill, E. Effects of disulfiram on positron emission tomography and neuropsychological studies in severe chronic alcoholism. *Alcohol. Clin. Exp. Res.* **1996**, *20*, 1456–1461. [CrossRef] [PubMed]
126. Heinz, A.; Siessmeier, T.; Wrase, J.; Buchholz, H.G.; Gründer, G.; Kumakura, Y.; Cumming, P.; Schreckenberger, M.; Smolka, M.N.; Rösch, F. Correlation of alcohol craving with striatal dopamine synthesis capacity and D2/3 receptor availability: A combined [^{18}F] DOPA and [^{18}F] DMFP PET study in detoxified alcoholic patients. *Am. J. Psychiatry* **2005**, *162*, 1515–1520. [CrossRef]
127. Rominger, A.; Cumming, P.; Xiong, G.; Koller, G.; Böning, G.; Wulff, M.; Zwergal, A.; Förster, S.; Reilhac, A.; Munk, O. [^{18}F] fallypride PET measurement of striatal and extrastriatal dopamine D2/3 receptor availability in recently abstinent alcoholics. *Addict. Biol.* **2012**, *17*, 490–503. [CrossRef] [PubMed]
128. Repo, E.; Kuikka, J.T.; Bergström, K.A.; Karhu, J.; Hiltunen, J.; Tiihonen, J.; Kuikka, J. Dopamine transporter and D2-receptor density in late-onset alcoholism. *Psychopharmacology* **1999**, *147*, 314–318. [CrossRef]
129. Narendran, R.; Mason, N.S.; Paris, J.; Himes, M.L.; Douaihy, A.B.; Frankle, W.G. Decreased prefrontal cortical dopamine transmission in alcoholism. *Am. J. Psychiatry* **2014**, *171*, 881–888. [CrossRef] [PubMed]

130. Erritzoe, D.; Tziortzi, A.; Bargiela, D.; Colasanti, A.; Searle, G.E.; Gunn, R.N.; Beaver, J.D.; Waldman, A.; Nutt, D.J.; Bani, M. In vivo imaging of cerebral dopamine D3 receptors in alcoholism. *Neuropsychopharmacology* **2014**, *39*, 1703–1712. [CrossRef]

131. Murphy, A.; Nestor, L.J.; McGonigle, J.; Paterson, L.; Boyapati, V.; Ersche, K.D.; Flechais, R.; Kuchibatla, S.; Metastasio, A.; Orban, C. Acute D3 antagonist GSK598809 selectively enhances neural response during monetary reward anticipation in drug and alcohol dependence. *Neuropsychopharmacology* **2017**, *42*, 1049–1057. [CrossRef] [PubMed]

132. Zafar, R.; Wall, M.; Erritzoe, D.; Nutt, D. Role of D3 Receptor antagonism in alcohol dependence: A combined PET/MRI investigation. Manuscript in prep.

133. Martinez, D.; Gil, R.; Slifstein, M.; Hwang, D.-R.; Huang, Y.; Perez, A.; Kegeles, L.; Talbot, P.; Evans, S.; Krystal, J. Alcohol dependence is associated with blunted dopamine transmission in the ventral striatum. *Biol. Psychiatry* **2005**, *58*, 779–786. [CrossRef]

134. Begleiter, H.; Porjesz, B. What is inherited in the predisposition toward alcoholism? A proposed model. *Alcohol. Clin. Exp. Res.* **1999**, *23*, 1125–1135. [CrossRef] [PubMed]

135. Abi-Dargham, A.; Krystal, J.H.; Anjilvel, S.; Scanley, B.E.; Zoghbi, S.; Baldwin, R.M.; Rajeevan, N.; Ellis, S.; Petrakis, I.L.; Seibyl, J.P. Alterations of benzodiazepine receptors in type II alcoholic subjects measured with SPECT and [^{123}I] iomazenil. *Am. J. Psychiatry* **1998**, *155*, 1550–1555. [CrossRef]

136. Lingford-Hughes, A.R.; Acton, P.; Gacinovic, S.; Suckling, J.; Busatto, G.; Boddington, S.; Bullmore, E.; Woodruff, P.; Costa, D.; Pilowsky, L. Reduced levels of GABA-benzodiazepine receptor in alcohol dependency in the absence of grey matter atrophy. *Br. J. Psychiatry* **1998**, *173*, 116–122. [CrossRef] [PubMed]

137. Lingford-Hughes, A.; Wilson, S.; Cunningham, V.J.; Feeney, A.; Stevenson, B.; Brooks, D.; Nutt, D.J. GABA-benzodiazepine receptor function in alcohol dependence: A combined 11 C-flumazenil PET and pharmacodynamic study. *Psychopharmacology* **2005**, *180*, 595–606. [CrossRef]

138. Lingford-Hughes, A.; Reid, A.G.; Myers, J.; Feeney, A.; Hammers, A.; Taylor, L.G.; Rosso, L.; Turkheimer, F.; Brooks, D.J.; Grasby, P. A [^{11}C] Ro15 4513 PET study suggests that alcohol dependence in man is associated with reduced α5 benzodiazepine receptors in limbic regions. *J. Psychopharmacol.* **2012**, *26*, 273–281. [CrossRef] [PubMed]

139. Volkow, N.D.; Wang, G.J.; Begleiter, H.; Hitzemann, R.; Pappas, N.; Burr, G.; Pascani, K.; Wong, C.; Fowler, J.S.; Wolf, A.P. Regional brain metabolic response to lorazepam in subjects at risk for alcoholism. *Alcohol. Clin. Exp. Res.* **1995**, *19*, 510–516. [CrossRef] [PubMed]

140. Weerts, E.M.; Wand, G.S.; Kuwabara, H.; Munro, C.A.; Dannals, R.F.; Hilton, J.; Frost, J.J.; McCaul, M.E. Positron emission tomography imaging of mu-and delta-opioid receptor binding in alcohol-dependent and healthy control subjects. *Alcohol. Clin. Exp. Res.* **2011**, *35*, 2162–2173. [CrossRef]

141. Mitchell, J.M.; O'Neil, J.P.; Janabi, M.; Marks, S.M.; Jagust, W.J.; Fields, H.L. Alcohol consumption induces endogenous opioid release in the human orbitofrontal cortex and nucleus accumbens. *Sci. Transl. Med.* **2012**, *4*, 116ra116. [CrossRef] [PubMed]

142. Heinz, A.; Reimold, M.; Wrase, J.; Hermann, D.; Croissant, B.; Mundle, G.; Dohmen, B.M.; Braus, D.H.; Schumann, G.; Machulla, H.-J. Correlation of stable elevations in striatal μ-opioid receptor availability in detoxified alcoholic patients with alcohol craving: A positron emission tomography study using carbon 11–labeled carfentanil. *Arch. Gen. Psychiatry* **2005**, *62*, 57–64. [CrossRef]

143. Hermann, D.; Hirth, N.; Reimold, M.; Batra, A.; Smolka, M.N.; Hoffmann, S.; Kiefer, F.; Noori, H.R.; Sommer, W.H.; Reischl, G. Low μ-opioid receptor status in alcohol dependence identified by combined positron emission tomography and post-mortem brain analysis. *Neuropsychopharmacology* **2017**, *42*, 606–614. [CrossRef]

144. Mann, K.; Torup, L.; Sørensen, P.; Gual, A.; Swift, R.; Walker, B.; van den Brink, W. Nalmefene for the management of alcohol dependence: Review on its pharmacology, mechanism of action and meta-analysis on its clinical efficacy. *Eur. Neuropsychopharmacol.* **2016**, *26*, 1941–1949. [CrossRef]

145. Parsons, L.H.; Hurd, Y.L. Endocannabinoid signalling in reward and addiction. *Nat. Rev. Neurosci.* **2015**, *16*, 579–594. [CrossRef] [PubMed]

146. Neumeister, A.; Normandin, M.D.; Murrough, J.W.; Henry, S.; Bailey, C.R.; Luckenbaugh, D.A.; Tuit, K.; Zheng, M.Q.; Galatzer-Levy, I.R.; Sinha, R. Positron emission tomography shows elevated cannabinoid CB 1 receptor binding in men with alcohol dependence. *Alcohol. Clin. Exp. Res.* **2012**, *36*, 2104–2109. [CrossRef]

147. Hirvonen, J.; Zanotti-Fregonara, P.; Umhau, J.C.; George, D.T.; Rallis-Frutos, D.; Lyoo, C.H.; Li, C.-T.; Hines, C.S.; Sun, H.; Terry, G.E. Reduced cannabinoid CB 1 receptor binding in alcohol dependence measured with positron emission tomography. *Mol. Psychiatry* **2013**, *18*, 916–921. [CrossRef]

148. Brown, A.K.; George, D.T.; Fujita, M.; Liow, J.S.; Ichise, M.; Hibbeln, J.; Ghose, S.; Sangare, J.; Hommer, D.; Innis, R.B. PET [^{11}C] DASB imaging of serotonin transporters in patients with alcoholism. *Alcohol. Clin. Exp. Res.* **2007**, *31*, 28–32. [CrossRef] [PubMed]

149. Szabo, Z.; Owonikoko, T.; Peyrot, M.; Varga, J.; Mathews, W.B.; Ravert, H.T.; Dannals, R.F.; Wand, G. Positron emission tomography imaging of the serotonin transporter in subjects with a history of alcoholism. *Biol. Psychiatry* **2004**, *55*, 766–771. [CrossRef] [PubMed]

150. Martinez, D.; Slifstein, M.; Gil, R.; Hwang, D.-R.; Huang, Y.; Perez, A.; Frankle, W.G.; Laruelle, M.; Krystal, J.; Abi-Dargham, A. Positron emission tomography imaging of the serotonin transporter and 5-HT1A receptor in alcohol dependence. *Biol. Psychiatry* **2009**, *65*, 175–180. [CrossRef] [PubMed]

151. Kumar, J.; Ismail, Z.; Hatta, N.H.; Baharuddin, N.; Hapidin, H.; Get Bee, Y.-T.; Yap, E.; Pakri Mohamed, R.M. Alcohol addiction-metabotropic glutamate receptor subtype 5 and its Ligands: How They All Come Together? *Curr. Drug Targets* **2018**, *19*, 907–915. [CrossRef] [PubMed]

152. Akkus, F.; Mihov, Y.; Treyer, V.; Ametamey, S.M.; Johayem, A.; Senn, S.; Rösner, S.; Buck, A.; Hasler, G. Metabotropic glutamate receptor 5 binding in male patients with alcohol use disorder. *Transl. Psychiatry* **2018**, *8*, 1–8. [CrossRef]
153. Erol, A.; Karpyak, V.M. Sex and gender-related differences in alcohol use and its consequences: Contemporary knowledge and future research considerations. *Drug Alcohol Depend.* **2015**, *156*, 1–13. [CrossRef]
154. Verplaetse, T.L.; Cosgrove, K.P.; Tanabe, J.; McKee, S.A. Sex/gender differences in brain function and structure in alcohol use: A narrative review of neuroimaging findings over the last 10 years. *J. Neurosci. Res.* **2021**, *99*, 309–323. [CrossRef]
155. McCaul, M.E.; Roach, D.; Hasin, D.S.; Weisner, C.; Chang, G.; Sinha, R. Alcohol and women: A brief overview. *Alcohol. Clin. Exp. Res.* **2019**, *43*, 774. [CrossRef]
156. Fairbanks, J.; Umbreit, A.; Kolla, B.P.; Karpyak, V.M.; Schneekloth, T.D.; Loukianova, L.L.; Sinha, S. Evidence-based pharmacotherapies for alcohol use disorder: Clinical pearls. In *Mayo Clinic Proceedings*; Elsevier: Amsterdam, The Netherlands, 2020; pp. 1964–1977.

nutrients

Review

Alcohol's Impact on the Fetus

Svetlana Popova [1,2,3,4,*], Danijela Dozet [1,4], Kevin Shield [1,2], Jürgen Rehm [1,2,4,5,6,7] and Larry Burd [8]

[1] Institute for Mental Health Policy Research, Centre for Addiction and Mental Health, 33 Ursula Franklin Street, Toronto, ON M5S 2S1, Canada; danijela.dozet@camh.ca (D.D.); kevin.shield@camh.ca (K.S.); jtrehm@gmail.com (J.R.)

[2] Dalla Lana School of Public Health, University of Toronto, 155 College Street, Toronto, ON M5T 3M7, Canada

[3] Factor-Inwentash Faculty of Social Work, University of Toronto, 246 Bloor Street W, Toronto, ON M5S 1V4, Canada

[4] Institute of Medical Science, Faculty of Medicine, Medical Sciences Building, University of Toronto, 1 King's College Circle, Toronto, ON M5S 1A8, Canada

[5] Institute of Clinical Psychology and Psychotherapy & Center of Clinical Epidemiology and Longitudinal Studies, Technische Universität Dresden, Chemnitzer Street 46, 01187 Dresden, Germany

[6] Department of Psychiatry, University of Toronto, 250 College Street, Toronto, ON M5T 1R8, Canada

[7] Department of International Health Projects, Institute for Leadership and Health Management, I.M. Sechenov First Moscow State Medical University, Trubetskaya Street, 8, b. 2, 119992 Moscow, Russia

[8] Department of Pediatrics, University of North Dakota School of Medicine and Health Sciences, 1301 N Columbia Rd., Grand Forks, ND 58202, USA; larry.burd@ndus.edu

[*] Correspondence: lana.popova@camh.ca; Tel.: +1-(416)-535-8501 (ext. 34558)

Abstract: Background: Alcohol is a teratogen and prenatal exposure may adversely impact the developing fetus, increasing risk for negative outcomes, including Fetal Alcohol Spectrum Disorder (FASD). Global trends of increasing alcohol use among women of childbearing age due to economic development, changing gender roles, increased availability of alcohol, peer pressure and social acceptability of women's alcohol use may put an increasing number of pregnancies at risk for prenatal alcohol exposure (PAE). This risk has been exacerbated by the ongoing COVID-19 pandemic in some countries. Method: This literature review presents an overview on the epidemiology of alcohol use among childbearing age and pregnant women and FASD by World Health Organization regions; impact of PAE on fetal health, including FASD; associated comorbidities; and social outcomes. Results/Conclusion: The impact of alcohol on fetal health and social outcomes later in life is enormous, placing a huge economic burden on countries. Prevention of prenatal alcohol exposure and early identification of affected individuals should be a global public health priority.

Keywords: alcohol; fetal; fetal alcohol spectrum disorder; morbidity; mortality; pregnancy

Citation: Popova, S.; Dozet, D.; Shield, K.; Rehm, J.; Burd, L. Alcohol's Impact on the Fetus. *Nutrients* **2021**, *13*, 3452. https://doi.org/10.3390/nu13103452

Academic Editor: Peter Anderson

Received: 9 August 2021
Accepted: 27 September 2021
Published: 29 September 2021

1. Introduction

Alcohol consumption can harm not only the person consuming alcohol, but also to their family, friends and community. Alcohol use during pregnancy is one of the examples of "harm to others"—to the developing fetus. Alcohol has been a well-established teratogen for many years and may adversely impact the developing fetus, increasing risk for many adverse outcomes, including Fetal Alcohol Spectrum Disorder (FASD).

Alcohol use is increasing at an alarming rate among women globally, particularly among women of childbearing age (15–49 years) and pregnant women [1–5]. Alcohol consumption has increased in terms of the prevalence of current drinkers [4], and in the volume and frequency of drinking among drinkers [5], as well as the lowering of the age of initiation for alcohol use [3]. Though a global assessment of alcohol use among pregnant women during the pandemic has yet to be conducted, there is some evidence to suggest that the COVID-19 pandemic has been associated with increased alcohol use among women of childbearing age in some countries [6–8]. Higher than usual rates of alcohol use, combined with unplanned pregnancies, places many pregnancies at increased risk of alcohol exposure and therefore, many infants at risk of harm [9]. Changes to substance use during the pandemic may be mediated by mental health status; for example, pregnant women's psychological distress in the United States was predictive of the number of substances they used [10], whereas a Canadian study found no such association [11]. This article will overview epidemiology of alcohol use among childbearing and pregnant women, FASD in general and special sub-populations as well as the impact of alcohol on the fetus.

1.1. Alcohol Use among Childbearing-Aged Women

Based on data from the World Health Organization's (WHO) 2018 Global Information System on Alcohol and Health [4], global adult per capita consumption of alcohol (APC) among women of childbearing age was estimated to be 2.3 L in 2016. Worldwide, APC among women of childbearing age varies, ranging from as low as 0.1 L in the Eastern Mediterranean Region to as high as 4.6 L in the European Region. Countries with high-income economies reported the greatest APC (4.5 L), followed by upper-middle-income economies (2.7 L), lower-middle-income economies (1.5 L), and low-income economies (1.2 L) (see Table 1).

Table 1. Adult per capita consumption and the prevalence of current drinking, former drinking, lifetime abstention, and heavy episodic drinking among women of childbearing age (15 to 49 years of age) in 2016.

Regions	Per Capita Consumption (Litres)	Prevalence (%)			
		Current Drinking	Former Drinking	Lifetime Abstention	Heavy Episodic Drinking
Global	2.3	32.1	56.7	11.3	8.7
WHO regions					
African	2.1	21.7	69.4	9.0	7.6
America	3.1	42.9	25.0	32.0	10.5
Eastern Mediterranean	0.1	1.3	97.8	0.9	0.1
European	4.6	53.9	30.3	15.9	18.7
South-East Asia	1.4	22.5	70.5	7.0	5.2
Western Pacific	2.8	42.7	49.6	7.6	10.7
World Bank regions					
Low-income economies	1.2	17.4	73.8	8.8	5.5
Lower-middle-income economies	1.5	20.6	71.3	8.1	5.0
Upper-middle-income economies	2.7	37.6	50.7	11.8	10.4
High-income economies	4.5	60.7	18.4	20.9	17.3

Source: The Global Information System on Alcohol and Health [12].

Globally, in 2016, 32.1% of women of childbearing age consumed alcohol. The prevalence of current drinking among childbearing-aged women was lowest in the Eastern Mediterranean Region (1.3%) and highest in the European Region (53.9%) [4].

Worldwide, approximately 8.7% of women of childbearing age engaged in heavy episodic drinking (HED), defined as consuming at least 60 g of pure alcohol on at least one occasion in the past 30 days [4]. The definition of what constitutes a standard drink varies: in the United States, a standard drink contains 14 g of pure alcohol, but in Europe, a standard drink contains approximately 10 g of alcohol.

Among women of childbearing age who drink, the prevalence of HED was lowest in the Eastern Mediterranean Region (0.1%) and highest in the European Region (18.7%). Among childbearing-aged women who drink, the prevalence of HED was highest in countries with high-income economies (60.7% and 17.3%, respectively), followed by upper-middle-income economies (37.6% and 10.4%, respectively), lower-middle-income economies (20.6% and 5.0%, respectively), and low-income economies (17.4% and 5.5%, respectively).

The ratio of men's APC to women's APC in 2016 was estimated to be 4.6 globally; highest in the Eastern Mediterranean Region (10.8) and lowest in the European Region (3.8). Assuming a prospective trend of convergence of gendered alcohol use [13], consumption by women is expected to continue to increase in those regions where currently there exists a large discrepancy in the amount of alcohol consumed by men versus women.

1.2. Alcohol Use during Pregnancy

Many countries provide information to healthcare practitioners and the public regarding the detrimental effects of alcohol consumption during pregnancy, often in the form of clinical guidelines. Examples include guidelines produced by Australia [14], Denmark [15], Canada [16], France [17], the United States [18], and the WHO's 2014 guidelines for the management of substance use during pregnancy [19]. Currently, no level of alcohol exposure during pregnancy is known to be safe [20], as even relatively low levels of alcohol use can substantially increase the risk of FASD [21]. Despite these public health efforts, approximately 10% of women worldwide continue to consume alcohol during pregnancy [22,23].

The WHO European Region (EUR) was estimated to have the highest prevalence of alcohol use during pregnancy (25.2%) [22] (see Table 2). This is not surprising; according to the latest Global Status Report on Alcohol and Health [4], the European Region ranked highest in nearly all major alcohol indicators, including prevalence of consumption, levels of consumption, rates of alcohol use, HED, and Alcohol Use Disorders (AUDs).

Table 2. Global prevalence of any amount of alcohol use and binge drinking (4 or more drinks on a single occasion) during pregnancy, and of FAS and FASD among the general population, by WHO Region in 2012, and corresponding 95% confidence intervals.

Region	Alcohol Use (Any Amount) during Pregnancy (%) [a]	Binge Drinking during Pregnancy (%) [b]	FAS (per 10,000)	FASD (per 10,000) [c]
Globally	9.8 (8.9, 11.1)	-	9.4 (9.4, 23.3)	77.3 (49.0, 116.1)
WHO Regions				
African	10.0 (8.5, 11.8)	0.1 (0.1, 6.1)	8.9 (8.9, 21.5)	78.3 (53.6, 107.1)
America	11.2 (9.4, 12.6)	0.1 (0.1, 5.6)	11.0 (11.0, 24.0)	87.9 (63.7, 132.4)
Eastern Mediterranean	0.2 (0.1, 0.9)	-	0.2 (0.2, 0.9)	1.3 (0.9, 4.5)
European	25.2 (21.6, 29.6)	0.0 (0.0, 5.3)	24.7 (24.7, 54.2)	198.2 (140.9, 280.0)
South-East Asia	1.8 (0.9, 5.1)	-	1.3 (1.3, 8.1)	14.1 (6.4, 53.1)
Western Pacific	8.6 (4.5, 11.6)	0.0 (0.0, 3.5)	7.7 (7.7, 19.4)	67.4 (45.4, 116.6)

Source: Popova et al., 2017 [23]. [a] The prevalence of any amount of alcohol use during pregnancy is inclusive of the prevalence of binge drinking during pregnancy. [b] It was not possible to estimate the prevalence of binge drinking during pregnancy for the Eastern-Mediterranean or South-East Asia Regions due to a lack of available data for countries in these regions; therefore, the global prevalence could not be estimated. [c] The prevalence of FASD includes the prevalence of FAS. © Canadian Science Publishing.

Additionally, the five countries with the highest prevalence of alcohol use among pregnant women were in the European Region: Ireland (60.4%), Belarus (46.6%), Denmark (45.8%), the United Kingdom (41.3%), and Russia (36.5%) [22,23]. Among these five countries alone, there are an annual 902,180 prenatally alcohol-exposed births, 69,395 of which have FASD. In total, there are approximately 1,249,110 cases with FASD among those aged 0–18 years in these five countries alone (see Table 3).

Table 3. Estimated annual number of cases of FASD in the five countries (Ireland, Belarus, Denmark, the United Kingdom, and Russia) reporting the highest prevalence of alcohol use during pregnancy in the world in 2016.

Country	Estimated Number of Cases with PAE (%) *	Annual Births (Prenatally Alcohol Exposed Births)	Annual Number of Cases with FASD *	Number of Cases with FASD from Birth up to Age 18
Ireland	60.4	55,959 (33,799)	2599	46,782
Belarus	46.6	107,930 (50,295)	3868	69,624
Denmark	45.8	61,167 (28,014)	2154	38,772
United Kingdom	41.3	640,370 (264,472)	20,344	366,192
Russia	36.5	1,440,000 (525,600)	40,430	727,775
Total cases with FASD		2,305,426 (902,180)	69,395	1,249,110

* Based on the assumption that one in every 13 alcohol-exposed pregnancies will result in FASD in any given year [22–24].

The lowest prevalence of alcohol use during pregnancy was estimated to be in the Eastern-Mediterranean Region (50 times lower than the global average) followed by the South-East Asia Region (five times lower than the global average); in these regions, a large proportion of the population abstains from alcohol use, especially within the female population [19].

In addition to high rates of alcohol consumption during pregnancy, in 40% of the countries examined, over one-quarter of pregnant women who consumed alcohol engaged in binge drinking [23,24]. This is alarming, as binge drinking is the most detrimental pattern of drinking during pregnancy and increases the risk of FASD. The highest prevalence of binge drinking during pregnancy was estimated to be in the African Region (3.1%), while the lowest prevalence of binge drinking during pregnancy was estimated to be in the Western-Pacific Region (1.8%) (see Table 2).

Among women who drank any amount of alcohol during pregnancy, the proportion who engaged in binge drinking was estimated to range from 10.7% in the European Region to 31.0% in the African Region [23]. The five countries with the highest estimated prevalence of binge drinking during pregnancy were Paraguay (13.9%), Moldova (10.6%), Ireland (10.5%), Lithuania (10.5%), and the Czech Republic (9.4%) [23].

These estimates are representative of the general populations of the respective countries; however, prevalence rates of alcohol use during pregnancy have been reported to be much higher among some sub-populations. For example, among pregnant Inuit women in northern Quebec (Canada), alcohol use was reported to be 60.5%, which is more than ten times higher than the estimate (10%) for the general population of pregnant women in Canada [25].

2. Effects of Alcohol Use on Fetus

Alcohol is a teratogen that can readily cross the placenta, resulting in damage to the developing embryo and fetus. Studies (both animal and clinical) have demonstrated that ethanol diffuses through the placenta and distributes rapidly into the fetal compartment, accumulating in the amniotic fluid [26]. This reservoir causes greater fetal exposure to ethanol and is intensified by fetal swallowing, caused by the fetal kidneys excreting xenobiotics into the amniotic fluid, which is then swallowed by the fetus [27]. Alcohol has a prolonged effect on the fetus due to amniotic accumulation, reduced concentrations of fetal metabolic enzymes, and reduced elimination, which results in damage to the developing embryo and fetus.

The identification of the range of adverse outcomes from prenatal exposure to alcohol is still an emerging field. Thus far, fetal exposure to alcohol is an established risk factor for a number of adverse outcomes, including stillbirth [28], spontaneous abortion [29], premature birth [30–32], intrauterine growth retardation [32,33], low birthweight [32,34] and FASD [35].

Cellular responses resulting from prenatal exposure to the teratogen alcohol is an area of considerable interest to a wide range of researchers and clinicians. Currently, alcohol is known to be teratogenic for all organ systems of the developing fetus [36]. Alcohol seems to be especially harmful to the developing nervous system. Some of these mechanisms, include increased oxidative stress to the central nervous system (CNS) [37], impaired angiogenesis and neurogenesis that occurs during brain development in utero [38], increased cell death in various brain structures [39], as well as disruptions to the endocrine system [40], to gene expressions [41] and to prostaglandin synthesis [42]. Prenatal alcohol exposure dysregulates amino acid homeostasis and induces excitatory neurotoxicity, increasing taurine levels as a neuroprotective response [43].

PAE is generally associated with reduced overall brain volume in exposed individuals, which can be observed at birth and later in life. Specifically, individuals with PAE may have reduced gray and white matter in the cerebrum and cerebellum, and reduced gray matter in brain structures including the amygdala, hippocampus, putamen, caudate, thalamus and pallidum [44]. Poorer cognitive and behavioural outcomes are associated with larger brain volume decreases from PAE; however, rates of brain volume growth are similar in early childhood and there are abnormal brain-behaviour connections displayed in children with PAE, suggesting there is a well-defined, long window of opportunity for intervention [45]. The asymmetry of the cerebral cortex thickness throughout development is preserved in individuals with PAE despite their thinner cortex and reduced globally reduced volume, suggesting that typical brain developmental patterns do occur in individuals with PAE and that deficits in cognitive performance are more related to the teratogenic exposure in utero [44].

In addition to the substantial cognitive, neurological, and behavioural deficits resulting from damage to the central nervous system, PAE can also lead to organ defects found in the liver, kidney, and heart, as well as disruptions in the gastrointestinal and endocrine systems [46,47]. Consistent and high levels of alcohol intake on a daily basis during pregnancy are associated with significant fetal impairments in gross and fine motor function observed in childhood [48]. It is theorized that PAE programs the fetal HPA axis to have increased tone, affecting stress response and regulation during baseline and stressful conditions throughout life [40]. This increased tone may be linked to increased reactivity in adverse life events and may represent an increased vulnerability to depression, as HPA dysregulation is commonly found in individuals with depression [49]. There is also preliminary evidence to suggest that PAE can adversely impact immune function later in life as a result of changes in cytokines and lymphocytes that are related to atopic allergy and infection outcomes [50]. While more research is needed in this area, this early finding is important given the relationship between stress reactivity and immune function, suggesting that individuals with PAE are even more susceptible to various comorbidities.

Outcomes from alcohol exposure are modified by a large number of effect modifiers, which include timing and amount of alcohol consumption (dose-dependent), maternal/fetal polysubstance exposure (smoking, other drugs, prescription medications (both prescribed and diverted), nutritional deficiencies, exposure to environmental stressors (e.g., domestic violence, poverty, homelessness), comorbid mental disorders (e.g., depression, anxiety), and peer pressure from friends, families and especially partners' substance use patterns during pregnancy [51–53]. Some patterns of co-exposure are observed to be more detrimental, especially drinking alcohol and nicotine or cannabis exposure [54–56]. It has been shown that prenatal exposures to alcohol and nicotine are more potent influences on developing fetal brain structures than cocaine [57].

A growing body of evidence supports epigenetic etiology in the development of FASD due to ethanol-related changes in gene expression (without changes in the DNA sequence), particularly from 3 to 8 weeks of gestation [58]. The timing and pattern of prenatal alcohol exposure affects the alcohol-induced alterations; for example, binge drinking in the first trimester can affect cell proliferation, whereas cell migration/differentiation is affected in the second, and cellular networking is affected by binges in the third trimester [59].

Exposure to alcohol can affect biochemical reactions in the methionine–homocysteine cycle which is linked to epigenetic changes including abnormal DNA methylation [58] and induces oxidative stress which leads to DNA damage [60] affecting a fetus's cell proliferation and differentiation. Interestingly, chronic paternal alcohol use is associated with epigenetic dysregulation in neonates as well [61]. Furthermore, rodent studies have found that prenatal alcohol exposure affects gene transcription in adolescence [62].

2.1. Fetal Alcohol Spectrum Disorder

The most dramatic manifestation of PAE is Fetal Alcohol Spectrum Disorder (FASD). FASD describes a wide range of effects, these are broadly clustered into three categories: (1) cognitive impairments including central nervous system damage, congenital anomalies, prenatal or postnatal growth restrictions; (2) characteristic dysmorphic facial features; and (3) deficits in cognitive, behavioral, emotional, and adaptive functioning. Since the first description of FAS by Jones and Smith in 1973 [63,64], the terminology and the diagnostic guidelines have changed numerous times. Depending on the classification system, FASD includes several alcohol-related diagnoses, such as Fetal Alcohol Syndrome (FAS), partial FAS (pFAS), alcohol-related neurodevelopmental disorder (ARND), and Alcohol-Related Birth Defects (ARBD) [65,66]. The Canadian framework utilizes a single designation of FASD as a diagnostic term, with the specification of the presence or absence of the characteristic sentinel facial features [67]. In addition, the category of neurodevelopmental disorder associated with prenatal alcohol exposure (ND-PAE) is included in the Diagnostic and Statistical Manual of Mental Disorders, Fifth Edition (DSM-5; [68]). Although the criteria for FASD diagnoses have been described thoroughly, the diagnosis of FASD remains challenging, and the specific assessment techniques used to make the definitive diagnosis are still debated, especially for ARND. It is important to note that FASD captures only a modest proportion of damage to the developing fetus from alcohol exposure, so not every child with PAE will be diagnosed with FASD [22].

The risk of developing FASD is related to the patterns (the quantity, frequency, duration) of alcohol consumption, as well as other factors. It has been well documented in both animal and human studies that larger quantities of PAE and sustained drinking throughout all trimesters of pregnancy result in greater physical and cognitive deficits, and the appearance of facial features in FAS [69]. Binge drinking is associated with the highest risk of fetal damage [52]. However, many studies show that low and moderate drinking during pregnancy is also associated with fetal damage [20,21].

Research studies show that many other risk factors can also influence development of FASD, which include a smaller body profile of mother (height, weight, and body mass index [BMI]), maternal age, low socioeconomic status and smoking [52], a poor nutritional status (deficit of riboflavin, calcium, and zinc; [70], genetic polymorphisms [71,72], and paternal alcohol consumption [73].

Several categories of biomarkers of PAE and FASD have been proposed. These include biomarkers demonstrating adverse outcomes for Fetal Alcohol Syndrome including anatomical markers (e.g., facial abnormalities; micrognathia); biomarkers of neurological impairments (e.g., reduced memory retention), developmental biomarkers (e.g., delayed motor learning) and neurobehavioral biomarkers (e.g., deficits in executive functioning) [74]. For example, lower serum concentrations of insulin growth factors IGF-I and IGF-II have been found to be a reliable biomarker of prenatal alcohol exposure in children [75]. Certain imaging studies are proposed to be biomarkers for brain differences associated with FASD, including the use of computerized tomography, magnetic resonance imaging and positron emission tomography [74]. In addition, a number of biomarkers for PAE present at birth are available based on meconium and cord blood [74,76,77].

FASD has a very broad phenotype and is further complicated by high rates of comorbidity. More than 400 disease conditions, spanning across 18 of 22 chapters of the ICD-10 have been reported to co-occur in people diagnosed with FASD [36]. The most prevalent conditions that occur in individuals with FASD are found within the International

Statistical Classification of Diseases (ICD-10) chapters of "Congenital malformations, deformities, and chromosomal abnormalities" (43%) and "Mental and behavioral disorders" (18%) [78]. Some comorbid conditions that are highly prevalent among people with FASD include abnormal results of function studies of the peripheral nervous system and special senses, conduct disorder, chronic serous otitis media, receptive and expressive language disorders, and visual, developmental, cognitive, mental, and behavioral impairments, with prevalence estimates of these conditions ranging from 50% to 91% [36]. FASD, as indicated by the sheer number of conditions found to co-occur in this population, is a multifaceted spectrum of disorders, affecting multiple organs and systems.

The complexity and chronicity of FASD affects both the affected individual and their family. In many cases, people with FASD require lifelong support from a wide range of support services, including health care, social assistance, and remedial education. Accordingly, it has been shown that FASD has a substantial economic impact on any society [79–81]. In Canada, the annual cost attributable to FASD was estimated to be between CAD1.3 billion and CAD2.3 billion [81]. In North America, the lifetime cost for a complex case of FASD has been estimated to be more than CAD1 million [82]. Individuals from all socio-economic and ethnic backgrounds are affected by FASD [83]. The deficits expressed by individuals with FASD vary broadly in severity and type. Moreover, FASD is intergenerational and familial. Children with an affected sibling are at a higher risk of having FASD [84].

2.2. Changes over the Lifespan of People with Fetal Alcohol Spectrum Disorder

The neurodevelopmental impairments associated with PAE resulting in FASD lead to other disabilities later in life. These disabilities include, but are not limited to, academic failure, substance abuse, mental health problems, frequent contact with law enforcement, and inability to live independently and obtain and/or maintain employment—all of which have lifelong implications [69]. These generally are referred to as secondary disabilities, which increase the demands and costs on existing service systems. This includes the health care system with increased demand for services [85], educational systems with need for special services for these children and adolescents, corrections systems beginning in early adolescence and extending across adult life, mental health systems across the lifespan, social services and child protection services and developmental disabilities.

2.3. Prevalence of Fetal Alcohol Spectrum Disorder

Globally, the prevalence rates of FASD among the general population were estimated to be 77.3 per 10,000 people [22,86].The prevalence of FASD in the general population was estimated to be highest in the European Region, at 198.2 per 10,000 people, and lowest in the Eastern Mediterranean Region, at 1.3 per 10,000 people (Table 2; [22,86]), respectively. These estimates are in line with the above-reported prevalence of alcohol use during pregnancy. The five countries with highest prevalence of FASD per 1,000 were South Africa (111.1), Croatia (53.3), Ireland (47.5), Italy (45.0), and Belarus (36.6) [86]. The five countries with the lowest prevalence of FASD (<0.1 per 1,000) were Oman, United Arab Emirates, Saudi Arabia, Qatar, and Kuwait [86].

Recent multi-site active case ascertainment studies in the United States have estimated the prevalence of FASD to be 5% among students in grade one [21,87–89]. In Canada, the population-based prevalence of FASD among elementary school children was estimated to be between 2% and 3% [90]. These studies found that nearly all of these students in the United States and all of these students in Canada had not been previously diagnosed with FASD.

The prevalence of FASD in the Western Cape Province of South Africa, a region known for wine production and a high prevalence of binge drinking among women of childbearing age, has been reported to be 135.1 to 207.5 per 1000 (13.5–20.8%) among first grade students—one of the highest FASD prevalence rates in the world [91].

A systematic review and meta-analysis revealed that FASD prevalence is much higher in certain sub-populations such as children in care, correctional, or special education programs, specialized clinical populations, and Aboriginal populations, compared to the general population [83]. For example, compared to the general population, the prevalence of FASD among children in care was 32 times higher in the United States and 40 times higher in Chile, the prevalence of FASD among adults in the Canadian correctional system was 19 times higher, and the prevalence of FASD among special education populations in Chile was over 10 times higher [83].

FASD prevalence rates reported in individual studies are extremely alarming. For instance, the prevalence of FASD was 62% among children with intellectual disabilities in care in Chile [92], over 52% among adoptees from Eastern Europe [93], and approximately 40% among children residing in orphanages in Lithuania [94]. The highest prevalence estimates of Fetal Alcohol Syndrome (FAS), which ranged between 46% and 68%, were reported in Russian orphanages for children with developmental abnormalities [95]. Additionally, the prevalence of FASD among youth in correctional services was over 23% in Canada [96], and over 14% among psychiatric care populations in the United States [97].

These findings demonstrate the large service and cost burdens of FASD across various systems of care and reflect a substantial global health problem [81]. It is important to provide caregiver education, diagnosis-informed interventions, lifelong services and long-term risk reduction to individuals with FASD to reduce the prevalence of secondary disabilities. Appropriate interventions, as well as diagnostic and support services, must be available to individuals with FASD from an early age in order to decrease their chances of becoming involved with the legal system, whether as a victim or as an offender. FASD is a huge risk factor for incarceration. It was estimated based on available findings that, on any given day in a specific year, children and adolescents with FASD are 19 times more likely to be incarcerated compared to children and adolescents without FASD [98]. Lastly, high prevalence rates of FASD among special education and specialized clinical populations are not surprising given that individuals with FASD are at an increased risk of having learning difficulties and mental health problems, and of experiencing developmental delays [36].

2.4. Fetal Alcohol Spectrum Disorder: Global Implications

Globally, there are 130 million live births each year. In 2016, 32.1% of women of childbearing age globally consumed alcohol in the past year [4] and 10% of women consumed alcohol during pregnancy [22]. These data demonstrate that every year, 13,000,000 pregnancies are alcohol-exposed in the world. Based on a crude estimate of exposure outcomes from a meta-analysis, an estimated one in every 13 alcohol-exposed pregnancies will result in FASD in one specific year [24]. This broad estimate suggests that out of the 13 million alcohol-exposed pregnancies, 1 million of these pregnancies will result in a baby born with FASD. To put this in perspective, this results in 2739 new cases of FASD each day, or 114 every hour, or 2 per minute. The cumulative impact is astonishing. Globally, the number of children with FASD aged from birth through 18 years of age is 18,000,000. We have previously noted that only a small number of individuals with FASD have been diagnosed: perhaps only one of every 800 to 1000 cases [99]. FASD is especially underdiagnosed in certain age groups, as tens of millions of adults and elderly people remain undiagnosed.

These are conservative estimates. Since a large portion of pregnancies globally are unplanned (44%) [100], many of the 32% of women of childbearing age who already consume alcohol are at risk of having an alcohol-exposed pregnancy in the early stages, before pregnancy recognition typically occurs. Alcohol use during pregnancy is also widely underreported [101,102] so the number of pregnancies with prenatal alcohol exposure (PAE) could be much higher. This would also increase the estimated number of births of children with FASD.

In addition, the number of comorbid disorders found to co-occur in individuals with FASD may also account for the lower prevalence estimates of FASD (i.e., underdiagnoses), likely due to the shadowing that might occur by the other disease conditions. Often, clinicians diagnose the immediate condition that has brought the individual in to seek medical treatment (e.g., problems with concentration), rather than taking into consideration the potential associations and underlying causes of the condition or illness (e.g., PAE).

2.5. Fetal Alcohol Spectrum Disorder and Mortality

One of the main neglected manifestations of FASD is premature mortality. The majority of people who may have met the diagnostic criteria for FASD die before they get a chance to be diagnosed. Deaths from stillbirth, sudden infant death syndrome and infectious illness are examples [103–105].

People diagnosed with FASD have a five-fold increase in risk for a premature death compared to people without FASD [105]. Since this risk is familial, their siblings similarly have a higher risk of mortality, irrespective of FASD diagnosis [106]. Having children with FASD is an important risk factor in maternal mortality [107,108]. In one case–control study in North Dakota (United States), mothers of children with FASD had a 44.82-fold increase in mortality risk compared to control mothers of the same age [108]. Among the deaths of birth mothers of children with FASD, 87% were among women under the age of 50 years, and among the causes of death for these cases, 67% was attributed to cancer, accidents, or classified as alcohol-related [108].

2.6. Fetal Alcohol Spectrum Disorder and Parenting

The complexity of parenting is increased for parents of children with FASD due to multiple issues, including sleep and eating disorders, fine and gross motor delays, toileting, speech and language disorders, need for medication(s), ongoing travel (often years) for therapies, learning and behavior impairments at school, and unavailability of childcare for children with complex behavioral impairments as well as of respite care services [109–112]. The complexity of parenting increases during adolescence, as young adult life with FASD may be especially complex. Early recognition of FASD and early emphasis on prevention of these problems and secondary disabilities are seen to be most useful in decreasing demands on parents [109].

3. Treatment Options

To date, there is no known treatment to reverse alcohol-induced damage to the fetus; however, many animal and human research studies have explored treatment options to reverse or prevent the mechanisms of alcohol teratogenicity on fetus [113]. For instance, animal studies have demonstrated ethanol-induced oxidative stress reduction by administering vitamins C and E [114–117], astaxanthin [117,118], curcumin [119,120] and resveratrol [121,122]. Doses of Vitamin E were found to reduce and even prevent Pukinje cell loss in rats [123,124], and when paired with beta carotene, doses of Vitamin C can also reduce or ameliorate hippocampal cell loss [124,125]. In zebrafish embryos, doses of Vitamin A were found to reverse small eye and body length defects, but were unable to reverse heart edema and other physical anomalies following PAE [124,126,127]. It was also shown that zinc can decrease physical abnormalities when administered at the time of ethanol exposure in mice [124,128,129]. Administering food supplements with folic acid, selenium, DHA, L-glutamine, boricacid, and choline [124,130] to pregnant women with risk of developing a child with FASD, also demonstrated that it can reduce the severity of ethanol-induced toxicity [124,131]. Recent studies show that low maternal iron reserves during pregnancy are associated with more prevalent FASD features where there is PAE; this is especially important as women of reproductive age who are heavy users of alcohol tend to have low iron stores [132]. Furthermore, while there is a breadth of evidence using animal models, clinical trials are needed to study interventions and supplementation options for humans [117].

4. Limitations

The current review has several strengths, including an overview of the most up-to-date epidemiological data on alcohol use in pregnancy and FASD in conjunction with what is known about the effects of alcohol on the fetus, including individual and population-level impacts. These data must be understood in light of several existing limitations, however. First and foremost, alcohol use during pregnancy is underreported and in addition, FASD is largely undiagnosed. Some countries may not have available data on alcohol use during pregnancy (e.g., countries not included in WHO Member States) and though the global estimate is 10%, this varies widely across countries. Furthermore, studies on maternal alcohol consumption are often not based on representative sampling strategies, and often do not include data on alcohol use patterns, pregnancy planning, or timing of pregnancy recognition. The current review presents meta-analytic estimates of FASD prevalence in general and sub-populations based on active case ascertainment (ACA) studies, which are the gold standard; however, case definitions and diagnostic criteria for FASD vary widely across studies and in different countries.

5. Conclusions

The detrimental effects of alcohol on a fetus leading to preventable chronic disabilities should be recognized and addressed as a global public health issue. The increasing incidence and prevalence of PAE and FASD at alarming rates is a public health exigency which demands strategic and timely interventions for both pregnant and childbearing-aged women who consume alcohol, and for their offspring who may be at risk of PAE. Prevention initiatives aimed at reducing alcohol use prior to and during pregnancy should be implemented worldwide.

There must be recognition that FASD is not restricted solely to disadvantaged groups, but rather that it may occur throughout society, regardless of socio-economic status, education, or ethnicity [90]. Therefore, efforts must be made to better educate the general population (adult women and men, children, and teenagers) about the risks of alcohol use (especially binge drinking and frequent drinking) during pregnancy.

Appropriate screening for alcohol use in all women of childbearing age, in combination with health promotion before conception, contraceptive counselling, brief interventions, and referrals to substance abuse programs where appropriate, should become a routine standard of care. Research has shown that easy access to substance use programs utilizing effective treatment of identified cases of alcohol dependence or AUDs among pregnant women could reduce the risk of PAE and having children with FASD [133,134]. Collaborative research projects and maternal health promotion programs (e.g., the Women-Infant-Child (WIC) program or the Parent Child Assistance Program (PCAP)) can be funded to help identify mothers who are drinking during pregnancy and therefore infants and young children with PAE who are at risk for FASD. Such programs have been shown to be cost-effective, and similarly, it has been estimated that preventing one case of FASD incurs only 3% of the costs it would require to provide support services to individuals with FASD [135].

Screening for PAE could lead to close monitoring of a child's development, facilitate early diagnosis, and the implementation of timely interventions, if necessary. It would also prevent the occurrence of secondary disabilities later in life and the occurrence and/or recurrence of prenatal and postnatal alcohol exposure within families at risk for FASD. Improved access to FASD diagnosis also is an opportunity to identify mothers who not only have maternal risk factors for FASD but also have children with FASD. This is a priority population for the prevention of alcohol exposure in future pregnancies. Among people diagnosed with FASD, specific efforts must be made to prevent school failure, victimization, incarceration, and the development of substance use disorders. Effective and consistent execution of universal screening can assist all countries in the development of surveillance systems to monitor the prevalence of alcohol use during pregnancy and of FASD, over time, in order to identify vulnerable populations, develop resources, and evaluate the effectiveness of prevention strategies.

To conclude, the negative health outcomes of alcohol exposure on the fetus, including FASD, are preventable and the most effective measure is to completely avoid consuming any type of alcohol during the entire pregnancy and while trying to become pregnant. All countries should implement effective and cost-effective population-based policy options that aim to reduce alcohol use among populations, including women of childbearing age and pregnant women [136,137], which ultimately will decrease the incidence of FASD and other negative health outcomes of children and their mothers.

Author Contributions: Conceptualization, S.P. and L.B.; methodology, S.P., L.B. and K.S.; writing—original draft preparation, S.P. and L.B.; writing—review and editing, S.P., D.D., L.B., K.S. and J.R.; visualization, S.P.; supervision, S.P. All authors have read and agreed to the published version of the manuscript.

Funding: This research received no external funding.

Institutional Review Board Statement: Not applicable.

Informed Consent Statement: Not applicable.

Conflicts of Interest: The authors declare no conflict of interest.

References

1. Slade, T.; Chapman, C.; Swift, W.; Keyes, K.; Tonks, Z.; Teesson, M. Birth cohort trends in the global epidemiology of alcohol use and alcohol-related harms in men and women: Systematic review and metaregression. *BMJ Open* **2016**, *6*, e011827. [CrossRef] [PubMed]
2. Denny, C.H.; Acero, C.S.; Terplan, M.; Kim, S.Y. Trends in alcohol use among pregnant women in the U.S., 2011–2018. *Am. J. Prev. Med.* **2020**, *59*, 768–769. [CrossRef] [PubMed]
3. Geels, L.M.; Vink, J.M.; van Beek, J.H.D.A.; Bartels, M.; Willemsen, G.; Boomsma, D.I. Increases in alcohol consumption in women and elderly groups: Evidence from an epidemiological study. *BMC Public Health* **2013**, *13*, 207. [CrossRef] [PubMed]
4. World Health Organization. *Global Status Report on Alcohol and Health*; World Health Organization: Geneva, Switzerland, 2018.
5. Manthey, J.; Shield, K.D.; Rylett, M.; Hasan, O.S.M.; Probst, C.; Rehm, J. Global alcohol exposure between 1990 and 2017 and forecasts until 2030: A modelling study. *Lancet* **2019**, *393*, 2493–2502. [CrossRef]
6. NANOS Research. *COVID-19 and Increased Alcohol Consumption: NANOS Poll Summary Report*; Canadian Centre on Substance Use and Addiction: Ottawa, ON, Canada, 2020.
7. Naughton, F.; Ward, E.; Khondoker, M.; Belderson, P.; Marie Minihane, A.; Dainty, J.; Hanson, S.; Holland, R.; Brown, T.; Notley, C. Health behaviour change during the UK COVID-19 lockdown: Findings from the first wave of the C-19 health behaviour and well-being daily tracker study. *Br. J. Health Psychol.* **2021**, *26*, 624–643. [CrossRef]
8. Pollard, M.S.; Tucker, J.S.; Green, J.H.D. Changes in adult alcohol use and consequences during the COVID-19 pandemic in the US. *JAMA Netw. Open* **2020**, *3*, e2022942. [CrossRef]
9. Sher, J. Fetal alcohol spectrum disorders: Preventing collateral damage from COVID-19. *Lancet Public Health* **2020**, *5*, e424. [CrossRef]
10. Smith, C.L.; Waters, S.F.; Spellacy, D.; Burduli, E.; Brooks, O.; Carty, C.L.; Ranjo, S.; McPherson, S.; Barbosa-Leiker, C. Substance use and mental health in pregnant women during the COVID-19 pandemic. *J. Reprod. Infant Psychol.* **2021**, 1–14. [CrossRef]
11. Kar, P.; Tomfohr-Madsen, L.; Giesbrecht, G.; Bagshawe, M.; Lebel, C. Alcohol and substance use in pregnancy during the COVID-19 pandemic. *Drug Alcohol Depend.* **2021**, *225*, 108760. [CrossRef]
12. World Health Organization. *Global Information System on Alcohol and Health*; World Health Organization: Geneva, Switzerland, 2018.
13. Bloomfield, K.; Gmel, G.; Neve, R.; Mustonen, H. Investigating gender convergence in alcohol consumption in Finland, Germany, The Netherlands, and Switzerland: A repeated survey analysis. *Subst. Abuse* **2001**, *22*, 39–53. [CrossRef]
14. National Health and Medical Research Council. *Australian Guidelines to Reduce Health Risks from Drinking Alcohol*; National Health and Medical Research Council: Canberra, Australia, 2009.
15. Danish National Board of Health. *Healthy Habits before, during and after Pregnancy*, 1st English ed.; Danish National Board of Health: Copenhagen, Denmark, 2010. (translated from 2nd Danish edition).
16. Carson, N.; Leach, L.; Murphy, K.J. A re-examination of Montreal Cognitive Assessment (MoCA) cutoff scores. *Int. J. Geriatr. Psychiatry* **2018**, *33*, 379–388. [CrossRef]
17. French Ministry of Youth Affairs and Sports. *Health Comes with Eating and Moving: The Food Guide for Pregnancy*; French Ministry of Youth Affairs and Sports: Saint-Yrieix-la-Perche, France, 2015.
18. Centers for Disease Control and Prevention. *Advisory on Alcohol Use during Pregnancy. A 2005 Message to Women from the US Surgeon General*; Centers for Disease Control and Prevention: Atlanta, GA, USA, 2005.

19. World Health Organization. *Guidelines for the Identification and Management of Substance Use and Substance Use Disorders in Pregnancy*; World Health Organization: Geneva, Switzerland, 2014.

20. May, P.A.; Blankenship, J.; Marais, A.; Gossage, J.P.; Kalberg, W.O.; Joubert, B.; Cloete, M.; Barnard, R.; de Vries, M.; Hasken, J.; et al. Maternal alcohol consumption producing fetal alcohol spectrum disorders (FASD): Quantity, frequency, and timing of drinking. *Drug Alcohol Depend.* **2013**, *133*, 502–512. [CrossRef]

21. Chambers, C.D.; Coles, C.; Kable, J.; Akshoomoff, N.; Xu, R.; Zellner, J.A.; Honerkamp-Smith, G.; Manning, M.A.; Adam, M.P.; Jones, K.L. Fetal alcohol spectrum disorders in a pacific Southwest city: Maternal and child characteristics. *Alcohol. Clin. Exp. Res.* **2019**, *43*, 2578–2590. [CrossRef]

22. Popova, S.; Lange, S.; Probst, C.; Gmel, G.; Rehm, J. Estimation of national, regional, and global prevalence of alcohol use during pregnancy and fetal alcohol syndrome: A systematic review and meta-analysis. *Lancet Glob. Health* **2017**, *5*, e290–e299. [CrossRef]

23. Popova, S.; Lange, S.; Probst, C.; Gmel, G.; Rehm, J. Global prevalence of alcohol use and binge drinking during pregnancy, and fetal alcohol spectrum disorder. *Biochem. Cell Biol.* **2018**, *96*, 237–240. [CrossRef]

24. Lange, S.; Probst, C.; Rehm, J.; Popova, S. Prevalence of binge drinking during pregnancy by country and World Health Organization region: Systematic review and meta-analysis. *Reprod. Toxicol.* **2017**, *73*, 214–221. [CrossRef] [PubMed]

25. Fraser, S.L.; Muckle, G.; Abdous, B.B.; Jacobson, J.L.; Jacobson, S.W. Effects of binge drinking on infant growth and development in an Inuit sample. *Alcohol* **2012**, *46*, 277–283. [CrossRef] [PubMed]

26. Brien, J.F.; Loomis, C.W.; Tranmer, J.; McGrath, M. Disposition of ethanol in human maternal venous blood and amniotic fluid. *Am. J. Obstet. Gynecol.* **1983**, *146*, 181–186. [CrossRef]

27. Underwood, M.A.; Gilbert, W.M.; Sherman, M.P. Amniotic fluid: Not just fetal urine anymore. *J. Perinatol.* **2005**, *25*, 341–348. [CrossRef] [PubMed]

28. Kesmodel, U.; Wisborg, K.; Olsen, S.F.; Henriksen, T.B.; Secher, N.J. Moderate alcohol intake during pregnancy and the risk of stillbirth and death in the first year of life. *Am. J. Epidemiol.* **2002**, *155*, 305–312. [CrossRef]

29. Henriksen, T.B.; Hjollund, N.H.; Jensen, T.K.; Bonde, J.P.; Andersson, A.-M.; Kolstad, H.; Ernst, E.; Giwercman, A.; Skakkebæk, N.E.; Olsen, J. Alcohol consumption at the time of conception and spontaneous abortion. *Am. J. Epidemiol.* **2004**, *160*, 661–667. [CrossRef]

30. Albertsen, K.; Andersen, A.-M.N.; Olsen, J.; Grønbæk, M. Alcohol consumption during pregnancy and the risk of preterm delivery. *Am. J. Epidemiol.* **2004**, *159*, 155–161. [CrossRef]

31. Kesmodel, U.; Olsen, S.F.; Secher, N.J. Does alcohol increase the risk of preterm delivery? *Epidemiology* **2000**, 512–518. [CrossRef] [PubMed]

32. Patra, J.; Bakker, R.; Irving, H.; Jaddoe, V.W.; Malini, S.; Rehm, J. Dose–response relationship between alcohol consumption before and during pregnancy and the risks of low birthweight, preterm birth and small for gestational age (SGA)—A systematic review and meta-analyses. *BJOG Int.J. Obstet. Gynaecol.* **2011**, *118*, 1411–1421. [CrossRef] [PubMed]

33. Yang, Q.; Witkiewicz, B.B.; Olney, R.S.; Liu, Y.; Davis, M.; Khoury, M.J.; Correa, A.; Erickson, J.D. A case-control study of maternal alcohol consumption and intrauterine growth retardation. *Ann. Epidemiol.* **2001**, *11*, 497–503. [CrossRef]

34. O'Callaghan, F.V.; O'Callaghan, M.; Najman, J.M.; Williams, G.M.; Bor, W. Maternal alcohol consumption during pregnancy and physical outcomes up to 5 years of age: A longitudinal study. *Early Hum. Dev.* **2003**, *71*, 137–148. [CrossRef]

35. Stratton, K.R.; Howe, C.J.; Battaglia, F.C. *Fetal Alcohol Syndrome: Diagnosis, Epidemiology, Prevention, and Treatment*; National Academy Press: Washington, DC, USA, 1996.

36. Popova, S.; Lange, S.; Shield, K.; Mihic, A.; Chudley, A.E.; Mukherjee, R.A.S.; Bekmuradov, D.; Rehm, J. Comorbidity of fetal alcohol spectrum disorder: A systematic review and meta-analysis. *Lancet* **2016**, *387*, 978–987. [CrossRef]

37. Ornoy, A. Embryonic oxidative stress as a mechanism of teratogenesis with special emphasis on diabetic embryopathy. *Reprod. Toxicol.* **2007**, *24*, 31–41. [CrossRef] [PubMed]

38. Jégou, S.; El Ghazi, F.; de Lendeu, P.K.; Marret, S.; Laudenbach, V.; Uguen, A.; Marcorelles, P.; Roy, V.; Laquerrière, A.; Gonzalez, B.J. Prenatal alcohol exposure affects vasculature development in the neonatal brain. *Ann. Neurol.* **2012**, *72*, 952–960. [CrossRef]

39. Cartwright, M.M.; Smith, S.M. Increased cell death and reduced neural crest cell numbers in ethanol-exposed Embryos: Partial basis for the fetal alcohol syndrome phenotype. *Alcohol. Clin. Exp. Res.* **1995**, *19*, 378–386. [CrossRef]

40. Haley, D.W.; Handmaker, N.S.; Lowe, J. Infant stress reactivity and prenatal alcohol exposure. *Alcohol. Clin. Exp. Res.* **2006**, *30*, 2055–2064. [CrossRef]

41. Vangipuram, S.D.; Grever, W.E.; Parker, G.C.; Lyman, W.D. Ethanol increases fetal human neurosphere size and alters adhesion molecule gene expression. *Alcohol. Clin. Exp. Res.* **2008**, *32*, 339–347. [CrossRef]

42. Randall, C.L.; Anton, R.F. Aspirin reduces alcohol-induced prenatal mortality and malformations in mice. *Alcohol. Clin. Exp. Res.* **1984**, *8*, 513–515. [CrossRef]

43. Lunde-Young, R.; Davis-Anderson, K.; Naik, V.; Nemec, M.; Wu, G.; Ramadoss, J. Regional dysregulation of taurine and related amino acids in the fetal rat brain following gestational alcohol exposure. *Alcohol* **2018**, *66*, 27–33. [CrossRef] [PubMed]

44. Zhou, D.; Rasmussen, C.; Pei, J.; Andrew, G.; Reynolds, J.N.; Beaulieu, C. Preserved cortical asymmetry despite thinner cortex in children and adolescents with prenatal alcohol exposure and associated conditions: Cortical asymmetry in prenatal alcohol exposure. *Hum. Brain Mapp.* **2018**, *39*, 72–88. [CrossRef] [PubMed]

45. Gautam, P.; Lebel, C.; Narr, K.L.; Mattson, S.N.; May, P.A.; Adnams, C.M.; Riley, E.P.; Jones, K.L.; Kan, E.C.; Sowell, E.R. Volume changes and brain-behavior relationships in white matter and subcortical gray matter in children with prenatal alcohol exposure. *Hum. Brain Mapp.* **2015**, *36*, 2318–2329. [CrossRef] [PubMed]

46. Lee, X.A.; Mostafaie, Y. Acute alcohol exposure in early gestation programmes sex-specific insulin resistance in offspring: A shifting tide in prenatal alcohol exposure models. *J. Physiol.* **2020**, *598*, 1807–1808. [CrossRef] [PubMed]

47. Caputo, C.; Wood, E.; Jabbour, L. Impact of fetal alcohol exposure on body systems: A systematic review: Impact of fetal alcohol exposure on body systems. *Birth Defects Res.* **2016**, *108*, 174–180. [CrossRef]

48. Bay, B.; Kesmodel, U.S. Prenatal alcohol exposure—A systematic review of the effects on child motor function. *Acta Obstet. Gynecol. Scand.* **2011**, *90*, 210–226. [CrossRef] [PubMed]

49. Weinberg, J.; Sliwowska, J.H.; Lan, N.; Hellemans, K.G.C. Prenatal alcohol exposure: Foetal programming, the hypothalamic-pituitary-adrenal axis and sex differences in outcome. *J. Neuroendocrinol.* **2008**, *20*, 470–488. [CrossRef] [PubMed]

50. Reid, N.; Moritz, K.M.; Akison, L.K. Adverse health outcomes associated with fetal alcohol exposure: A systematic review focused on immune-related outcomes. *Pediatr. Allergy Immunol.* **2019**, *30*, 698–707. [CrossRef]

51. Eberhart, J.K.; Parnell, S.E. The genetics of fetal alcohol spectrum disorders. *Alcohol. Clin. Exp. Res.* **2016**, *40*, 1154–1165. [CrossRef]

52. May, P.A.; Gossage, J.P. Maternal risk factors for fetal alcohol spectrum disorders: Not as simple as it might seem. *Alcohol Res. Health* **2011**, *34*, 15–26.

53. Day, J.; Savani, S.; Krempley, B.D.; Nguyen, M.; Kitlinska, J.B. Influence of paternal preconception exposures on their offspring: Through epigenetics to phenotype. *Am. J. Stem Cells* **2016**, *5*, 11–18. [PubMed]

54. Odendaal, H.J.; Steyn, D.W.; Elliott, A.; Burd, L. Combined effects of cigarette smoking and alcohol consumption on perinatal outcome. *Gynecol. Obstet. Investig.* **2009**, *67*, 1–8. [CrossRef]

55. Shuffrey, L.C.; Myers, M.M.; Isler, J.R.; Lucchini, M.; Sania, A.; Pini, N.; Nugent, J.D.; Condon, C.; Ochoa, T.; Brink, L.; et al. Association between prenatal exposure to alcohol and tobacco and neonatal brain activity: Results from the safe passage study. *JAMA Netw. Open* **2020**, *3*, e204714. [CrossRef]

56. Boa-Amponsem, O.; Zhang, C.; Mukhopadhyay, S.; Ardrey, I.; Cole, G.J. Ethanol and cannabinoids interact to alter behavior in a zebrafish fetal alcohol spectrum disorder model. *Birth Defects Res.* **2019**, *111*, 775–788. [CrossRef] [PubMed]

57. Levitt, P.; Stanwood, G.D.; Thompson, B.L. Prenatal exposure to drugs: Effects on brain development and implications for policy and education. *Nat. Rev. Neurosci.* **2009**, *10*, 303–312. [CrossRef]

58. Kobor, M.S.; Weinberg, J. Focus on: Epigenetics and fetal alcohol spectrum disorders. *Alcohol Res. Health* **2011**, *34*, 29–37.

59. Kleiber, M.L.; Mantha, K.; Stringer, R.L.; Singh, S.M. Neurodevelopmental alcohol exposure elicits long-term changes to gene expression that alter distinct molecular pathways dependent on timing of exposure. *J. Neurodev. Disord.* **2013**, *5*, 6. [CrossRef]

60. Bhatia, S.; Drake, D.M.; Miller, L.; Wells, P.G. Oxidative stress and DNA damage in the mechanism of fetal alcohol spectrum disorders. *Birth Defects Res.* **2019**, *111*, 714–748. [CrossRef] [PubMed]

61. Sánchez-Alvarez, R.; Gayen, S.; Vadigepalli, R.; Anni, H. Ethanol diverts early neuronal differentiation trajectory of embryonic stem cells by disrupting the balance of lineage specifiers. *PLoS ONE* **2013**, *8*, e63794. [CrossRef]

62. Perkins, A.; Lehmann, C.; Lawrence, R.C.; Kelly, S.J. Alcohol exposure during development: Impact on the epigenome. *Int. J. Dev. Neurosci.* **2013**, *31*, 391–397. [CrossRef]

63. Jones, K.; Smith, D. Recognition of the fetal alcohol syndrome in early infancy. *Lancet* **1973**, *302*, 999–1001. [CrossRef]

64. Jones, K.; Smith, D.; Ulleland, C.; Streissguth, A. Pattern of malformation in offspring of chronic alcoholic mothers. *Lancet* **1973**, *301*, 1267–1271. [CrossRef]

65. Chudley, A.E.; Conry, J.; Cook, J.L.; Loock, C.; Rosales, T.; LeBlanc, N. Fetal alcohol spectrum disorder: Canadian guidelines for diagnosis. *Can. Med. Assoc. J. CMAJ* **2005**, *172*, S1–S21. [CrossRef] [PubMed]

66. Hoyme, H.E.; Kalberg, W.O.; Elliott, A.J.; Blankenship, J.; Buckley, D.; Marais, A.-S.; Manning, M.A.; Robinson, L.K.; Adam, M.P.; Abdul-Rahman, O.; et al. Updated clinical guidelines for diagnosing fetal alcohol spectrum disorders. *Pediatrics* **2016**, *138*, e20154256. [CrossRef]

67. Cook, J.L.; Green, C.R.; Lilley, C.M.; Anderson, S.M.; Baldwin, M.E.; Chudley, A.E.; Conry, J.L.; LeBlanc, N.; Loock, C.A.; Lutke, J.; et al. Fetal alcohol spectrum disorder: A guideline for diagnosis across the lifespan. *Can. Med. Assoc. J. CMAJ* **2016**, *188*, 191–197. [CrossRef] [PubMed]

68. American Psychiatric Association. *Diagnostic and Statistical Manual of Mental Disorders: DSM-5*, 5th ed.; American Psychiatric Association: Arlington, VA, USA, 2013.

69. Streissguth, A.P.; Barr, H.M.; Kogan, J.; Bookstein, F.L. *Understanding the Occurrence of Secondary Disabilities in Clients with Fetal Alcohol Syndrome (FAS) and Fetal Alcohol Effects (FAE). Final Report to the Centers for Disease Control and Prevention (CDC)*; 96-06; University of Washington, Fetal Alcohol & Drug Unit: Seattle, WA, USA, 1996.

70. Keen, C.L.; Uriu-Adams, J.Y.; Skalny, A.; Grabeklis, A.; Grabeklis, S.; Green, K.; Yevtushok, L.; Wertelecki, W.W.; Chambers, C.D. The plausibility of maternal nutritional status being a contributing factor to the risk for fetal alcohol spectrum disorders: The potential influence of zinc status as an example. *BioFactors* **2010**, *36*, 125–135. [CrossRef]

71. Johnson, C.; Drgon, T.; Liu, Q.-R.; Walther, D.; Edenberg, H.; Rice, J.; Foroud, T.; Uhl, G.R. Pooled association genome scanning for alcohol dependence using 104,268 SNPs: Validation and use to identify alcoholism vulnerability loci in unrelated individuals from the collaborative study on the genetics of alcoholism. *Am. J. Med. Gen. Part B Neuropsychiat. Gen.* **2006**, *141B*, 844–853. [CrossRef]

72. Stoler, J.M.; Ryan, L.M.; Holmes, L.B. Alcohol dehydrogenase 2 genotypes, maternal alcohol use, and infant outcome. *J. Pediatr.* **2002**, *141*, 780–785. [CrossRef]

73. Ouko, L.A.; Shantikumar, K.; Knezovich, J.; Haycock, P.; Schnugh, D.J.; Ramsay, M. Effect of alcohol consumption on CpG methylation in the differentially methylated regions of H19 and IG-DMR in male gametes—Implications for fetal alcohol spectrum disorders. *Alcohol. Clin. Exp. Res.* **2009**, *33*, 1615–1627. [CrossRef]

74. Chabenne, A.; Moon, C.; Ojo, C.; Khogali, A.; Nepal, B.; Sharma, S. Biomarkers in fetal alcohol syndrome. *Biomark. Genom. Med.* **2014**, *6*, 12–22. [CrossRef]

75. Andreu-Fernández, V.; Bastons-Compta, A.; Navarro-Tapia, E.; Sailer, S.; Garcia-Algar, O. Serum concentrations of IGF-I/IGF-II as biomarkers of alcohol damage during foetal development and diagnostic markers of foetal alcohol syndrome. *Sci. Rep.* **2019**, *9*, 1562. [CrossRef] [PubMed]

76. Himes, S.K.; Dukes, K.A.; Tripp, T.; Petersen, J.M.; Raffo, C.; Burd, L.; Odendaal, H.; Elliott, A.J.; Hereld, D.; Signore, C.; et al. Clinical sensitivity and specificity of meconium fatty acid ethyl ester, ethyl glucuronide, and ethyl sulfate for detecting maternal drinking during pregnancy. *Clin. Chem.* **2015**, *61*, 523–532. [CrossRef] [PubMed]

77. Howlett, H.; Abernethy, S.; Brown, N.W.; Rankin, J.; Gray, W.K. How strong is the evidence for using blood biomarkers alone to screen for alcohol consumption during pregnancy? A systematic review. *Eur. J. Obstet.Gynecol. Reprod. Biol.* **2017**, *213*, 45–52. [CrossRef] [PubMed]

78. World Health Organization. *International Statistical Classification of Diseases and Related Health Problems*, 5th ed.; 10th Revision; World Health Organization: Geneva, Switzerland, 2016.

79. Popova, S.; Lange, S.; Burd, L.; Chudley, A.E.; Clarren, S.K.; Rehm, J. Cost of fetal alcohol spectrum disorder diagnosis in Canada. *PLoS ONE* **2013**, *8*, e60434. [CrossRef] [PubMed]

80. Popova, S.; Lange, S.; Burd, L.; Urbanoski, K.; Rehm, J. Cost of specialized addiction treatment of clients with fetal alcohol spectrum disorder in Canada. *BMC Public Health* **2013**, *13*, 570. [CrossRef]

81. Popova, S.; Lange, S.; Burd, L.; Rehm, J. The economic burden of fetal alcohol spectrum disorder in Canada in 2013. *Alcohol Alcohol.* **2016**, *51*, 367–375. [CrossRef]

82. Popova, S.; Stade, B.; Bekmuradov, D.; Lange, S.; Rehm, J. What do we know about the economic impact of fetal alcohol spectrum disorder? A systematic literature review. *Alcohol Alcohol.* **2011**, *46*, 490–497. [CrossRef]

83. Popova, S.; Lange, S.; Shield, K.; Burd, L.; Rehm, J. Prevalence of fetal alcohol spectrum disorder among special subpopulations: A systematic review and meta-analysis. *Addiction* **2019**, *114*, 1150–1172. [CrossRef]

84. Abel, E.L. Fetal alcohol syndrome in families. *Neurotoxicol. Teratol.* **1988**, *10*, 1–2. [CrossRef]

85. Oh, S.S.; Kim, Y.J.; Jang, S.-I.; Park, S.; Nam, C.M.; Park, E.-C. Hospitalizations and mortality among patients with fetal alcohol spectrum disorders: A prospective study. *Sci. Rep.* **2020**, *10*, 19512. [CrossRef] [PubMed]

86. Lange, S.; Probst, C.; Gmel, G.; Rehm, J.; Burd, L.; Popova, S. Global prevalence of fetal alcohol spectrum disorder among children and youth: A systematic review and meta-analysis. *JAMA Pediatr.* **2017**, *171*, 948–956. [CrossRef] [PubMed]

87. May, P.A.; Hasken, J.M.; Baete, A.; Russo, J.; Elliott, A.J.; Kalberg, W.O.; Buckley, D.; Brooks, M.; Ortega, M.A.; Hedrick, D.M.; et al. Fetal alcohol spectrum disorders in a Midwestern city: Child characteristics, maternal risk traits, and prevalence. *Alcohol. Clin. Exp. Res.* **2020**, *44*, 919–938. [CrossRef] [PubMed]

88. May, P.A.; Hasken, J.M.; Bozeman, R.; Jones, J.V.; Burns, M.K.; Goodover, J.; Kalberg, W.O.; Buckley, D.; Brooks, M.; Ortega, M.A.; et al. Fetal alcohol spectrum disorders in a rocky mountain region city: Child characteristics, maternal risk traits, and prevalence. *Alcohol. Clin. Exp. Res.* **2020**, *44*, 900–918. [CrossRef] [PubMed]

89. May, P.A.; Hasken, J.M.; Stegall, J.M.; Mastro, H.A.; Kalberg, W.O.; Buckley, D.; Brooks, M.; Hedrick, D.M.; Ortega, M.A.; Elliott, A.J.; et al. Fetal alcohol spectrum disorders in a southeastern county of the United States: Child characteristics and maternal risk traits. *Alcohol. Clin. Exp. Res.* **2020**, *44*, 939–959. [CrossRef]

90. Popova, S.; Lange, S.; Poznyak, V.; Chudley, A.E.; Shield, K.D.; Reynolds, J.N.; Murray, M.; Rehm, J. Population-based prevalence of fetal alcohol spectrum disorder in Canada. *BMC Public Health* **2019**, *19*, 845. [CrossRef]

91. May, P.A.; Blankenship, J.; Marais, A.S.; Gossage, J.P.; Kalberg, W.O.; Barnard, R.; de Vries, M.; Robinson, L.K.; Adnams, C.M.; Buckley, D. Approaching the prevalence of the full spectrum of fetal alcohol spectrum disorders in a south African population-based study. *Alcohol. Clin. Exp. Res.* **2013**, *37*, 818–830. [CrossRef]

92. Mena, M.; Navarrete, P.; Avila, P.; Bedregal, P.; Berríos, X. Alcohol drinking in parents and its relation with intellectual score of their children. *Rev. Med. Chile* **1993**, *121*, 98–105.

93. Landgren, M.; Svensson, L.; Strömland, K.; Grönlund, M.A. Prenatal alcohol exposure and neurodevelopmental disorders in children adopted from eastern Europe. *Pediatrics* **2010**, *125*, e1178–e1185. [CrossRef]

94. Kuzmenkovienė, E.; Prasauskienė, A.; Endzinienė, M. The prevalence of fetal alcohol spectrum disorders and concomitant disorders among orphanage children in Lithuania. *J. Popul. Ther. Clin. Pharmacol.* **2012**, *19*, e423.

95. Legonkova, S.V. *Kliniko-Funktcionalnaia Kharakteristika Fetalnogo Alkogolnogo Sindroma u Detei Rannego Vozrasta [Clinical and Functional Characteristics of Fetal Alcohol Syndrome in Early Childhood]*; St. Petersburg's State Paediatric Medical Academy: St. Petersburg, Russia, 2011.

96. Fast, D.K.; Conry, J.; Loock, C.A. Identifying fetal alcohol syndrome among youth in the criminal justice system. *J. Dev. Behav. Pediatr.* **1999**, *20*, 370–372. [CrossRef]

97. Bell, C.C.; Chimata, R. Prevalence of neurodevelopmental disorders among low-income African Americans at a clinic on Chicago's South Side. *Psychiatr. Serv.* **2015**, *66*, 539–542. [CrossRef]

98. Popova, S.; Lange, S.; Bekmuradov, D.; Mihic, A.; Rehm, J. Fetal alcohol spectrum disorder prevalence estimates in correctional systems: A systematic literature review. *Can. J. Public Health* **2011**, *102*, 336–340. [CrossRef] [PubMed]

99. Popova, S.; Dozet, D.; Burd, L. Fetal alcohol spectrum disorder: Can we change the future? *Alcohol. Clin.Exp.Res.* **2020**, *44*, 815–819. [CrossRef] [PubMed]

100. Bearak, J.; Popinchalk, A.; Alkema, L.; Sedgh, G. Global, regional, and subregional trends in unintended pregnancy and its outcomes from 1990 to 2014: Estimates from a Bayesian hierarchical model. *Lancet Global Health* **2018**, *6*, e380–e389. [CrossRef]

101. Lange, S.; Shield, K.; Koren, G.; Rehm, J.; Popova, S. A comparison of the prevalence of prenatal alcohol exposure obtained via maternal self-reports versus meconium testing: A systematic literature review and meta-analysis. *BMC Pregnancy Childbirth* **2014**, *14*, 127. [CrossRef] [PubMed]

102. Chiandetti, A.; Hernandez, G.; Mercadal-Hally, M.; Alvarez, A.; Andreu-Fernandez, V.; Navarro-Tapia, E.; Bastons-Compta, A.; Garcia-Algar, O. Prevalence of prenatal exposure to substances of abuse: Questionnaire versus biomarkers. *Reprod. Health* **2017**, *14*, 137. [CrossRef] [PubMed]

103. Schaff, E.; Moreno, M.; Foster, K.; Klug, M.G.; Burd, L. What do we know about prevalence and management of intoxicated women during labor and delivery? *Glob. Pediatr. Health* **2019**, *6*, 2333794X19894799. [CrossRef] [PubMed]

104. Elliott, A.J.; Kinney, H.C.; Haynes, R.L.; Dempers, J.D.; Wright, C.; Fifer, W.P.; Angal, J.; Boyd, T.K.; Burd, L.; Burger, E.; et al. Concurrent prenatal drinking and smoking increases risk for SIDS: Safe passage study report. *EClinicalMedicine* **2020**, *19*, 100247. [CrossRef]

105. Burd, L.; Martsolf, J.; Klug, M. Children with fetal alcohol syndrome in North Dakota: A case control study utilizing birth certificate data. *Addict. Biol.* **1996**, *1*, 181–189. [CrossRef]

106. Burd, L.; Klug, M.; Martsolf, J. Increased sibling mortality in children with fetal alcohol syndrome. *Addict. Biol.* **2004**, *9*, 179–186. [CrossRef]

107. Schwartz, M.; Hart, B.; Weyrauch, D.; Benson, P.; Klug, M.G.; Burd, L. The hidden face of fetal alcohol spectrum disorder. *Curr. Women's Health Rev.* **2017**, *13*, 96–102. [CrossRef]

108. Li, Q.; Fisher, W.W.; Peng, C.-Z.; Williams, A.D.; Burd, L. Fetal alcohol spectrum disorders: A population based study of premature mortality rates in the mothers. *Matern. Child Health J.* **2012**, *16*, 1332–1337. [CrossRef]

109. Petrenko, C.L.M.; Alto, M.E.; Hart, A.R.; Freeze, S.M.; Cole, L.L. "I'm doing my part, I just need help from the community": Intervention implications of foster and adoptive parents' experiences raising children and young adults with FASD. *J. Fam. Nurs.* **2019**, *25*, 314–347. [CrossRef]

110. Rutman, D.; van Bibber, M. Parenting with fetal alcohol spectrum disorder. *Int. J. Mental Health Addict.* **2010**, *8*, 351–361. [CrossRef]

111. Jirikowic, T.; Olson, H.C.; Astley, S. Parenting stress and sensory processing: Children with fetal alcohol spectrum disorders. *Occup. Particip. Health* **2012**, *32*, 160–168. [CrossRef]

112. Hanlon-Dearman, A.; Chen, M.L.; Olson, H.C. Understanding and managing sleep disruption in children with fetal alcohol spectrum disorder. *Biochem. Cell Biol.* **2018**, *96*, 267–274. [CrossRef] [PubMed]

113. Joya, X.; Garcia-Algar, O.; Salat-Batlle, J.; Pujades, C.; Vall, O. Advances in the development of novel antioxidant therapies as an approach for fetal alcohol syndrome prevention. *Birth Defects Res. Part A Clin. Mol. Teratol.* **2015**, *103*, 163–177. [CrossRef] [PubMed]

114. Peng, Y.; Kwok, K.; Yang, P.; Ng, S.; Liu, J.; Wong, O.; He, M.; Kung, H.; Lin, M. Ascorbic acid inhibits ROS production, NF-κB activation and prevents ethanol-induced growth retardation and microencephaly. *Neuropharmacology* **2005**, *48*, 426–434. [CrossRef] [PubMed]

115. Wentzel, P.; Rydberg, U.; Eriksson, U.J. Antioxidative treatment diminishes ethanol-induced congenital malformations in the rat. *Alcohol. Clin. Exp. Res.* **2006**, *30*, 1752–1760. [CrossRef]

116. Shirpoor, A.; Nemati, S.; Ansari, M.H.K.; Ilkhanizadeh, B. The protective effect of vitamin E against prenatal and early postnatal ethanol treatment-induced heart abnormality in rats: A 3-month follow-up study. *Int. Immunopharmacol.* **2015**, *26*, 72–79. [CrossRef]

117. Zhang, Y.; Wang, H.; Li, Y.; Peng, Y. A review of interventions against fetal alcohol spectrum disorder targeting oxidative stress. *Int. J. Dev. Neurosci.* **2018**, *71*, 140–145. [CrossRef]

118. Zheng, D.; Li, Y.; He, L.; Tang, Y.; Li, X.; Shen, Q.; Yin, D.; Peng, Y. The protective effect of astaxanthin on fetal alcohol spectrum disorder in mice. *Neuropharmacology* **2014**, *84*, 13–18. [CrossRef]

119. Tiwari, V.; Chopra, K. Protective effect of curcumin against chronic alcohol-induced cognitive deficits and neuroinflammation in the adult rat brain. *Neuroscience* **2013**, *244*, 147–158. [CrossRef] [PubMed]

120. Cantacorps, L.; Montagud-Romero, S.; Valverde, O. Curcumin treatment attenuates alcohol-induced alterations in a mouse model of foetal alcohol spectrum disorders. *Prog. Neuro-Psychopharmacol. Biol. Psychiatry* **2020**, *100*, 109899. [CrossRef]

121. Yuan, H.; Zhang, W.; Li, H.; Chen, C.; Liu, H.; Li, Z. Neuroprotective effects of resveratrol on embryonic dorsal root ganglion neurons with neurotoxicity induced by ethanol. *Food Chem. Toxicol.* **2013**, *55*, 192–201. [CrossRef] [PubMed]

122. Luo, G.; Huang, B.; Qiu, X.; Xiao, L.; Wang, N.; Gao, Q.; Yang, W.; Hao, L. Resveratrol attenuates excessive ethanol exposure induced insulin resistance in rats via improving NAD+/NADH ratio. *Mol. Nutr. Food Res.* **2017**, *61*, 1700087. [CrossRef] [PubMed]

123. Heaton, M.B.; Mitchell, J.J.; Paiva, M. Amelioration of ethanol-induced neurotoxicity in the neonatal rat central nervous system by antioxidant therapy. *Alcohol. Clin. Exp. Res.* **2000**, *24*, 512–518. [CrossRef]
124. Young, J.K.; Giesbrecht, H.E.; Eskin, M.N.; Aliani, M.; Suh, M. Nutrition implications for fetal alcohol spectrum disorder. *Adv. Nutr.* **2014**, *5*, 675–692. [CrossRef] [PubMed]
125. Mitchell, J.J.; Paiva, M.; Heaton, M.B. The antioxidants vitamin E and beta-carotene protect against ethanol-induced neurotoxicity in embryonic rat hippocampal cultures. *Alcohol* **1999**, *17*, 163–168. [CrossRef]
126. Marrs, J.A.; Clendenon, S.G.; Ratcliffe, D.R.; Fielding, S.M.; Liu, Q.; Bosron, W.F. Zebrafish fetal alcohol syndrome model: Effects of ethanol are rescued by retinoic acid supplement. *Alcohol* **2010**, *44*, 707–715. [CrossRef]
127. Sarmah, S.; Marrs, J.A. Complex cardiac defects after ethanol exposure during discrete cardiogenic events in zebrafish: Prevention with folic acid. *Dev. Dyn.* **2013**, *242*, 1184–1201. [CrossRef] [PubMed]
128. Carey, L.C.; Coyle, P.; Philcox, J.C.; Rofe, A.M. Zinc Supplementation at the time of ethanol exposure ameliorates teratogenicity in mice. *Alcohol. Clin. Exp. Res.* **2003**, *27*, 107–110. [CrossRef]
129. Summers, B.L.; Rofe, A.M.; Coyle, P. Dietary zinc supplementation throughout pregnancy protects against fetal dysmorphology and improves postnatal survival after prenatal ethanol exposure in mice. *Alcohol. Clin. Exp. Res.* **2009**, *33*, 591–600. [CrossRef]
130. Jacobson, S.W.; Carter, R.C.; Molteno, C.D.; Stanton, M.E.; Herbert, J.S.; Lindinger, N.M.; Lewis, C.E.; Dodge, N.C.; Hoyme, H.E.; Zeisel, S.H.; et al. Efficacy of maternal choline supplementation during pregnancy in mitigating adverse effects of prenatal alcohol exposure on growth and cognitive function: A randomized, double-blind, placebo-controlled clinical trial. *Alcohol. Clin. Exp. Res.* **2018**, *42*, 1327–1341. [CrossRef]
131. Yanaguita, M.Y.; Gutierrez, C.M.; Ribeiro, C.N.M.; Lima, G.A.; Machado, H.R.; Peres, L.C. Pregnancy outcome in ethanol-treated mice with folic acid supplementation in saccharose. *Child's Nerv. System* **2008**, *24*, 99–104. [CrossRef]
132. Helfrich, K.K.; Saini, N.; Kling, P.J.; Smith, S.M. Maternal iron nutriture as a critical modulator of fetal alcohol spectrum disorder risk in alcohol-exposed pregnancies. *Biochem. Cell Biol.* **2018**, *96*, 204–212. [CrossRef]
133. Grant, T.M.; Ernst, C.C.; Streissguth, A.; Stark, K. Preventing alcohol and drug exposed births in Washington State: Intervention findings from three parent-child assistance program sites. *Am. J. Drug Alcohol Abuse* **2005**, *31*, 471–490. [CrossRef]
134. Thanh, N.X.; Jonsson, E.; Moffatt, J.; Dennett, L.; Chuck, A.W.; Birchard, S. An economic evaluation of the parent–child assistance program for preventing fetal alcohol spectrum disorder in Alberta, Canada. *Adm. Policy Mental Health Mental Health Serv. Res.* **2015**, *42*, 10–18. [CrossRef] [PubMed]
135. Greenmyer, J.R.; Popova, S.; Klug, M.G.; Burd, L. Fetal alcohol spectrum disorder: A systematic review of the cost of and savings from prevention in the United States and Canada. *Addiction* **2020**, *115*, 409–417. [CrossRef] [PubMed]
136. World Health Organization. *Global Strategy to Reduce the Harmful Use of Alcohol*; World Health Organization: Geneva, Switzerland, 2010.
137. World Health Organization. *Global Action Plan. for the Prevention and Control. of NCDs 2013–2020*; World Health Organization: Geneva, Switzerland, 2013.

Review

Alcohol's Impact on the Cardiovascular System

Michael Roerecke [1,2,3]

1   Centre for Addiction and Mental Health (CAMH), Institute for Mental Health Policy Research, 33 Ursula Franklin Street, Toronto, ON M5T 2S1, Canada; michael.roerecke@camh.ca

2   Campbell Family Mental Health Research Institute, CAMH, 33 Ursula Franklin Street, Toronto, ON M5T 2S1, Canada

3   Dalla Lana School of Public Health, University of Toronto, 155 College Street, Toronto, ON M5T 1P8, Canada

Abstract: Alcohol consumption has been shown to have complex, and sometimes paradoxical, associations with cardiovascular diseases (CVDs). Several hundred epidemiological studies on this topic have been published in recent decades. In this narrative review, the epidemiological evidence will be examined for the associations between alcohol consumption, including average alcohol consumption, drinking patterns, and alcohol use disorders, and CVDs, including ischaemic heart disease, stroke, hypertension, atrial fibrillation, cardiomyopathy, and heart failure. Methodological shortcomings, such as exposure classification and measurement, reference groups, and confounding variables (measured or unmeasured) are discussed. Based on systematic reviews and meta-analyses, the evidence seems to indicate non-linear relationships with many CVDs. Large-scale longitudinal epidemiological studies with multiple detailed exposure and outcome measurements, and the extensive assessment of genetic and confounding variables, are necessary to elucidate these associations further. Conflicting associations depending on the exposure measurement and CVD outcome are hard to reconcile, and make clinical and public health recommendations difficult. Furthermore, the impact of alcohol on other health outcomes needs to be taken into account. For people who drink alcohol, the less alcohol consumed the better.

Keywords: alcohol drinking; binge drinking; cardiovascular diseases; ischaemic heart disease; hypertension; stroke; review

Citation: Roerecke, M. Alcohol's Impact on the Cardiovascular System. *Nutrients* **2021**, *13*, 3419. https://doi.org/10.3390/nu13103419

Academic Editor: Peter Anderson

Received: 10 August 2021
Accepted: 24 September 2021
Published: 28 September 2021

1. Introduction

Alcohol is one of the most important risk factors for disease and mortality globally [1]. The relationship between alcohol consumption and cardiovascular diseases (CVDs) is complex, and hundreds if not thousands of individual research reports have been published. Due to the potential beneficial effects of alcohol consumption on some CVD outcomes, the relationship between alcohol consumption and CVDs, in particular ischaemic heart disease (IHD), is controversial and highly debated [2–7].

Diseases under the umbrella of CVDs are differential in their aetiology. Therefore, based on major reviews and meta-analyses, this review is divided into the major CVD sub-categories: IHD, ischaemic stroke (IS), haemorrhagic stroke (HS), hypertension, atrial fibrillation (AF), cardiomyopathy, and heart failure.

Alcohol consumption is multi-dimensional, and there is no agreement in the literature on how to label different levels of alcohol consumption. What is considered low, moderate, and heavy alcohol consumption varies widely. Even the amount of alcohol in a 'standard' drink varies considerably [8]. For example, in the UK, one standard unit is 8 g pure alcohol (half a pint); in the US, it is 14 g per standard drink; and in Canada, it is 13.6 g. Episodic heavy drinking, sometimes called 'binge' drinking, is not consistently defined [9–12]. In the US, episodic heavy drinking occasions are defined as alcohol consumption that brings the blood alcohol concentration to at least 0.08% (or 0.08 g of alcohol per deciliter), corresponding to $\geq$5 US standard drinks per occasion in men and $\geq$4 standard drinks in women, in about 2 h [13]. Accordingly, in this narrative review, in order to standardize the exposure, alcohol intake is referred to in grams of pure alcohol based on reported conversion factors. The following search terms were used in Medline and Embase: (cardiovascular diseases or cardiac diseases or stroke or heart diseases or heart failure or cardiac myopathy or cardiac arrhythmia or hypertensive heart disease or hypertension or high blood pressure or elevated blood pressure or resistant hypertension).mp AND (alcohol consumption.mp. or exp alcohol consumption) AND (systematic reviews and meta-analysis).mp.

2. Ischaemic Heart Disease

Alcohol consumption and IHD are both highly prevalent in high-income countries. Many systematic reviews and meta-analyses [5,14–20] and numerous individual studies have been published in recent decades on the relationship between alcohol consumption and IHD, or myocardial infarction, the main subcategory of IHD. This relationship and its implications remain controversial due to a lack of long-term randomized controlled trials with CVD endpoints. Most meta-analyses of epidemiological data on the topic have found a J-shaped or sometimes inverse relationship between average alcohol consumption and IHD, with lifetime abstainers showing a higher risk compared to current 'moderate' drinkers (various amounts of alcohol are used to define these drinking groups), and then an uptake of the risk curve to similar or higher risks compared to those seen for heavier drinkers. Oftentimes, whether or not a J-curve or an inverse or U-shaped relationship is observed depends on the range of alcohol consumption reported in an individual study and the specific IHD endpoint considered (fatal or non-fatal).

The J-shaped risk relationship has been found in both sexes and for IHD morbidity and mortality [16,21]. In a meta-analysis comprising 957,684 participants and 38,627 events, a J-shaped curve in relation to lifetime abstainers was observed in women for both fatal and non-fatal IHD outcomes, and an inverse relationship was observed in men with non-fatal IHD events [16]. Using only studies fully stratified by sex and endpoint, the nadir was found at 32 g per day for IHD mortality in men, 69 g per day for IHD morbidity in men, 11 g per day for IHD mortality in women, and 14 g per day for IHD morbidity in women. The evidence suggests that the type of alcoholic beverage does not play a role in the shape of the relationship. A meta-analysis [22] of fatal or non-fatal CVD events showed that a J-shaped association was observed for the consumption of wine, an inverse relationship for beer consumption, and a negative association for spirits.

IHD mortality among men who drink 60 or more g of pure alcohol on average per day has been found to be similar to that of lifetime abstainers [18]. Among women, such drinking levels are rarely observed in typical epidemiological studies. The risk from alcohol consumption is typically higher in women for the same amount of alcohol consumption compared to men, due to body fat distribution, body size, and alcohol solubility [23–25]. Nevertheless, both men and women with alcohol use disorders, who oftentimes, but not always, drink very heavily, have been associated with some of the highest mortality risks for IHD (RR = 1.62; 95% CI: 1.34 to 1.95 in men; RR = 2.09; 95% CI: 1.28 to 3.41 in women compared to the general population) [17].

Due to the large heterogeneity observed in meta-analyses of alcohol consumption and CVD outcomes, it is clear that not all drinking is associated with a lower risk for IHD. For example, data from Russia consistently show a detrimental association with IHD outcomes; however, it should be noted that the most prevalent drinking pattern in Russia at the time of these studies was infrequent heavy drinking rather than low amounts more frequently. Both drinking patterns would result in the same magnitude of average alcohol intake over the week [26,27].

Perhaps the most compelling observational evidence for a beneficial association between average alcohol consumption and IHD comes from an individual-participant analysis of nearly 600,000 current drinkers of the Emerging Risk Factors Collaboration, EPIC-CVD, and the UK Biobank cohorts [28]. In an inverse relationship, the hazard ratio for an increase in 100 g pure alcohol per week in comparison to >0 to <50 g per week was

0.94 (95% CI: 0.91–0.97) for myocardial infarction (14,539 events). However, except for myocardial infarction, the risk (per 100 g per week increase in consumption) for stroke (HR = 1.14, 1.10–1.17), coronary disease other than myocardial infarction (1.06, 1.00–1.11), heart failure (1.09, 1.03–1.15), and fatal hypertensive disease (1.24, 1.15–1.33) increased in a linear fashion. The shape of the relationship between average alcohol consumption and myocardial infarction was J-shaped for fatal myocardial infarction (2748 events) and non-fatal coronary disease excluding myocardial infarction (6000 events), and inverse for non-fatal myocardial infarction (11,706 events). The data point for the highest consumption was 300 g per week, which translates to about 25 standard drinks of 12 g each per week, or 3.57 standard drinks per day on average. The analyses were adjusted for age, sex, smoking, and history of diabetes. Due to the strong association with MI, the nadir (i.e., the lowest risk) for overall CVD events (39,018 events) was at an alcohol consumption level of 100 g per week.

The threat of unmeasured confounding variables and other sources of bias [7] is not unique to the alcohol–IHD relationship. Early on, the sick-quitter hypothesis [29] was widely thought to be the cause of the J-shaped curve reported in many studies. Meta-analyses have shown that former drinkers are associated with a higher risk for IHD mortality than lifetime or long-term abstainers (RR = 1.25; 95% CI: 1.15–1.36) [30]. No association was found for both sexes for IHD morbidity. The sick-quitter hypothesis has been systematically evaluated, and meta-analyses have shown that even when lifetime abstainers are the reference group, thereby eliminating the sick-quitter effect, or, in other words, former drinking bias, a lower risk for people reporting alcohol consumption up to 30 g per day without irregular heavy drinking episodes remained [5].

Aside from the issue of reference groups, heavy episodic drinking, i.e., drinking about five standard drinks for men and four for women on one occasion or within two hours based on some definitions [13], seems to be an effect modifier for the relationship between average alcohol consumption and IHD. A meta-analysis [5] found an RR = 1.75 (95% CI: 1.36–2.25) for people who drink up to 30 g on average per day, but who actually have a drinking pattern characterized by less frequent drinking and mostly heavy drinking episodes compared to drinkers without such a drinking pattern. This increased risk seems to negate any lower risk for IHD found in people who drink up to 30 g on average per day without heavy drinking episodes. Thus, the risk was similar for lifetime abstainers and people who consume alcohol mostly in heavy drinking episodes. However, the concept of heavy episodic drinking, at least regarding the CVD effects of alcohol consumption, is not clearly defined, and different studies use different thresholds for heavy episodic drinking. These vary from three drinks per drinking day to six or even eight drinks per drinking day [31,32]. For an overview of mechanisms, please see [33,34].

Adjustment for possible confounders, some of which may lie in the pathway of CVD development and could be considered mediators, remains an issue in alcohol epidemiology [35]. An analysis of individual-level data from eight cohort studies showed that adjustment for age; year of baseline; smoking; body mass index; education; physical activity; energy intake; intake of polyunsaturated fat, monounsaturated fat, saturated fat, fiber, and cholesterol; and study design did not explain the J-shaped association [36].

A J-shaped or inverse association has also been reported in patients with CVD. In a meta-analysis of 11 cohorts published in 2014, an inverse risk relationship between average alcohol consumption and IHD in patients with hypertension was reported [37]. Similar associations have been reported among people with diabetes and non-fatal myocardial infarction [38–42]. A recent large-scale study from the UK reported a J-curve for most CVD outcomes in patients with CVD [43].

The pathways by which alcohol consumption may exert a beneficial effect on ischaemic diseases are not well understood. In the absence of long-term randomized controlled trials on CVD endpoints, 44 intervention studies on surrogate biomarkers for CVD were summarized in a meta-analysis in 2011 [44]. The results showed a substantial dose–response relationship for high-density lipoprotein-C with (in comparison to no alcohol consumption):

mean difference: 0.072 mmol per L (95% CI: 0.024–0.119) for 12.5–29.9 g per day; mean difference: 0.103 mmol per L (95% CI: 0.065–0.141) for 30–60 g per day; mean difference: 0.141 mmol per L (95% CI: 0.042–0.240) for >60 g per day. The effect on fibrinogen levels was −0.20 g per L (95% CI: −0.29 to −0.11), and for Adiponectin, 0.56 mg per L (95% CI: 0.39–0.72). Alcohol consumption did not substantially change the levels of total cholesterol, low density lipoprotein cholesterol, triglycerides, Lp(a) lipoprotein, C-reactive protein, interleukin 6, or tumour necrosis factor α. Analyses stratified by type of alcoholic beverage were similar to analyses of all alcoholic beverages combined.

Due to the limitations of typical epidemiological studies, other types of study design, such as Mendelian randomization studies using an instrumental variable approach, sought to answer questions about the causality of the lower risk of low-level alcohol drinkers. However, the use of such an approach [45,46], which depends on several assumptions that are not easily met in a complex relationship, such as between alcohol consumption patterns and CVD risk, is highly debated [47–50].

3. Hypertension

Several meta-analyses have been published over the last two decades that summarize the relationship between average alcohol consumption and incidence of hypertension [51,52,54–58]. While older reviews (e.g., [54,58]) found a small but significantly lower risk in women who reported very small amounts of alcohol intake, more recent meta-analyses with more data did not find such an association. In particular, in a meta-analysis of 361,254 participants from 20 cohort studies (125,907 men and 235,347 women) with 90,160 incident cases of hypertension, the risk compared to abstainers was elevated for any amount of alcohol consumption in men, and in women, there was no increased risk for up to 24 g per day (RR = 0.94; 95% CI: 0.88–1.01), with an increased risk beyond this level of consumption [57]. The risk to former drinkers was similar to lifetime abstainers (RR = 1.03; 95% CI: 0.89–1.20). Among men, the risk increased to 1.68 (95% CI: 1.31–2.14) for drinking 60 g per day on average. No such data were available for women. The difference in risk for up to 24 g per day for women compared to men was significant (RR = 0.79; 95% CI: 0.67–0.93). One possible explanation for this difference could be more detrimental drinking patterns among men, which typically includes more binge drinking episodes. Heavy episodic drinking elevates blood pressure and, subsequentially, the risk for hypertension [73,74].

The relationship between alcohol consumption and blood pressure and hypertension has to be seen as causal and reversible, with experimental evidence showing that a reduction in alcohol consumption leads to a reduction in both systolic and diastolic blood pressure in a dose–response relationship with effects of clinical importance [75]. The reduction in systolic blood pressure is sizable with a mean difference of −5.50 mmHg (95% CI: −6.70 to −4.30) for people who drink 72 g per day on average and reduce their consumption by about 50%. There was no discernible difference for drinkers of up to 24 g per day in comparison to abstainers; however, data were sparse. For a discussion of mechanisms, please see [74,76].

4. Stroke

There are two major stroke subtypes with differing aetiologies: IS (based on ischaemic disease processes) and HS (based on haemorrhagic processes, i.e., bleeding processes). Due to higher prevalence, IS typically drives investigations of total stroke. With similarities in aetiology, one would expect IS to show a similar relationship with alcohol consumption in comparison to IHD. Indeed, several earlier [52,60] and more recent [59,62] meta-analyses have shown that the association between average alcohol consumption and IS follows a J-curve. In a meta-analysis of 27 prospective cohort studies with 3824 IS cases (2216 men and 1608 women) compared to abstainers, the risk for IS was below one for up to 24 g per day on average (RR = 0.90 (95% CI: 0.85–0.95) for <12 g; RR = 0.92 (95% CI: 0.87–0.97) for 12–24 g per day), and increased for alcohol consumption >24 g per day [59]. In contrast, the analysis of the Emerging Risk Factors Collaboration, EPIC-CVD, and the UK Biobank cohorts [28] showed an increased risk for both fatal and non-fatal total stroke based on average alcohol consumption.

The risk for intracerebral and subarachnoid HS increased with every drink, and the consumption of >48 g per day resulted in an RR = 1.67 (95% CI: 1.25–2.23) for intracerebral stroke and 1.82 (95% CI: 1.18–2.82) for subarachnoid HS [59,61].

Several studies have reported an elevated risk for both IS and HS from heavy episodic drinking [77–79]. One study showed that the risk increased with a higher frequency of heavy episodic drinking [78]. Alcohol consumption is also a trigger for stroke events. The higher the alcohol consumption within 24 h or one week, the higher the risk for IS or HS [53,80].

5. AF and Cardiomyopathy

Several meta-analyses have investigated the risk of AF in relation to alcohol consumption [63–66]. In a meta-analysis of seven cohort studies with 12,554 cases of AF, in comparison to non-drinkers, the risk for AF was elevated in all drinking groups, even when heavy episodic drinkers were excluded from the analysis. The pooled linear risk increase was 1.08 (95% CI: 1.06–1.10) for each 12 g per day increase in average alcohol consumption. More recently, using data from 249,496 participants, a meta-analysis concluded that there was no increase in risk for AF for consumption of 6–7 drinks (10–12 g per drink) per week [63]. Another large cohort study of 403,281 participants from the UK Biobank with 21,312 incident cases of AF reported a J-shaped relationship for average alcohol consumption, with people drinking 56 g per week or less having the lowest risk in comparison to lifetime abstainers [67]. A beverage-specific analysis showed that the J-shaped curve was found in wine drinkers, but not in beer or spirit drinkers [67]. A recent randomized controlled trial indicates that a reduction in drinking is associated with a lower recurrence of AF [81].

While the exact amounts remain unknown, alcohol consumption (either regular or irregular), can cause, aside from hypertension, structural damage to the heart muscle and arrythmias [68]. Cardiomyopathy, which is characterized by ventricle dilatation, hypertrophy, and dysfunction, can be caused by alcohol consumption and its metabolites, both of which have a direct toxic effect on the heart muscles [68]. A quantification of a potential dose–response relationship has not been possible to date; however, it seems that the consumption of >80 g per day leads to a substantially increased risk [69]. Few studies have been conducted among women, for which less alcohol consumption per day and a short duration of such consumption over several years have a similar effect compared to men. It has been estimated that 1–40% of alcohol use disorder patients have cardiomyopathy, or, conversely, that 23–47% of patients with dilated cardiomyopathy have, in fact, alcoholic cardiomyopathy [69].

6. Heart Failure

CVD categories, such as IHD, hypertension, and cardiomyopathy, increase the risk of heart failure. Three meta-analyses largely came to similar conclusions [70–72]. In the most recent meta-analysis published in 2018, which consisted of 355,804 participants with 13,738 cases of HF based on 13 cohort studies, it was shown that the dose–response relationship between alcohol consumption and HF was curvilinear [71]. Compared to non-drinkers, the risk for 1–84 g per week, 85–168 g per week, 168–336 g per week, and >336 g per week were RR = 0.86 (95% CI: 0.81–0.90), 0.88 (0.77–1.01), 0.91 (0.80–1.04), and 1.16 (0.92–1.47), respectively. Based on eight studies, the meta-analysis concluded that former drinkers are at a higher risk for AF compared to lifetime abstainers (RR = 1.22; 95% CI: 1.11–1.33) [71]. Due to data limitations, the role of sex and other potential effect modifiers remains unclear.

7. Conclusions

Epidemiological studies indicate a complex relationship between various dimensions of alcohol consumption (i.e., life course drinking patterns) and CVD outcomes. Indeed, substantial heterogeneity is evident. Most epidemiological studies to date have relied on a single measurement of alcohol intake at baseline. It is assumed that the self-reported drinking levels, preferably including drinking patterns, remains the same before and after the baseline measurement. For many people this is clearly not the case, and even lifetime abstainers are hard to identify [82].

Does some alcohol consumption protect some people against ischaemic diseases to some degree? Epidemiological data, as outlined in this review, suggest that this is the case (Table 1). For example, a J-shaped relationship emerges for average alcohol consumption and IHD and IS. On the other hand, the relationship with incident hypertension, which is a potent risk factor for most if not all CVDs, is quite different between men and women, with an increased risk for any amount of alcohol consumption in men. While potential sources of bias, such as the reference group, i.e., separating lifetime abstainers, former drinkers, and heavy episodic drinkers, have been systematically investigated for the relationship between alcohol and IHD, their impact on other CVD outcomes remains less clear. While there is a lack of large-scale randomized studies on the long-term effect of alcohol consumption on various CVD endpoints, short-term clinical trial data indicate a sizable effect of alcohol consumption on HDL-C and fibrinogen. However, the heterogeneity found in epidemiological studies points to more than just biological differences. Socioeconomic status, for example, might influence the impact of alcohol on CVD [83]. More research is necessary to advance knowledge on this topic.

Table 1. Shape of the relation between alcohol consumption and CVD categories based on current evidence syntheses.

CVD Category	Shape of the Relationship	Meta-Analyses and Systematic Reviews
Ischaemic heart disease	J-shaped, modified by episodic heavy drinking	[5,14–22,28,30,37,43,44,51–53]
Hypertension	Any consumption detrimental in men, detrimental beyond 24 g in women; reduction in drinkers >24 g/day lowers blood pressure	[51,52,54–58]
Stroke	Ischaemic stroke: J-shaped, possibly modified by episodic heavy drinkingHaemorrhagic stroke: detrimental, possibly modified by episodic heavy drinking	[21,28,51,52,59–62]
Atrial fibrillation	Detrimental beyond 60 g/week	[63–67]
Cardiomyopathy	Consumption of >80 g/day leads to a substantially increased risk	[68,69]
Heart failure	J-shaped	[63,70–72]

It should also be noted that due to the limitations of alcohol-epidemiological studies, the beneficial associations tend to be overestimated. Furthermore, potential beneficial effects of non-heavy alcohol consumption on CVD endpoints, as described in this review, have already been observed at very low levels, such as 100 g pure alcohol per week, which, at the lower end, translates to about 1 drink every other day. As such, most drinkers should drink less. Recommending drinking as a primary or secondary prevention measure for CVDs, which comes up occasionally in the literature, should be discouraged due to the substantial risks of any alcohol consumption for many health outcomes. Alcohol is a carcinogen, neuro-toxin, hepato-toxin, and psychoactive drug.

Funding: This research received no external funding.

Institutional Review Board Statement: Not applicable, as the study did not involve humans or animals.

Informed Consent Statement: Not applicable.

Data Availability Statement: Not applicable.

Conflicts of Interest: The author declares no conflict of interest.

References

1. World Health Organization. *Global Status Report on Alcohol and Health*; World Health Organization: Geneva, Switzerland, 2018.
2. O'Keefe, J.H.; Bhatti, S.K.; Bajwa, A.; DiNicolantonio, J.J.; Lavie, C.J. Alcohol and cardiovascular health: The dose makes the poison . . . or the remedy. *Mayo Clin. Proc.* **2014**, *89*, 382–393. [CrossRef]
3. Chikritzhs, T.N.; Naimi, T.S.; Stockwell, T.R.; Liang, W. Mendelian randomisation meta-analysis sheds doubt on protective associations between 'moderate' Alcohol. consumption and coronary heart disease. *Evid. Based Med.* **2015**, *20*, 38. [CrossRef]
4. Rehm, J.; Gmel Sr, G.E.; Gmel, G.; Hasan, O.S.M.; Imtiaz, S.; Popova, S.; Probst, C.; Roerecke, M.; Room, R.; Samokhvalov, A.V.; et al. The relationship between different dimensions of Alcohol. use and the burden of disease—An update. *Addiction* **2017**, *112*, 968–1001. [CrossRef] [PubMed]
5. Roerecke, M.; Rehm, J. Alcohol consumption, drinking patterns, and ischemic heart disease: A narrative review of meta-analyses and a systematic review and meta-analysis of the impact of heavy drinking occasions on risk for moderate drinkers. *BMC Med.* **2014**, *12*, 182. [CrossRef] [PubMed]
6. O'Keefe, E.L.; DiNicolantonio, J.J.; O'Keefe, J.H.; Lavie, C.J. Alcohol and CV Health: Jekyll and Hyde J-Curves. *Prog. Cardiovasc. Dis.* **2018**, *61*, 68–75. [CrossRef]
7. Naimi, T.S.; Stockwell, T.; Zhao, J.; Xuan, Z.; Dangardt, F.; Saitz, R.; Liang, W.; Chikritzhs, T. Selection biases in observational studies affect associations between 'moderate' Alcohol. consumption and mortality. *Addiction* **2017**, *112*, 207–214. [CrossRef]
8. Kalinowski, A.; Humphreys, K. Governmental standard drink definitions and low-risk Alcohol. consumption guidelines in 37 countries. *Addiction* **2016**, *111*, 1293–1298. [CrossRef]
9. Dawson, D.A. Defining risk drinking. *Alcohol. Res. Health J. Natl. Inst. Alcohol. Abuse Alcohol.* **2011**, *34*, 144–156.
10. Gmel, G.; Kuntsche, E.; Rehm, J. Risky single-occasion drinking: Bingeing is not bingeing. *Addiction* **2011**, *106*, 1037–1045. [CrossRef] [PubMed]
11. Dawson, D.A.; Grant, B.F. The "gray area" of consumption between moderate and risk drinking. *J. Stud. Alcohol Drugs* **2011**, *72*, 453–458. [CrossRef] [PubMed]
12. Biagioni, N.; Pettigrew, S.; Jones, S.C.; Stafford, J.; Daube, M.; Chikritzhs, T. Defining binge drinking: Young drinkers' perceptions of risky Alcohol. consumption. *Public Health* **2017**, *152*, 55–57. [CrossRef] [PubMed]
13. National Institute on Alcohol. Abuse and Alcoholism (NIAAA). NIAAA Council approves definition of binge drinking. *NIAAA Newsl.* **2004**, *3*, 3.
14. Corrao, G.; Rubbiati, L.; Bagnardi, V.; Zambon, A.; Poikolainen, K. Alcohol and coronary heart disease: A meta-analysis. *Addiction* **2000**, *95*, 1505–1523. [CrossRef] [PubMed]
15. Roerecke, M.; Rehm, J. Irregular heavy drinking occasions and risk of ischemic heart disease: A systematic review and meta-analysis. *Am. J. Epidemiol.* **2010**, *171*, 633–644. [CrossRef] [PubMed]
16. Roerecke, M.; Rehm, J. The cardioprotective association of average Alcohol. consumption and ischaemic heart disease: A systematic review and meta-analysis. *Addiction* **2012**, *107*, 1246–1260. [CrossRef]
17. Roerecke, M.; Rehm, J. Cause-specific mortality risk in Alcohol. use disorder treatment patients: A systematic review and meta-analysis. *Int. J. Epidemiol.* **2014**, *43*, 906–919. [CrossRef] [PubMed]
18. Roerecke, M.; Rehm, J. Chronic heavy drinking and ischaemic heart disease: A systematic review and meta-analysis. *Open Heart* **2014**, *1*, e000135. [CrossRef]
19. Maclure, M. Demonstration of deductive meta-analysis: Ethanol intake and risk of myocardial infarction. *Epidemiol. Rev.* **1993**, *15*, 328–351. [CrossRef]
20. Zhao, J.; Stockwell, T.; Roemer, A.; Naimi, T.; Chikritzhs, T. Alcohol Consumption and Mortality From Coronary Heart Disease: An Updated Meta-Analysis of Cohort Studies. *J. Stud. Alcohol. Drugs* **2017**, *78*, 375–386. [CrossRef]
21. Ronksley, P.E.; Brien, S.E.; Turner, B.J.; Mukamal, K.J.; Ghali, W.A. Association of Alcohol. consumption with selected cardiovascular disease outcomes: A systematic review and meta-analysis. *BMJ* **2011**, *342*, d671. [CrossRef]
22. Costanzo, S.; Di Castelnuovo, A.; Donati, M.B.; Iacoviello, L.; de Gaetano, G. Wine, beer or spirit drinking in relation to fatal and non-fatal cardiovascular events: A meta-analysis. *Eur. J. Epidemiol.* **2011**, *26*, 833–850. [CrossRef]
23. Piano, M.R.; Thur, L.A.; Hwang, C.L.; Phillips, S.A. Effects of Alcohol. on the Cardiovascular System in Women. *Alcohol. Res.* **2020**, *40*, 12. [CrossRef] [PubMed]
24. Kwo, P.Y.; Ramchandani, V.A.; O'Connor, S.; Amann, D.; Carr, L.G.; Sandrasegaran, K.; Kopecky, K.K.; Li, T.K. Gender differences in Alcohol. metabolism: Relationship to liver volume and effect of adjusting for body mass. *Gastroenterology* **1998**, *115*, 1552–1557. [CrossRef]
25. Frezza, M.; di Padova, C.; Pozzato, G.; Terpin, M.; Baraona, E.; Lieber, C.S. High blood Alcohol. levels in women. The role of decreased gastric Alcohol. dehydrogenase activity and first-pass metabolism. *N. Engl. J. Med.* **1990**, *322*, 95–99. [CrossRef] [PubMed]

26. Zaridze, D.; Brennan, P.; Boreham, J.; Boroda, A.; Karpov, R.; Lazarev, A.; Konobeevskaya, I.; Igitov, V.; Terechova, T.; Boffetta, P.; et al. Alcohol and cause-specific mortality in Russia: A retrospective case-control study of 48,557 adult deaths. *Lancet* **2009**, *373*, 2201–2214. [CrossRef]

27. Malyutina, S.; Bobak, M.; Kurilovitch, S.; Gafarov, V.; Simonova, G.; Nikitin, Y.; Marmot, M. Relation between heavy and binge drinking and all-cause and cardiovascular mortality in Novosibirsk, Russia: A prospective cohort study. *Lancet* **2002**, *360*, 1448–1454. [CrossRef]

28. Wood, A.M.; Kaptoge, S.; Butterworth, A.S.; Willeit, P.; Warnakula, S.; Bolton, T.; Paige, E.; Paul, D.S.; Sweeting, M.; Burgess, S.; et al. Risk thresholds for Alcohol. consumption: Combined analysis of individual-participant data for 599,912 current drinkers in 83 prospective studies. *Lancet* **2018**, *391*, 1513–1523. [CrossRef]

29. Shaper, A.G.; Wannamethee, G.; Walker, M. Alcohol and mortality in british men: Explaining the u-shaped curve. *Lancet* **1988**, *332*, 1267–1273. [CrossRef]

30. Roerecke, M.; Rehm, J. Ischemic heart disease mortality and morbidity rates in former drinkers: A meta-analysis. *Am. J. Epidemiol.* **2011**, *173*, 245–258. [CrossRef] [PubMed]

31. Kuntsche, E.; Kuntsche, S.; Thrul, J.; Gmel, G. Binge drinking: Health impact, prevalence, correlates and interventions. *Psychol. Health* **2017**, *32*, 976–1017. [CrossRef] [PubMed]

32. Maurage, P.; Lannoy, S.; Mange, J.; Grynberg, D.; Beaunieux, H.; Banovic, I.; Gierski, F.; Naassila, M. What We Talk About When We Talk About Binge Drinking: Towards an Integrated Conceptualization and Evaluation. *Alcohol Alcohol.* **2020**, *55*, 468–479. [CrossRef]

33. Piano, M.R.; Mazzuco, A.; Kang, M.; Phillips, S.A. Cardiovascular Consequences of Binge Drinking: An Integrative Review with Implications for Advocacy, Policy, and Research. *Alcohol. Clin. Exp. Res.* **2017**, *41*, 487–496. [CrossRef]

34. Molina, P.E.; Nelson, S. Binge Drinking's Effects on the Body. *Alcohol. Res.* **2018**, *39*, 99–109.

35. Wallach, J.D.; Serghiou, S.; Chu, L.; Egilman, A.C.; Vasiliou, V.; Ross, J.S.; Ioannidis, J.P.A. Evaluation of confounding in epidemiologic studies assessing Alcohol. consumption on the risk of ischemic heart disease. *BMC Med. Res. Methodol.* **2020**, *20*, 64. [CrossRef] [PubMed]

36. Hvidtfeldt, U.A.; Tolstrup, J.S.; Jakobsen, M.U.; Heitmann, B.L.; Grønbaek, M.; O'Reilly, E.; Bälter, K.; Goldbourt, U.; Hallmans, G.; Knekt, P.; et al. Alcohol intake and risk of coronary heart disease in younger, middle-aged, and older adults. *Circulation* **2010**, *121*, 1589–1597. [CrossRef] [PubMed]

37. Huang, C.; Zhan, J.; Liu, Y.J.; Li, D.J.; Wang, S.Q.; He, Q.Q. Association between Alcohol. consumption and risk of cardiovascular disease and all-cause mortality in patients with hypertension: A meta-analysis of prospective cohort studies. *Mayo Clin. Proc.* **2014**, *89*, 1201–1210. [CrossRef] [PubMed]

38. Pai, J.K.; Mukamal, K.J.; Rimm, E.B. Long-term Alcohol. consumption in relation to all-cause and cardiovascular mortality among survivors of myocardial infarction: The Health Professionals Follow-up Study. *Eur. Heart J.* **2012**, *33*, 1598–1605. [CrossRef]

39. Jackson, V.A.; Sesso, H.D.; Buring, J.E.; Gaziano, J.M. Alcohol consumption and mortality in men with preexisting cerebrovascular disease. *Arch. Intern. Med.* **2003**, *163*, 1189–1193. [CrossRef]

40. Tanasescu, M.; Hu, F.B.; Willett, W.C.; Stampfer, M.J.; Rimm, E.B. Alcohol consumption and risk of coronary heart disease among men with type 2 diabetes mellitus. *J. Am. Coll. Cardiol.* **2001**, *38*, 1836–1842. [CrossRef]

41. Solomon, C.G.; Hu, F.B.; Stampfer, M.J.; Colditz, G.A.; Speizer, F.E.; Rimm, E.B.; Willett, W.C.; Manson, J.E. Moderate Alcohol. consumption and risk of coronary heart disease among women with type 2 diabetes mellitus. *Circulation* **2000**, *102*, 494–499. [CrossRef]

42. Beulens, J.W.; Algra, A.; Soedamah-Muthu, S.S.; Visseren, F.L.; Grobbee, D.E.; van der Graaf, Y. Alcohol consumption and risk of recurrent cardiovascular events and mortality in patients with clinically manifest vascular disease and diabetes mellitus: The Second Manifestations of ARTerial (SMART) disease study. *Atherosclerosis* **2010**, *212*, 281–286. [CrossRef]

43. Ding, C.; O'Neill, D.; Bell, S.; Stamatakis, E.; Britton, A. Association of Alcohol. consumption with morbidity and mortality in patients with cardiovascular disease: Original data and meta-analysis of 48,423 men and women. *BMC Med.* **2021**, *19*, 167. [CrossRef] [PubMed]

44. Brien, S.E.; Ronksley, P.E.; Turner, B.J.; Mukamal, K.J.; Ghali, W.A. Effect of Alcohol. consumption on biological markers associated with risk of coronary heart disease: Systematic review and meta-analysis of interventional studies. *BMJ* **2011**, *342*, d636. [CrossRef]

45. Holmes, M.V.; Dale, C.E.; Zuccolo, L.; Silverwood, R.J.; Guo, Y.; Ye, Z.; Prieto-Merino, D.; Dehghan, A.; Trompet, S.; Wong, A.; et al. Association between Alcohol and cardiovascular disease: Mendelian randomisation analysis based on individual participant data. *BMJ* **2014**, *349*, g4164. [CrossRef] [PubMed]

46. Millwood, I.Y.; Walters, R.G.; Mei, X.W.; Guo, Y.; Yang, L.; Bian, Z.; Bennett, D.A.; Chen, Y.; Dong, C.; Hu, R.; et al. Conventional and genetic evidence on Alcohol and vascular disease aetiology: A prospective study of 500,000 men and women in China. *Lancet* **2019**, *393*, 1831–1842. [CrossRef]

47. Mukamal, K.J.; Rimm, E.B.; Stampfer, M.J. Reply to: Mendel's laws, Mendelian randomization and causal inference in observational data: Substantive and nomenclatural issues. *Eur. J. Epidemiol.* **2020**, *35*, 725–726. [CrossRef] [PubMed]

48. Mukamal, K.J.; Stampfer, M.J.; Rimm, E.B. Genetic instrumental variable analysis: Time to call mendelian randomization what it is. The example of Alcohol and cardiovascular disease. *Eur. J. Epidemiol.* **2020**, *35*, 93–97. [CrossRef]

49. Davey Smith, G.; Holmes, M.V.; Davies, N.M.; Ebrahim, S. Mendel's laws, Mendelian randomization and causal inference in observational data: Substantive and nomenclatural issues. *Eur. J. Epidemiol.* **2020**, *35*, 99–111. [CrossRef]
50. Gmel, G. Beneficial effects of moderate Alcohol use—A case for Occam's razor? *Addiction* **2017**, *112*, 215–217. [CrossRef]
51. Corrao, G.; Bagnardi, V.; Zambon, A.; Arico, S. Exploring the dose-response relationship between Alcohol. consumption and the risk of several alcohol-related conditions: A meta-analysis. *Addiction* **1999**, *94*, 1551–1573. [CrossRef]
52. Corrao, G.; Bagnardi, V.; Zambon, A.; La Vecchia, C. A meta-analysis of Alcohol. consumption and the risk of 15 diseases. *Prev. Med.* **2004**, *38*, 613–619. [CrossRef]
53. Mostofsky, E.; Chahal, H.S.; Mukamal, K.J.; Rimm, E.B.; Mittleman, M.A. Alcohol and Immediate Risk of Cardiovascular Events: A Systematic Review and Dose-Response Meta-Analysis. *Circulation* **2016**, *133*, 979–987. [CrossRef] [PubMed]
54. Briasoulis, A.; Agarwal, V.; Messerli, F.H. Alcohol consumption and the risk of hypertension in men and women: A systematic review and meta-analysis. *J. Clin. Hypertens.* **2012**, *14*, 792–798. [CrossRef] [PubMed]
55. Jung, M.H.; Shin, E.S.; Ihm, S.H.; Jung, J.G.; Lee, H.Y.; Kim, C.H. The effect of Alcohol. dose on the development of hypertension in Asian and Western men: Systematic review and meta-analysis. *Korean J. Intern. Med.* **2020**, *35*, 906–916. [CrossRef] [PubMed]
56. Liu, F.; Liu, Y.; Sun, X.; Yin, Z.; Li, H.; Deng, K.; Zhao, Y.; Wang, B.; Ren, Y.; Liu, X.; et al. Race- and sex-specific association between Alcohol. consumption and hypertension in 22 cohort studies: A systematic review and meta-analysis. *Nutr. Metab. Cardiovasc. Dis.* **2020**, *30*, 1249–1259. [CrossRef] [PubMed]
57. Roerecke, M.; Tobe, S.W.; Kaczorowski, J.; Bacon, S.L.; Vafaei, A.; Hasan, O.S.M.; Krishnan, R.J.; Raifu, A.O.; Rehm, J. Sex-Specific Associations Between Alcohol. Consumption and Incidence of Hypertension: A Systematic Review and Meta-Analysis of Cohort Studies. *J. Am. Heart Assoc.* **2018**, *7*. [CrossRef] [PubMed]
58. Taylor, B.; Irving, H.M.; Baliunas, D.; Roerecke, M.; Patra, J.; Mohapatra, S.; Rehm, J. Alcohol and hypertension: Gender differences in dose–response relationships determined through systematic review and meta-analysis. *Addiction* **2009**, *104*, 1981–1990. [CrossRef]
59. Larsson, S.C.; Wallin, A.; Wolk, A.; Markus, H.S. Differing association of Alcohol. consumption with different stroke types: A systematic review and meta-analysis. *BMC Med.* **2016**, *14*, 178. [CrossRef]
60. Patra, J.; Taylor, B.; Irving, H.; Roerecke, M.; Baliunas, D.; Mohapatra, S.; Rehm, J. Alcohol consumption and the risk of morbidity and mortality for different stroke types—A systematic review and meta-analysis. *BMC Public Health* **2010**, *10*, 258. [CrossRef]
61. Yao, X.; Zhang, K.; Bian, J.; Chen, G. Alcohol consumption and risk of subarachnoid hemorrhage: A meta-analysis of 14 observational studies. *Biomed. Rep.* **2016**, *5*, 428–436. [CrossRef]
62. Zhang, C.; Qin, Y.-Y.; Chen, Q.; Jiang, H.; Chen, X.-Z.; Xu, C.-L.; Mao, P.-J.; He, J.; Zhou, Y.-H. Alcohol intake and risk of stroke: A dose–response meta-analysis of prospective studies. *Int. J. Cardiol.* **2014**, *174*, 669–677. [CrossRef]
63. Gallagher, C.; Hendriks, J.M.L.; Elliott, A.D.; Wong, C.X.; Rangnekar, G.; Middeldorp, M.E.; Mahajan, R.; Lau, D.H.; Sanders, P. Alcohol and incident atrial fibrillation—A systematic review and meta-analysis. *Int. J. Cardiol.* **2017**, *246*, 46–52. [CrossRef]
64. Kodama, S.; Saito, K.; Tanaka, S.; Horikawa, C.; Saito, A.; Heianza, Y.; Anasako, Y.; Nishigaki, Y.; Yachi, Y.; Iida, K.T.; et al. Alcohol consumption and risk of atrial fibrillation: A meta-analysis. *J. Am. Coll. Cardiol.* **2011**, *57*, 427–436. [CrossRef]
65. Larsson, S.C.; Drca, N.; Wolk, A. Alcohol consumption and risk of atrial fibrillation: A prospective study and dose-response meta-analysis. *J. Am. Coll. Cardiol.* **2014**, *64*, 281–289. [CrossRef] [PubMed]
66. Samokhvalov, A.V.; Irving, H.M.; Rehm, J. Alcohol consumption as a risk factor for atrial fibrillation: A systematic review and meta-analysis. *Eur. J. Cardiovasc. Prev. Rehabil.* **2010**, *17*, 706–712. [CrossRef] [PubMed]
67. Tu, S.J.; Gallagher, C.; Elliott, A.D.; Linz, D.; Pitman, B.M.; Hendriks, J.M.L.; Lau, D.H.; Sanders, P.; Wong, C.X. Risk Thresholds for Total and Beverage-Specific Alcohol. Consumption and Incident Atrial Fibrillation. *JACC Clin. Electrophysiol.* **2021**. [CrossRef] [PubMed]
68. Fernández-Solà, J. The Effects of Ethanol on the Heart: Alcoholic Cardiomyopathy. *Nutrients* **2020**, *12*, 572. [CrossRef] [PubMed]
69. Rehm, J.; Hasan, O.S.M.; Imtiaz, S.; Neufeld, M. Quantifying the contribution of Alcohol. to cardiomyopathy: A systematic review. *Alcohol* **2017**, *61*, 9–15. [CrossRef] [PubMed]
70. Larsson, S.C.; Orsini, N.; Wolk, A. Alcohol consumption and risk of heart failure: A dose-response meta-analysis of prospective studies. *Eur. J. Heart Fail.* **2015**, *17*, 367–373. [CrossRef]
71. Larsson, S.C.; Wallin, A.; Wolk, A. Alcohol consumption and risk of heart failure: Meta-analysis of 13 prospective studies. *Clin. Nutr.* **2018**, *37*, 1247–1251. [CrossRef]
72. Padilla, H.; Gaziano, J.M.; Djoussé, L. Alcohol Consumption and Risk of Heart Failure: A Meta-Analysis. *Null* **2010**, *38*, 84–89. [CrossRef] [PubMed]
73. Piano, M.R.; Burke, L.; Kang, M.; Phillips, S.A. Effects of Repeated Binge Drinking on Blood Pressure Levels and Other Cardiovascular Health Metrics in Young Adults: National Health and Nutrition Examination Survey, 2011–2014. *J. Am. Heart Assoc.* **2018**, *7*, e008733. [CrossRef]
74. Puddey, I.B.; Rakic, V.; Dimmitt, S.B.; Beilin, L.J. Influence of pattern of drinking on cardiovascular disease and cardiovascular risk factors—A review. *Addiction* **1999**, *94*, 649–663. [CrossRef] [PubMed]
75. Roerecke, M.; Kaczorowski, J.; Tobe, S.W.; Gmel, G.; Hasan, O.S.M.; Rehm, J. The effect of a reduction in Alcohol. consumption on blood pressure: A systematic review and meta-analysis. *Lancet Public Health* **2017**, *2*, e108–e120. [CrossRef]
76. Puddey, I.B.; Mori, T.A.; Barden, A.E.; Beilin, L.J. Alcohol and Hypertension-New Insights and Lingering Controversies. *Curr. Hypertens. Rep.* **2019**, *21*, 79. [CrossRef] [PubMed]

77. Aigner, A.; Grittner, U.; Rolfs, A.; Norrving, B.; Siegerink, B.; Busch, M.A. Contribution of Established Stroke Risk Factors to the Burden of Stroke in Young Adults. *Stroke* **2017**, *48*, 1744–1751. [CrossRef]
78. Sull, J.W.; Yi, S.W.; Nam, C.M.; Ohrr, H. Binge drinking and mortality from all causes and cerebrovascular diseases in korean men and women: A Kangwha cohort study. *Stroke* **2009**, *40*, 2953–2958. [CrossRef]
79. Sundell, L.; Salomaa, V.; Vartiainen, E.; Poikolainen, K.; Laatikainen, T. Increased stroke risk is related to a binge-drinking habit. *Stroke* **2008**, *39*, 3179–3184. [CrossRef]
80. Guiraud, V.; Amor, M.B.; Mas, J.L.; Touzé, E. Triggers of ischemic stroke: A systematic review. *Stroke* **2010**, *41*, 2669–2677. [CrossRef]
81. Voskoboinik, A.; Kalman, J.M.; De Silva, A.; Nicholls, T.; Costello, B.; Nanayakkara, S.; Prabhu, S.; Stub, D.; Azzopardi, S.; Vizi, D.; et al. Alcohol Abstinence in Drinkers with Atrial Fibrillation. *N. Engl. J. Med.* **2020**, *382*, 20–28. [CrossRef]
82. Rehm, J.; Irving, H.; Ye, Y.; Kerr, W.C.; Bond, J.; Greenfield, T.K. Are lifetime abstainers the best control group in Alcohol. epidemiology? On the stability and validity of reported lifetime abstention. *Am. J. Epidemiol.* **2008**, *168*, 866–871. [CrossRef] [PubMed]
83. Degerud, E.; Ariansen, I.; Ystrom, E.; Graff-Iversen, S.; Høiseth, G.; Mørland, J.; Davey Smith, G.; Næss, Ø. Life course socioeconomic position, Alcohol. drinking patterns in midlife, and cardiovascular mortality: Analysis of Norwegian population-based health surveys. *PLoS Med.* **2018**, *15*, e1002476. [CrossRef] [PubMed]

Review

Alcohol Use and the Risk of Communicable Diseases

Neo K. Morojele [1,*], Sheela V. Shenoi [2,3], Paul A. Shuper [4,5,6,7], Ronald Scott Braithwaite [8] and Jürgen Rehm [4,5,9,10,11,12,13,14]

[1] Department of Psychology, University of Johannesburg, Johannesburg 2006, South Africa
[2] Section of Infectious Diseases, Department of Medicine, Yale University School of Medicine, New Haven, CT 06510, USA; sheela.shenoi@yale.edu
[3] Yale Institute for Global Health, Yale University, New Haven, CT 06520, USA
[4] Centre for Addiction and Mental Health, Institute for Mental Health Policy Research and Campbell Family Mental Health Research Institute, Toronto, ON M5S 2S1, Canada; paul.shuper@camh.ca (P.A.S.); jtrehm@gmail.com (J.R.)
[5] Dalla Lana School of Public Health, University of Toronto, Toronto, ON M5T 3M7, Canada
[6] Institute for Collaboration on Health, Intervention, and Policy, University of Connecticut, Storrs, CT 06269, USA
[7] Alcohol, Tobacco and Other Drug Research Unit, South African Medical Research Council, Pretoria 0001, South Africa
[8] Division of Comparative Effectiveness and Decision Science, Department of Population Health, NYU Grossman School of Medicine, New York University, New York, NY 10013, USA; Scott.Braithwaite@nyulangone.org
[9] Department of Psychiatry, University of Toronto, Toronto, ON M5T 1R8, Canada
[10] Center for Interdisciplinary Addiction Research (ZIS), Department of Psychiatry and Psychotherapy, University Medical Center Hamburg-Eppendorf (UKE), 20246 Hamburg, Germany
[11] Institute of Clinical Psychology and Psychotherapy, Technische Universität Dresden, 01187 Dresden, Germany
[12] Faculty of Medicine, Institute of Medical Science, University of Toronto, Toronto, ON M5S 1A8, Canada
[13] Program on Substance Abuse, Public Health Agency of Catalonia, 08005 Barcelona, Spain
[14] Department of International Health Projects, Institute for Leadership and Health Management, I.M. Sechenov First Moscow State Medical University (Sechenov University), 119991 Moscow, Russia
[*] Correspondence: nmorojele@uj.ac.za; Tel.: +27-11-559-3125

Citation: Morojele, N.K.; Shenoi, S.V.; Shuper, P.A.; Braithwaite, R.S.; Rehm, J. Alcohol Use and the Risk of Communicable Diseases. *Nutrients* **2021**, *13*, 3317. https://doi.org/10.3390/nu13103317

Academic Editor: Peter Henneman

Received: 30 July 2021
Accepted: 14 September 2021
Published: 23 September 2021

Publisher's Note: MDPI stays neutral with regard to jurisdictional claims in published maps and institutional affiliations.

Abstract: The body of knowledge on alcohol use and communicable diseases has been growing in recent years. Using a narrative review approach, this paper discusses alcohol's role in the acquisition of and treatment outcomes from four different communicable diseases: these include three conditions included in comparative risk assessments to date—Human Immunodeficiency Virus (HIV)/AIDS, tuberculosis (TB), and lower respiratory infections/pneumonia—as well as Severe Acute Respiratory Syndrome Coronavirus 2 (SARS-CoV-2) because of its recent and rapid ascension as a global health concern. Alcohol-attributable TB, HIV, and pneumonia combined were responsible for approximately 360,000 deaths and 13 million disability-adjusted life years lost (DALYs) in 2016, with alcohol-attributable TB deaths and DALYs predominating. There is strong evidence that alcohol is associated with increased incidence of and poorer treatment outcomes from HIV, TB, and pneumonia, via both behavioral and biological mechanisms. Preliminary studies suggest that heavy drinkers and those with alcohol use disorders are at increased risk of COVID-19 infection and severe illness. Aside from HIV research, limited research exists that can guide interventions for addressing alcohol-attributable TB and pneumonia or COVID-19. Implementation of effective individual-level interventions and alcohol control policies as a means of reducing the burden of communicable diseases is recommended.

Keywords: alcohol; communicable diseases; infectious diseases; HIV; tuberculosis; pneumonia; severe acute respiratory syndrome coronavirus 2

1. Introduction

Alcohol consumption was recognized as a risk factor for infectious lung diseases, such as pneumonia, as early as 1785, in Benjamin Rush's seminal work on the effects of spirits on the human body and mind [1]. However, the first global comparative risk assessment on alcohol use as a risk factor for disease burden and mortality, conducted in the last decade of the last century [2], did not include any effects of alcohol consumption on infectious disease. The impact of alcohol use on infectious disease outcomes only entered comparative risk assessments in the Global Burden of Disease Study and the World Health Organization's (WHO) Global Status Reports after 2010 (starting with [3]; for an overview of the reasoning to include it, see [4,5]).

We can only speculate as to why the strong association between alcohol use and infectious disease was overlooked in global risk assessments for such a long time. This oversight is all the more astonishing as the association is readily apparent in research and practice, for instance, by the high prevalence of people with alcohol use disorders (AUDs) in tuberculosis treatment [6]; or by the strong associations between alcohol use and HIV/AIDS in surveys or other empirical studies [7–12]. However, these associations were not judged to be necessarily causal, even when alcohol use was related to the incidence of HIV infection [13,14]. Another likely reason is that the impact of alcohol use was indirect, via behavioral and biological pathways which were impacted by many social and other factors, making it difficult to identify alcohol use as a necessary element in a multi-component process of causation [15].

For example, the impact of alcohol on HIV/AIDS is mainly mediated by the impact of alcohol use on decision-making, resulting in riskier sexual behaviors [16] and lower adherence to virus suppression therapies [17–20], which results in higher transmission of HIV and other sexually transmitted diseases [21–23]. It took strong experimental methodology to ascertain that the widely recognized associations between alcohol and HIV/AIDS had substantial causal components (for more details, see [16,17]). For other infectious disease outcomes, such as tuberculosis (TB), the toxic effects of heavy alcohol consumption on the immune system render the host more susceptible to TB disease, which is again an indirect effect ([24,25]; for systematic overviews on all mechanisms, see [6,26]).

Given the plethora of multi-component causal pathways involving alcohol and infectious diseases and the complexities required to elucidate them, additional evidence is likely to continue to emerge regarding the causal impact of alcohol use on infectious diseases. For example, there is an association between alcohol and other sexually transmitted infections [27], and the causal mechanism for the impact of alcohol on HIV infection seems to also apply to these other sexually transmitted infections [5]. However, the present review will be restricted to conditions that have been included in global comparative risk assessments to date (HIV/AIDS, tuberculosis, pneumonia) with one exception, COVID-19 infection, which has been included because of its recent and rapid ascension as a global health concern, even though it occurred after the last global comparative risk assessment was performed. All sections on disease outcomes discuss both behavioral and biological risk factors and are split into sections regarding incidence (Does alcohol use cause new infections with the disease?) and impact upon the course (How does alcohol use impact the course of disease?), and all sections also discuss different dimensions of alcohol consumption, in particular, irregular and heavy drinking occasions.

2. Alcohol and the Risk of Human Immunodeficiency Virus (HIV) and Acquired Immune Deficiency Syndrome (AIDS)

HIV persists as a global health issue. In 2020, there were an estimated 37.6 million people living with HIV, including 1.5 million newly infected individuals and 690,000 who died from AIDS-related illnesses [28]. Alcohol has been identified as a driver of this epidemic, facilitating HIV acquisition/transmission and disease progression through both behavioral and biological means.

2.1. Alcohol and HIV Acquisition/Transmission

2.1.1. Behavioral Mechanisms

Most HIV seroconversions result from sexual activity [29], and alcohol has been associated with a diminished likelihood of engagement in the behaviors necessary to prevent sexually based HIV acquisition/transmission. Consuming alcohol in sexual contexts can result in alcohol myopia [30], which entails an alcohol-induced constraint in cognitive capacity that causes a focus on risk-impelling cues (e.g., sexual arousal) and a disregard of risk-inhibiting cues (e.g., the prospect of HIV acquisition/transmission), thereby increasing the likelihood of condomless sex. This mechanism and corresponding alcohol–condomless sex association have been supported through a number of reviews and meta-analyses [7–12,14,31] as well as through controlled experiments that have provided evidence for the causal nature of this link [16,17,32–34].

More recently, HIV prevention efforts have emphasized biomedical approaches, which include HIV Pre-Exposure Prophylaxis (PrEP)—a medication taken daily by those living without HIV to prevent HIV acquisition [35,36]; and Treatment as Prevention (TasP)—which involves people living with HIV taking antiretroviral therapy (ART) to achieve viral suppression, thereby eliminating the possibility of viral transmission [21–23]. Despite their biomedical basis, these approaches are directly reliant on a behavior, namely adherence, which has been shown to be negatively associated with alcohol use [18–20,37–41]. A variety of underlying mechanisms for this association have been proposed, which, for the sake of conciseness, are presented below under "Alcohol and HIV Disease Progression". It is possible that long-acting formulations of PrEP and ART may be particularly well suited for HIV prevention in alcohol users because those formulations diminish the adherence burden. This hypothesis needs to be evaluated in future research.

2.1.2. Biological Mechanisms

Alcohol use can facilitate HIV acquisition/transmission by (1) decreasing host immune efficiencies among those living without HIV and (2) increasing viral replication among people living with HIV. Regarding the former, alcohol disrupts the physiology of the liver, causing a disturbance to non-specific innate and adaptive immune responses [42–45]. Both acute and chronic alcohol consumption can suppress the production of lymphocytes and cytokines [46–50], inhibit T-lymphocyte proliferation [51], and decrease or inhibit the production of CD4+ and CD8+ T-cells and natural killer cells [52,53], which, taken together, can result in immunodeficiency and autoimmunity, and increase host susceptibility to HIV infection [25,54,55]. These effects can be further exacerbated by liver diseases such as liver fibrosis and cirrhosis observed among those who chronically abuse alcohol [25,54–59]. Among people living with HIV, moderate to heavy alcohol consumption has been significantly associated with changes in vaginal flora, increased proinflammatory cytokines, and genital tract inflammation, which increase HIV shedding and replication, and, in turn, the likelihood of HIV transmission [60–63].

2.2. Alcohol Use and HIV Disease Progression

2.2.1. Behavioral Mechanisms

The successful treatment of HIV, which entails achieving viral suppression to halt disease progression, relies on enacting the behaviorally underpinned steps of the HIV care continuum that include HIV testing, linkage, and retention in HIV care, and ART initiation and adherence. Alcohol use has been associated with poor outcomes at all steps of the continuum [37,64–69], and some evidence suggestive of the *causal* role of alcohol use, particularly with respect to adherence, has been yielded [17–20]. Alcohol-HIV care continuum associations can result from a range of mechanisms, including alcohol-related stigmatization that prevents alcohol users from accessing HIV testing and care [70,71], and alcohol-derived diminished cognitive functioning that poses a challenge for ongoing adherence and clinic attendance [72,73]. Among individuals who are alcohol-dependent, the syndrome of dependence may shift priorities towards obtaining and consuming alcohol and away from health, self-care, and other concerns [70]. Finally, specific to ART adherence, some alcohol-consuming people living with HIV consciously and intentionally decide not to take their doses due to factors including the possession of beliefs surrounding toxic alcohol–ART interactions [74].

2.2.2. Biological Mechanisms

The role of alcohol in HIV disease progression is manifested through its effects on host liver and immunomodulation, resulting in increased activation of CD4+ T-cells and its subsequent depletion at mucosal sites [63], as well as inhibition and abnormalities of T and B lymphocytes and natural killer cells [31,75,76], all of which are necessary for the containment of HIV pathogens. Alcohol may also enhance HIV viral replication by increasing or altering the HIV-binding CXCR4 coreceptor [77,78]. Accordingly, among ART-naïve individuals, heavy drinking (vs. lower consumption) has been linked to higher CD8 cell counts and lower CD4 cell counts [79–81], and among those taking ART, it has been associated with reduced CD4 cell counts and higher log HIV RNA, even after controlling for adherence and age [80,82–84]. Relevant to this latter group, some ART medications are metabolized by the Cytochrome P450 enzyme pathway in the liver, which may be induced or inhibited by acute or chronic alcohol consumption [63,85,86]. This can affect the pharmacokinetics of some ART medications, resulting in either an increase or decrease of the available drug in plasma and causing drug toxicity or suboptimal control of the virus, respectively [63]. The effect of alcohol on ART can be further exacerbated by comorbidities, including drug dependence and Hepatitis C coinfection [17,63,87–89].

2.3. Addressing the Intersection of Alcohol Use and HIV

Alcohol use is closely intertwined with the persistent HIV epidemic. HIV prevention- and treatment-related outcomes can be improved by addressing alcohol use through behavioral [90], pharmacological [91], and policy/structural-level interventions [92,93]. Tailoring and targeting these interventions to meet the unique needs of diverse populations affected by HIV may further enhance their effectiveness and help reduce the global HIV burden [94].

3. Alcohol Use and the Risk of Tuberculosis

Tuberculosis (TB) is the leading cause of infectious death globally, surpassing HIV/AIDS and among the top 10 causes of death worldwide [95]. In 2019, 10 million people became ill with tuberculosis, and 1.4 million people died [95]. TB is caused by *Mycobacterium tuberculosis*, transmitted when affected individuals cough droplet nuclei containing the bacteria into the air, which is subsequently inhaled by others, causing latent infection and pulmonary and extrapulmonary disease. The WHO estimates that one-third of the global population is latently infected. Alcohol use is among the top modifiable risk factors for tuberculosis, with AUDs prevalent in 30% of patients with TB and 11.4% (9.3–13%) of TB mortality attributable to alcohol [95,96].

3.1. Behavioral Mechanisms

Alcohol use is well established as a risk factor for incident TB [6,97], responsible for 17% of incident TB globally [98]. Data suggest that alcohol use is associated with a 35% increased risk for developing active TB [98], while a systematic review of 21 studies demonstrated that only heavy alcohol use (defined as >40 g ethanol daily) or AUD provided a pooled risk of 3.50 (95% CI: 2.01–5.93) [6,99]. The mechanisms for the increased risk are not clearly delineated but are likely attributable to both biological and behavioral factors, the latter facilitated by close contact in crowded congregate settings [6,26]. Alcohol consumed in the context of social interactions, such as bars [6,26,99], has facilitated transmission and outbreaks of TB in institutionalized or service settings, including prisons [100] and among homeless populations [101] have been well documented.

Alcohol use is also an established risk factor for poor TB outcomes overall, including treatment failure, loss to follow up, and mortality for both drug-susceptible and drug-resistant TB [102]. This is predominantly attributed to behavioral mechanisms, notably poor adherence to TB treatment and poor retention in TB care [102–104]. Outcomes are worse with common comorbid conditions, including HIV, hepatitis C, substance use, and smoking [105–108]. Recognized limitations in studies assessing the relationship between alcohol and TB include poorly quantified alcohol consumption, a lack of alcohol standards across countries, and a lack of data on optimal screening for alcohol use among those receiving TB treatment.

Increased TB treatment failure and death independent of loss to follow-up suggests alcohol-related biological factors [102], with a novel study underway to gauge the role of alcohol use on outcomes controlling for adherence [109]. Interventions to reduce the impact of alcohol use on TB outcomes are scarce though emerging evidence suggests screening and intervention are feasible and promising to improve treatment completion and clinical outcomes [56,110–116].

Less is known about the role of alcohol in TB preventive therapy. Until recently, the only regimen available for prevention was 6–12 months of isoniazid with rare but recognized hepatic toxicity that can be exacerbated by alcohol use [117–121]. The balance between the benefits of preventing TB, with its individual and public health implications, against the risk of individual toxicity is currently being explored, considering the potential lower risk associated with shorter course regimens [122,123]. Data are needed to guide optimal screening and thresholds for alcohol use that halt TB preventive therapy; strategies to improve TB preventive therapy completion in the setting of alcohol use are being evaluated [124,125].

3.2. Biological Mechanisms

Pathophysiologically, data suggest multiple targets of alcohol use, including direct impairment of cell-mediated immunity [126], direct impact on the upper respiratory tract [127], indirect impact on adaptive immunity [128], and malnutrition [129,130], as pathways for increased susceptibility to TB.

Alcohol use complicates TB treatment for drug-susceptible and drug-resistant TB [102]. Frequent coinfection with hepatitis C and/or HIV increases the risk of hepatotoxicity [121,131,132]. Interactions between alcohol and anti-tuberculous medications are well established. Isoniazid, rifampin, and pyrazinamide, core agents of the first-line TB treatment regimen, uncommonly (~1–3%) cause hepatitis, which can be dose-related and reversible with cessation of the medications or may be due to hypersensitivity reaction [133]. Data are emerging on newer agents now available for drug-resistant TB [134]. In patients taking TB treatment who are at increased risk of hepatotoxicity, such as those with alcohol use or liver disease, closer monitoring is recommended, and non-hepatotoxic TB agents may be substituted.

3.3. Addressing the Intersection of Alcohol Use and TB

Alcohol use increases the risk of incident tuberculosis disease and risk of poor outcomes, primarily through behavioral mechanisms [98,99]. Data suggest alcohol use also impairs cell-mediated and adaptive immunity, though work remains to elucidate these mechanisms [126,127]. Additionally, there is a paucity of data on the thresholds for alcohol use that portend risk, implementation of screening and addressing alcohol use within TB programs, and the alcohol-related risk for latent TB and latent TB treatment. Emerging data suggest promising interventions that can improve TB outcomes.

4. Alcohol Use and the Risk of Lower Respiratory Infections (Pneumonia)

Pneumonia is the most important category of lower respiratory infections. Its most common type is bacterial pneumonia caused by the *Streptococcus pneumoniae*, but other forms may be viral or, rarely, caused by fungi or parasites [135]. In 2019, lower respiratory infections, the main category usually estimated in international statistics, caused about 2.5 million deaths globally (point estimate: 2,493,000; 95% confidence interval (CI): 2,268,000–2,736,000) and about 97 million disability years of life lost (DALYs; point estimate: 97,190,000; 95% CI: 84,871,000–113,083,000; all data are based on the 2019 Global Burden of Disease Study [136]. More than 80% of the lower respiratory infection deaths [136] and more than 90% of the DALYs lost were in low- and middle-income countries (LMIC) [136], with a clear gradient in age-adjusted rates by wealth: the higher the economic wealth, the lower the rate of lower respiratory infections. In total, 3.2% (95% CI: 1.6–6.0%) of the deaths and 1.8% of the DALYs (95% CI: (1.0–3.3%)) due to lower respiratory infections were attributable to alcohol, meaning they would not have occurred in a world without alcohol [137].

Alcohol use both impacts the etiology and the course of lower respiratory infections, most importantly in community-acquired infections. As with most infections, lower respiratory infections are more highly prevalent in crowded environments often inhabited by poor people. Additionally, within countries, pneumonia is associated with socioeconomic status, an indicator of wealth: the higher the socioeconomic status, the lower the prevalence of lower respiratory infections [138]. Alcohol contributes to these inequalities ([139,140]), especially via heavy drinking occasions [141]. Of course, factors other than crowding and alcohol use, which are associated with wealth at the individual and societal levels, also contribute to lower respiratory infection rates, such as tobacco smoking, undernutrition, indoor air pollution, and insufficient access to health care [96]. Most of these risk factors are known to interact with alcohol use [142].

The main impact of alcohol use on lower respiratory infections seems to be via the innate and the adaptive immune system [24,53,143–145]. There are a number of pathways leading to the weakening of various aspects of the immune system, with the key immune cells involved in combating pulmonary conditions being neutrophils, lymphocytes, alveolar macrophages, and the cells responsible for innate immune responses [44,55,145–149]. In addition, alcohol use is causally linked to more than 200 disease and injury outcomes (such as various types of cancer, stroke, liver cirrhosis, or traffic injury) which weaken the immune system and increase the risk for lower respiratory infections [138].

Although a number of studies on pathways have been conducted among people with AUDs, two dose–response meta-analyses found an almost linearly increasing risk with increasing average consumption of alcohol [150,151]. These two meta-analyses estimated that an average increase of one drink per day was associated with an increased risk of 8% (95% CI 6–9%) and 6% (95% CI 1–11%). As people with AUDs tend to have the highest average consumption [152], risk for lower respiratory infections is highest in this group. For, instance, in a cohort study of more than 12 million French hospital patients, the relative risk for hospitalization for pneumococcal pneumonia in patients with an AUD was 3.71 (95% 3.60–3.83; [153]). Other studies have found similar and higher risks [154–156].

The same mechanisms which lead to the incidence of lower respiratory infections also worsen its course. Clearly, the living conditions and behaviors, as well as alcohol-induced compromised immunity, hinder the healing process for people with lower respiratory infections [157]. Abstinence or at least an absence of heavy drinking occasions should thus be the norm during such infections, bearing in mind that for some people with AUD, abruptly abstaining may lead to alcohol withdrawal syndrome, which in itself may have severely negative effects [158].

5. Alcohol Use and the Risk of COVID-19

The ongoing global pandemic of coronavirus disease 2019 (COVID-19) is caused by the severe acute respiratory syndrome coronavirus 2 (SARS-CoV-2). Following the first cases in China late in 2019, which soon spread to other countries, the WHO declared it a Public Health Emergency of International Concern on 30 January 2020, and later—based on more than 118,000 cases in 114 countries and 4291 deaths—a pandemic on 11 March 2020 [159]. To date, on 10 July 2021, the estimates are that there have been more than 187 million COVID-19 infections and more than 4 million deaths [160], which are widely considered to be conservative estimates, as only direct cases with ascertainment are included (for a total estimate, see [161]). Alcohol use may play a role in both the incidence and the course of the disease [162], with both behavioral and biological pathways.

5.1. Behavioral Pathways

COVID-19–alcohol behavioral pathways hinge on the social drift hypotheses—the phenomenon that alcohol problems, especially in heavy drinkers and people with AUDs, are associated with long-term negative effects on the place of residence, involving an elevated likelihood of moving into or remaining in disadvantaged neighborhoods [163,164]. These environments hinder physical distancing and have been established as risk factors for COVID-19 infections and poor outcomes [165]. Independent of characteristics of disadvantaged neighborhoods, alcohol consumption has been shown to narrow physical distancing [166].

5.2. Biological Pathways

COVID-19 has only recently emerged as a pathogen, so there is less comprehensive and systematic knowledge about its relationship with alcohol use than there is for more established pathogens. The pathophysiology of COVID-19 is complex [167] but can be conceptually simplified as:

(1)   COVID-19 travels from the upper respiratory tract (highest transmission risk) to the lower respiratory tract (highest disease risk), causing pneumonia;

(2)   COVID-19 initiates innate and adaptive immune responses that are often maladaptive, leading to ineffective pathogen eradication combined with inflammation that causes host tissue damage;

(3)   Damage is concentrated not at the alveolus (i.e., the interface of air–blood oxygen exchange), as is typical of pneumonia, but instead at epithelial cells (i.e., cells lining the lower respiratory tract) and endothelial cells (i.e., cells lining blood vessels);

(4)   Endothelial damage occurs not only in the lungs but also systematically, leading to vasculitis (i.e., damaged small blood vessels) and thrombosis (i.e., blood clots), potentially causing multi-organ failure.

Accordingly, alcohol use may impact COVID-19 by facilitating some or all of these steps. Heavy alcohol consumption is a well-known risk factor for aspiration pneumonia, so it is likely that alcohol, if consumed heavily, leads to increased aspiration of the upper respiratory tract COVID-19 to the lower respiratory tract.

Heavy alcohol use weakens the innate and adaptive immune systems [168,169]. The processes have been described in other sections and are summarised in [126,128]. A recent network meta-analysis [170] explored the potential effects of alcohol use on inflammation, based on the fact that many COVID-19 patients present with fever in the early phase, with some progressing to a hyperinflammatory phase. This network meta-analysis demonstrated that alcohol exposure might augment COVID-19-induced inflammation by altering the activity of key inflammatory mediators (augmenting inflammatory effects and inhibiting the activity of anti-inflammatory mediators, including the glucocorticoid receptor). Finally, a large study showed genetically informative putative causal effects of alcohol use on worsening the course of COVID-19 [171]. However, the last study has not undergone peer review as of yet.

Finally, chronic heavy use of alcohol leads to frailty, arterial hypertension, and liver and other organ damage, rendering people more susceptible to COVID-related complications. Susceptibility to COVID is also enhanced if alcohol leads to obesity or co-occurring infectious diseases or arterial hypertension [172–175]. Pathways involving obesity, a known independent risk factor for COVID-19 [176], have received research interest, both from theoretical [177] and empirical perspectives [173].

5.3. Association with Alcohol Use or Heavy Alcohol Use/AUDs

Studies on the association between alcohol use and the incidence and severity of COVID-19 have yielded mixed results. While some studies have found associations, in particular for heavy drinkers [178–180] or people with AUDs [181], other studies have demonstrated that alcohol use per se was not necessarily associated with the incidence of COVID-19 or with a more severe course of the disease [182–184]. This is in line with the postulated pathways described above, which mainly report effects for heavy drinking and/or in people with AUDs (see also [185–187]).

6. Interventions for Preventing Transmission and Improving Treatment Outcomes of Alcohol-Attributable Diseases

The evidence reviewed above suggests that alcohol is a clear risk factor for the incidence of and poor treatment outcomes from HIV, TB, and pneumonia, with the evidence regarding its effects on COVID-19 still emerging. Alcohol-attributable TB, HIV, and pneumonia combined were responsible for approximately 360,000 deaths and 14 million DALYs in 2016 (Table 1), and alcohol-attributable TB deaths and DALYs far exceeded alcohol-attributable lower respiratory infection and HIV deaths and DALYs [137]. Given the observed role of alcohol use in these diseases, reductions in alcohol consumption should lead to reduced incidence of and improved disease outcomes, including fewer deaths, among those with these illnesses. Feasible and effective alcohol-reduction interventions must be prioritized, but how best to intervene has not been fully delineated. We discuss individual-level interventions followed by structural interventions (or alcohol control measures) that may prevent transmission and improve treatment outcomes of alcohol-attributable communicable diseases.

Table 1. Alcohol-attributable communicable diseases: 2016 estimates [137].

	Deaths (Thousands)	DALYs (Millions)
Tuberculosis	236.3 (74.6–456.6)	9.9 (3.2–18.6)
HIV/AIDS	30.4 (22.8–56.7)	1.7 (1.2–3.1)
Lower respiratory infections	95.2 (48.5–177.6)	2.3 (1.3–4.3)

6.1. Reducing the Incidence of Communicable Diseases

Individual-level approaches focusing on alcohol reduction in order to reduce the incidence of pneumonia and TB are relatively rare, whereas more studies focused on alcohol use reduction for preventing HIV transmission have been conducted. Within a systematic review of studies of interventions for reducing the incidence of TB, no studies that examined alcohol-reduction interventions for reducing the incidence of TB were found [188].

Similarly, despite the role of alcohol use in increasing the risk of pneumonia acquisition, alcohol use reduction seems to be missing as part of a number of texts providing recommendations for the prevention of pneumonia (e.g., [189,190]). On the other hand, one of the main preventative measures for pneumonia is vaccination, and commentators have recommended vaccinating individuals with an AUD in order to prevent (re-)infection with pneumonia [158]. Others (e.g., [191]) have suggested that clinicians should identify individuals who are at high risk of developing pneumonia as potential candidates for pneumonia vaccinations due to their possession of risk factors, including alcohol use, smoking, older age, and lower socioeconomic status, among a few others [191].

In terms of HIV, a number of systematic reviews of alcohol–HIV reduction interventions [8,192,193], mostly conducted in clinic or treatment settings, have shown that behavioral interventions can reduce alcohol use in sexual contexts and alcohol consumption among individuals at risk of alcohol-related HIV acquisition. A systematic review [8] of alcohol–HIV interventions targeting both alcohol and sexual risk behavior reduction among STI clinic and substance use treatment patients in Russia showed evidence of effectiveness in increasing condom use. Interventions in other settings, such as bars and communities, may also be ideal and feasible (e.g., [194–196]) but have yielded mixed results [194,195]. Secondary prevention, which entails TasP (discussed below), with high adherence to ART to bring about viral suppression, is particularly important yet problematic in people living with HIV who drink alcohol [94].

6.2. Improving Treatment Outcomes

Since alcohol use complicates the treatment of many communicable diseases, integration of alcohol use reduction counseling or screening and brief interventions into TB [197], HIV [94], or pneumonia [150] treatment services has been recommended. Similarly, screening for TB [197] or HIV among people with AUDs has also been recommended, as has the co-location of services [94]. However, the evidence base regarding the effectiveness of such approaches for all communicable disease categories of interest in the current report is fairly limited.

A few primary studies that have evaluated the efficacy of individual-level alcohol reduction interventions for improving TB treatment outcomes [56,114,116,198] have yielded disappointing results. In Russia, Shin et al. [198] found no differences between the TB and alcohol use outcomes of new TB patients with AUDs who received: (1) a brief counseling intervention (BCI) and treatment as usual; (2) naltrexone combined with brief behavioral compliance enhancement counseling (BBCET) (naltrexone adherence counseling); (3) BCI and naltrexone with BBCET and treatment as usual; and (4) treatment as usual—referral to a narcologist (namely, an addiction psychiatrist in the Russian system). One sub-group analysis revealed that among those with previous quit attempts (n = 111), the TB treatment outcome was better for the naltrexone group (92.3%) compared with the non-naltrexone group (75.9%). In a cluster RCT in South Africa, Peltzer et al. [116] found no effect for a two-session screening and brief intervention on TB and alcohol use outcomes among new TB patients who had Alcohol Use Disorder Identification Test (AUDIT) scores of ≥ 7 if they were women and ≥ 8 if they were men. More research on individual-level alcohol-reduction interventions among patients on TB treatment is needed.

Several recommendations regarding the treatment of patients with pneumonia who drink alcohol or have AUDs have been put forward. These include preventing further bouts of pneumonia by providing alcohol counseling [151] and pneumococcal vaccination [158]. Screening and brief interventions for AUDs among all patients undergoing treatment for pneumonia have also been recommended so that the clinician can be well informed about their patients' alcohol use and manage their pneumonia accordingly [150]. Assessment for potential alcohol withdrawal syndrome that may occur as a result of abstinence is also recommended as it can have serious and even fatal consequences if not managed appropriately [158]. Providing guidelines on screening for the risk of AUDs and alcohol withdrawal syndrome to TB treatment providers has also been recommended [199].

Efforts to improve treatment outcomes and improve secondary prevention for people living with HIV who drink alcohol require emphasizing linkage and retention in care, ART initiation, ART adherence, viral suppression, and condom use [94]. A recently published high-quality systematic review and meta-analysis involving 21 studies and 8461 people living with HIV, 69% of whom were on ART, has indicated (contrary to other findings, [200–202]) that individual-level behavioral interventions were effective in reducing the quantity and heavy consumption (but not alcohol use or alcohol use frequency), increasing condom use (but not affecting the number of sexual partners or a composite index of sexual risk), reducing viral load, and increasing ART adherence [90]. Interventions in which participants were recruited from clinics were most likely to be effective. As supplements to such interventions, additional approaches that have been recommended include the use of technology to deliver interventions, use of ultra-brief interventions, prevention of increased alcohol consumption or the development of AUDs, a focus on aging populations, addressing psychosocial comorbidities, and improving accessibility and convenience of HIV care [94]. Occasionally, health workers have stigmatizing attitudes or inadequate knowledge that can lead to inadvertent ART nonadherence among their patients [203].

Very clear guidelines are needed to enable health workers to provide appropriate and consistent messaging [204].

Pharmacological interventions for reducing alcohol use and improving treatment outcomes may be especially appropriate for those with communicable diseases [91]. Farhadian et al.'s systematic review, including seven studies, provided some evidence of naltrexone's effectiveness in reducing alcohol consumption and HIV viral load, but it did not affect ART adherence, CD4 cell count, or disease severity. However, as discussed above, a study in which naltrexone was used in combination with naltrexone adherence counseling, and in another group also behavioral counseling to reduce drinking and improve TB treatment outcomes among TB patients, did not yield positive results [198].

6.3. Alcohol Control Measures

Alcohol control policies that are aligned with the three alcohol "best buys"—increasing excise tax, bans or restrictions on alcohol advertising, and restricting the availability of alcohol [205]—are most effective for reducing population-level alcohol use. These measures can be expected to be effective for reducing the incidence of and morbidity and mortality due to alcohol-attributable TB, pneumonia, and HIV. There is some evidence that may provide support for such effects. For example, a study in the United States showed that longer sales hours at the state/district level were associated with high-risk sexual behaviors [92], which are associated with HIV transmission. However, implementation of effective alcohol control policies such as the best buys is relatively low around the world, particularly in lower- and middle-income countries [206–209], many of which have the highest disease burden with respect to many communicable diseases [136]. Implementation and enforcement of effective alcohol control policies as a means of reducing the burden of communicable diseases is recommended.

7. Discussion/Conclusions

Alcohol use is a clear risk factor for the incidence of and poor treatment outcomes from HIV, TB, and pneumonia. Emerging evidence suggests that heavy and chronic alcohol use is associated with an increased risk of acquisition of COVID-19 and more severe disease once infected, while evidence regarding the role of alcohol use per se in COVID-19 is mixed.

Alcohol's role in communicable diseases can be explained by both behavioral and biological mechanisms (Figure 1). Alcohol use increases susceptibility to infectious diseases through several immunologic mechanisms. Chronic or irregular heavy drinking leads to increased susceptibility to viral and bacterial infections, including mycobacterial infections and decreased response to vaccination. Chronic heavy drinking stimulates inflammation yet impairs neutrophil function in the innate (immediate) immune response and leads to loss of T cells and B cells in the adaptive (delayed or humoral) response. In contrast, moderate alcohol consumption seems to strengthen the response to infection, though the exact mechanisms of alcohol's mixed effects on the immune system, particularly on the adaptive immune response, remain under investigation [126,128]. In terms of behavioral mechanisms, most of alcohol's effects on disease acquisition result from impaired decision making (or impaired control), which gives rise to increased risk behaviors such as condomless sex. Furthermore, alcohol consumption is negatively associated with linkage to and retention in care and with medication adherence (particularly for HIV and TB treatment).

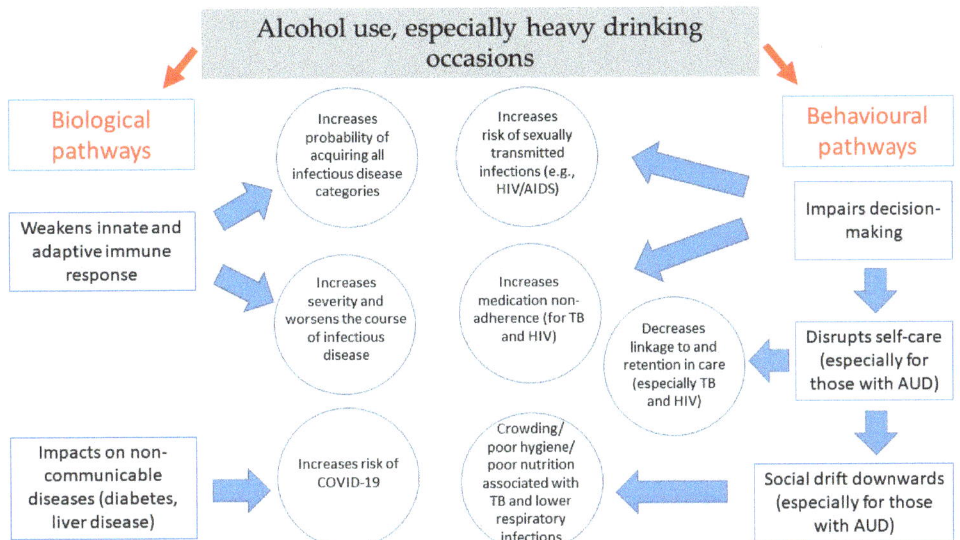

Figure 1. Key biological and behavioral mechanisms through which alcohol use is associated with communicable diseases.

There has been limited research that has identified effective interventions for addressing alcohol-attributable TB and pneumonia, suggesting an urgent need for research in these areas, while several effective interventions to address alcohol-attributable HIV infection have been determined. Implementation of effective individual-level interventions, as well as alcohol control measures as a means of reducing the burden of communicable diseases, are recommended.

Author Contributions: Conceptualization, N.K.M. and J.R.; methodology, N.K.M. and J.R.; writing—original draft preparation, N.K.M., J.R., S.V.S., P.A.S. and R.S.B.; writing—review and editing, N.K.M., J.R., S.V.S., P.A.S. and R.S.B.; project administration, N.K.M. and J.R. All authors have read and agreed to the published version of the manuscript.

Funding: This research was supported by funding from the US National Institute on Alcohol Abuse and Alcoholism (1R01AA028009) and the Canadian Institutes of Health Research, Institute of Neurosciences, and Mental Health and Addiction (CRISM Ontario Node grant No. SMN-13950). J.R. P.A.S.'s research related to this work is supported by funding from the NIAAA (5UH3AA026212-05), and his salary is supported in part by the Ontario HIV Treatment Network (OHTN). This work was also supported by funding from the NIAAA (R01AA027481) to S.V.S. and R.S.B. The funders had no role in the design of the study; in the collection, analyses, or interpretation of data; in the writing of the manuscript, or in the decision to publish the results.

Acknowledgments: The authors would like to acknowledge Mpho Mathebula for literature searching, Narges Joharchi for her work on the biologically focused HIV components, Christine Buchanan for manuscript preparation and formatting, and Astrid Otto for proofreading and copyediting the text.

Conflicts of Interest: The authors declare no conflict of interest.

References

1. Rush, B. *An Inquiry into the Effects of Ardent Spirits upon the Human Body and Mind: With an Account of the Means of Preventing, and of the Remedies for Curing Them*, 6th ed.; Cornelius Davis: New York, NY, USA, 1811; (originally published 1785).
2. Murray, C.J.L.; Lopez, A. Quantifying the burden of disease and injury attributable to ten major risk factors. In *The Global Burden of Disease: A Comprehensive Assessment of Mortality and Disability from Diseases, Injuries and Risk Factors in 1990 and Projected to 2020*; Murray, C.J.L., Lopez, A.D., Eds.; Harvard School of Public Health on Behalf of the World Health Organization and the World Bank: Boston, MA, USA, 1996; pp. 295–324.

3. Lim, S.S.; Vos, T.; Flaxman, A.D.; Danaei, G.; Shibuya, K.; Adair-Rohani, H.; Amann, M.; Anderson, H.R.; Andrews, K.G.; Aryee, M.; et al. A comparative risk assessment of burden of disease and injury attributable to 67 risk factors and risk factor clusters in 21 regions, 1990–2010: A systematic analysis for the Global Burden of Disease Study 2010. *Lancet* **2012**, *380*, 2224–2260. [CrossRef]
4. Rehm, J.; Baliunas, D.; Borges, G.L.; Graham, K.; Irving, H.; Kehoe, T.; Parry, C.D.; Patra, J.; Popova, S.; Poznyak, V. The relation between different dimensions of alcohol consumption and burden of disease: An overview. *Addiction* **2010**, *105*, 817–843. [CrossRef] [PubMed]
5. Rehm, J.; Gmel, G.E., Sr.; Gmel, G.; Hasan, O.S.M.; Imtiaz, S.; Popova, S.; Probst, C.; Roerecke, M.; Room, R.; Samokhvalov, A.V.; et al. The relationship between different dimensions of alcohol use and the burden of disease-an update. *Addiction* **2017**, *112*, 968–1001. [CrossRef] [PubMed]
6. Lönnroth, K.; Williams, B.G.; Stadlin, S.; Jaramillo, E.; Dye, C. Alcohol use as a risk factor for tuberculosis: A systematic review. *BMC Public Health* **2008**, *8*, 289. [CrossRef] [PubMed]
7. Kalichman, S.C.; Simbayi, L.C.; Kaufman, M.; Cain, D.; Jooste, S. Alcohol use and sexual risks for HIV/AIDS in sub-Saharan Africa: Systematic review of empirical findings. *Prev. Sci.* **2007**, *8*, 141–151. [CrossRef] [PubMed]
8. Lan, C.W.; Scott-Sheldon, L.A.; Carey, K.B.; Johnson, B.T.; Carey, M.P. Alcohol and sexual risk reduction interventions among people living in Russia: A systematic review and meta-analysis. *AIDS Behav.* **2014**, *18*, 1835–1846. [CrossRef] [PubMed]
9. Scott-Sheldon, L.A.; Walstrom, P.; Carey, K.B.; Johnson, B.T.; Carey, M.P. Alcohol use and sexual risk behaviors among individuals infected with HIV: A systematic review and meta-analysis 2012 to early 2013. *Curr. HIV/AIDS Rep.* **2013**, *10*, 314–323. [CrossRef]
10. Okoro, U.J.; Carey, K.B.; Johnson, B.T.; Carey, M.P.; Scott-Sheldon, L.A.J. Alcohol consumption, risky sexual behaviors, and HIV in Nigeria: A meta-analytic review. *Curr. Drug. Res. Rev.* **2019**, *11*, 92–110. [CrossRef]
11. Shuper, P.A.; Joharchi, N.; Irving, H.; Rehm, J. Alcohol as a correlate of unprotected sexual behavior among people living with HIV/AIDS: Review and meta-analysis. *AIDS Behav.* **2009**, *13*, 1021–1036. [CrossRef]
12. Przybyla, S.M.; Krawiec, G.; Godleski, S.A.; Crane, C.A. Meta-Analysis of alcohol and serodiscordant condomless sex among people living with HIV. *Arch. Sex. Behav.* **2018**, *47*, 1351–1366. [CrossRef]
13. Fisher, J.C.; Bang, H.; Kapiga, S.H. The association between HIV infection and alcohol use: A systematic review and meta-analysis of African studies. *Sex. Transm. Dis.* **2007**, *34*, 856–863. [CrossRef] [PubMed]
14. Baliunas, D.; Rehm, J.; Irving, H.; Shuper, P. Alcohol consumption and risk of incident human immunodeficiency virus infection: A meta-analysis. *Int. J. Public Health* **2010**, *55*, 159–166. [CrossRef] [PubMed]
15. Rothman, K.J.; Greenland, S.; Lash, T.L. *Modern Epidemiology*, 3rd ed.; Lippincott Williams & Wilkins: Philadelphia, PA, USA, 2008.
16. Scott-Sheldon, L.A.; Carey, K.B.; Cunningham, K.; Johnson, B.T.; Carey, M.P. Alcohol use predicts sexual decision-making: A systematic review and meta-analysis of the experimental literature. *AIDS Behav.* **2016**, *20* (Suppl. 1), S19–S39. [CrossRef]
17. Rehm, J.; Probst, C.; Shield, K.D.; Shuper, P.A. Does alcohol use have a causal effect on HIV incidence and disease progression? A review of the literature and a modeling strategy for quantifying the effect. *Popul. Health Metr.* **2017**, *15*, 4. [CrossRef]
18. Braithwaite, R.S.; McGinnis, K.A.; Conigliaro, J.; Maisto, S.A.; Crystal, S.; Day, N.; Cook, R.L.; Gordon, A.; Bridges, M.W.; Seiler, J.F.; et al. A temporal and dose-response association between alcohol consumption and medication adherence among veterans in care. *Alcohol. Clin. Exp. Res.* **2005**, *29*, 1190–1197. [CrossRef] [PubMed]
19. Parsons, J.T.; Rosof, E.; Mustanski, B. The temporal relationship between alcohol consumption and HIV-medication adherence: A multilevel model of direct and moderating effects. *Health Psychol.* **2008**, *27*, 628–637. [CrossRef] [PubMed]
20. Schensul, J.J.; Ha, T.; Schensul, S.; Sarna, A.; Bryant, K. Identifying the intersection of alcohol, adherence and sex in HIV positive men on ART treatment in India using an adapted timeline followback procedure. *AIDS Behav.* **2017**, *21*, 228–242. [CrossRef]
21. Cohen, M.S.; Chen, Y.Q.; McCauley, M.; Gamble, T.; Hosseinipour, M.C.; Kumarasamy, N.; Hakim, J.G.; Kumwenda, J.; Grinsztejn, B.; Pilotto, J.H.; et al. Antiretroviral therapy for the prevention of HIV-1 transmission. *N. Engl. J. Med.* **2016**, *375*, 830–839. [CrossRef]
22. Cohen, M.S.; Chen, Y.Q.; McCauley, M.; Gamble, T.; Hosseinipour, M.C.; Kumarasamy, N.; Hakim, J.G.; Kumwenda, J.; Grinsztejn, B.; Pilotto, J.H.; et al. Prevention of HIV-1 infection with early antiretroviral therapy. *N. Engl. J. Med.* **2011**, *365*, 493–505. [CrossRef]
23. Eisinger, R.W.; Dieffenbach, C.W.; Fauci, A.S. HIV Viral load and transmissibility of HIV infection: Undetectable equals untransmittable. *JAMA* **2019**, *321*, 451–452. [CrossRef]
24. Happel, K.I.; Nelson, S. Alcohol, immunosuppression, and the lung. *Proc. Am. Thorac. Soc.* **2005**, *2*, 428–432. [CrossRef] [PubMed]
25. Szabo, G. Alcohol's contribution to compromised immunity. *Alcohol. Health Res. World* **1997**, *21*, 30.
26. Rehm, J.; Samokhvalov, A.V.; Neuman, M.G.; Room, R.; Parry, C.; Lönnroth, K.; Patra, J.; Poznyak, V.; Popova, S. The association between alcohol use, alcohol use disorders and tuberculosis (TB). A systematic review. *BMC Public Health* **2009**, *9*, 1–12. [CrossRef] [PubMed]
27. Cook, R.L.; Clark, D.B. Is there an association between alcohol consumption and sexually transmitted diseases? A systematic review. *Sex. Transm. Dis.* **2005**, *32*, 156–164. [CrossRef] [PubMed]
28. UNAIDS. *Global HIV & AIDS Statistics—Fact Sheet*; UNAIDS: Geneva, Switzerland, 2021.
29. UNAIDS. *Seizing the Moment: Tackling Entrenched Inequalities to End Epidemics*; Joint United Nations Programme on HIV/AIDS: Geneva, Switzerland, 2020.
30. Steele, C.M.; Josephs, R.A. Alcohol myopia: Its prized and dangerous effects. *Am. Psychol.* **1990**, *45*, 921–933. [CrossRef] [PubMed]
31. Shuper, P.A.; Neuman, M.; Kanteres, F.; Baliunas, D.; Joharchi, N.; Rehm, J. Causal considerations on alcohol and HIV/AIDS: A systematic review. *Alcohol Alcohol.* **2010**, *45*, 159–166. [CrossRef] [PubMed]

32. Rehm, J.; Shield, K.D.; Joharchi, N.; Shuper, P. Alcohol consumption and the intention to engage in unprotected sex: Systematic review and meta-analysis of experimental studies. *Addiction* **2012**, *107*, 51–59. [CrossRef]

33. Shuper, P.A.; Joharchi, N.; Monti, P.M.; Loutfy, M.; Rehm, J. Acute alcohol consumption directly increases HIV transmission risk: A randomized controlled experiment. *J. Acquir. Immune Defic. Syndr.* **2017**, *76*, 493–500. [CrossRef]

34. Shuper, P.A.; Joharchi, N.; Rehm, J. Protocol for a controlled experiment to identify the causal role of acute alcohol consumption in condomless sex among HIV-positive MSM: Study procedures, ethical considerations, and implications for HIV prevention. *AIDS Behav.* **2016**, *20* (Suppl. 1), S173–S184. [CrossRef] [PubMed]

35. Baeten, J.M.; Donnell, D.; Ndase, P.; Mugo, N.R.; Campbell, J.D.; Wangisi, J.; Tappero, J.W.; Bukusi, E.A.; Cohen, C.R.; Katabira, E.; et al. Antiretroviral prophylaxis for HIV prevention in heterosexual men and women. *N. Engl. J. Med.* **2012**, *367*, 399–410. [CrossRef] [PubMed]

36. Grant, R.M.; Lama, J.R.; Anderson, P.L.; McMahan, V.; Liu, A.Y.; Vargas, L.; Goicochea, P.; Casapia, M.; Guanira-Carranza, J.V.; Ramirez-Cardich, M.E.; et al. Preexposure chemoprophylaxis for HIV prevention in men who have sex with men. *N. Engl. J. Med.* **2010**, *363*, 2587–2599. [CrossRef] [PubMed]

37. Hendershot, C.S.; Stoner, S.A.; Pantalone, D.W.; Simoni, J.M. Alcohol use and antiretroviral adherence: Review and meta-analysis. *J. Acquir. Immune Defic. Syndr.* **2009**, *52*, 180–202. [CrossRef] [PubMed]

38. Shuper, P.A.; Joharchi, N.; Bogoch, I.I.; Loutfy, M.; Crouzat, F.; El-Helou, P.; Knox, D.C.; Woodward, K.; Rehm, J. Alcohol consumption, substance use, and depression in relation to HIV Pre-Exposure Prophylaxis (PrEP) nonadherence among gay, bisexual, and other men-who-have-sex-with-men. *BMC Public Health* **2020**, *20*, 1782. [CrossRef] [PubMed]

39. Haberer, J.E.; Baeten, J.M.; Campbell, J.; Wangisi, J.; Katabira, E.; Ronald, A.; Tumwesigye, E.; Psaros, C.; Safren, S.A.; Ware, N.C.; et al. Adherence to antiretroviral prophylaxis for HIV prevention: A substudy cohort within a clinical trial of serodiscordant couples in East Africa. *PLoS Med.* **2013**, *10*, e1001511. [CrossRef] [PubMed]

40. Mugo, P.M.; Sanders, E.J.; Mutua, G.; van der Elst, E.; Anzala, O.; Barin, B.; Bangsberg, D.R.; Priddy, F.H.; Haberer, J.E. Understanding adherence to daily and intermittent regimens of oral HIV pre-exposure prophylaxis among men who have sex with men in Kenya. *AIDS Behav.* **2015**, *19*, 794–801. [CrossRef] [PubMed]

41. van der Elst, E.M.; Mbogua, J.; Operario, D.; Mutua, G.; Kuo, C.; Mugo, P.; Kanungi, J.; Singh, S.; Haberer, J.; Priddy, F.; et al. High acceptability of HIV pre-exposure prophylaxis but challenges in adherence and use: Qualitative insights from a phase I trial of intermittent and daily PrEP in at-risk populations in Kenya. *AIDS Behav.* **2013**, *17*, 2162–2172. [CrossRef] [PubMed]

42. Lau, A.H.; Szabo, G.; Thomson, A.W. Antigen-presenting cells under the influence of alcohol. *Trends Immunol.* **2009**, *30*, 13–22. [CrossRef]

43. McClain, C.J.; Shedlofsky, S.; Barve, S.; Hill, D.B. Cytokines and alcoholic liver disease. *Alcohol. Health Res. World* **1997**, *21*, 317–320.

44. Szabo, G.; Mandrekar, P. A recent perspective on alcohol, immunity, and host defense. *Alcohol. Clin. Exp. Res.* **2009**, *33*, 220–232. [CrossRef]

45. Friedman, H.; Newton, C.; Klein, T.W. Microbial infections, immunomodulation, and drugs of abuse. *Clin. Microbiol. Rev.* **2003**, *16*, 209–219. [CrossRef]

46. Friedman, H.; Pross, S.; Klein, T.W. Addictive drugs and their relationship with infectious diseases. *FEMS Immunol. Med. Microbiol.* **2006**, *47*, 330–342. [CrossRef]

47. Goral, J.; Choudhry, M.A.; Kovacs, E.J. Acute ethanol exposure inhibits macrophage IL-6 production: Role of p38 and ERK1/2 MAPK. *J. Leukoc. Biol.* **2004**, *75*, 553–559. [CrossRef] [PubMed]

48. Goral, J.; Kovacs, E.J. In vivo ethanol exposure down-regulates TLR2-, TLR4-, and TLR9-mediated macrophage inflammatory response by limiting p38 and ERK1/2 activation. *J. Immunol.* **2005**, *174*, 456–463. [CrossRef] [PubMed]

49. Pruett, S.B.; Schwab, C.; Zheng, Q.; Fan, R. Suppression of innate immunity by acute ethanol administration: A global perspective and a new mechanism beginning with inhibition of signaling through TLR3. *J. Immunol.* **2004**, *173*, 2715–2724. [CrossRef]

50. Brodie, C.; Domenico, J.; Gelfand, E.W. Ethanol inhibits early events in T-lymphocyte activation. *Clin. Immunol. Immunopathol.* **1994**, *70*, 129–136. [CrossRef] [PubMed]

51. Zhang, T.; Guo, C.J.; Douglas, S.D.; Metzger, D.S.; O'Brien, C.P.; Li, Y.; Wang, Y.J.; Wang, X.; Ho, W.Z. Alcohol suppresses IL-2-induced CC chemokine production by natural killer cells. *Alcohol. Clin. Exp. Res.* **2005**, *29*, 1559–1567. [CrossRef]

52. Shellito, J.E.; Olariu, R. Alcohol decreases T-lymphocyte migration into lung tissue in response to Pneumocystis carinii and depletes T-lymphocyte numbers in the spleens of mice. *Alcohol. Clin. Exp. Res.* **1998**, *22*, 658–663. [CrossRef]

53. Cook, R.T. Alcohol abuse, alcoholism, and damage to the immune system—A review. *Alcohol. Clin. Exp. Res.* **1998**, *22*, 1927–1942.

54. Szabo, G. Consequences of alcohol consumption on host defence. *Alcohol Alcohol.* **1999**, *34*, 830–841. [CrossRef]

55. Neuman, M.G. Cytokines—central factors in alcoholic liver disease. *Alcohol Res. Health* **2003**, *27*, 307–316.

56. Mathew, T.A.; Yanov, S.A.; Mazitov, R.; Mishustin, S.P.; Strelis, A.K.; Yanova, G.V.; Golubchikova, V.T.; Taran, D.V.; Golubkov, A.; Shields, A.L.; et al. Integration of alcohol use disorders identification and management in the tuberculosis programme in Tomsk Oblast, Russia. *Eur. J. Public Health* **2009**, *19*, 16–18. [CrossRef] [PubMed]

57. Neuman, M.G.; Sha, K.; Esguerra, R.; Zakhari, S.; Winkler, R.E.; Hilzenrat, N.; Wyse, J.; Cooper, C.L.; Seth, D.; Gorrell, M.D.; et al. Inflammation and repair in viral hepatitis C. *Dig. Dis. Sci.* **2008**, *53*, 1468–1487. [CrossRef]

58. Voiculescu, M.; Winkler, R.E.; Moscovici, M.; Neuman, M.G. Chemotherapies and targeted therapies in advanced hepatocellular carcinoma: From laboratory to clinic. *J. Gastrointestin Liver Dis.* **2008**, *17*, 315–322. [PubMed]

59. Alcohol-Induced hepatic fibrosis: Mechanisms. Proceedings of a satellite symposium. Hilton Head Island, South Carolina, USA. June 1998. *Alcohol. Clin. Exp. Res.* **1999**, *23*, 901–954.

60. Theall, K.P.; Amedee, A.; Clark, R.A.; Dumestre, J.; Kissinger, P. Alcohol consumption and HIV-1 vaginal RNA shedding among women. *J. Stud. Alcohol. Drugs* **2008**, *69*, 454–458. [CrossRef] [PubMed]

61. Coleman, J.S.; Hitti, J.; Bukusi, E.A.; Mwachari, C.; Muliro, A.; Nguti, R.; Gausman, R.; Jensen, S.; Patton, D.; Lockhart, D.; et al. Infectious correlates of HIV-1 shedding in the female upper and lower genital tracts. *AIDS* **2007**, *21*, 755–759. [CrossRef] [PubMed]

62. Rebbapragada, A.; Howe, K.; Wachihi, C.; Pettengell, C.; Sunderji, S.; Huibner, S.; Ball, T.B.; Plummer, F.A.; Jaoko, W.; Kaul, R. Bacterial vaginosis in HIV-infected women induces reversible alterations in the cervical immune environment. *J. Acquir. Immune Defic. Syndr.* **2008**, *49*, 520–522. [CrossRef]

63. Pandrea, I.; Happel, K.I.; Amedee, A.M.; Bagby, G.J.; Nelson, S. Alcohol's role in HIV transmission and disease progression. *Alcohol Res. Health* **2010**, *33*, 203–218.

64. Azar, M.M.; Springer, S.A.; Meyer, J.P.; Altice, F.L. A systematic review of the impact of alcohol use disorders on HIV treatment outcomes, adherence to antiretroviral therapy and health care utilization. *Drug Alcohol. Depend.* **2010**, *112*, 178–193. [CrossRef] [PubMed]

65. Vagenas, P.; Azar, M.M.; Copenhaver, M.M.; Springer, S.A.; Molina, P.E.; Altice, F.L. The impact of alcohol use and related disorders on the HIV continuum of care: A systematic review. *Curr. HIV/AIDS Rep.* **2015**, *12*, 421–436. [CrossRef] [PubMed]

66. Walter, A.W.; Lundgren, L.; Umez-Eronini, A.; Ritter, G.A. Alcohol use and HIV testing in a national sample of women. *AIDS Behav.* **2016**, *20* (Suppl. 1), S84–S96. [CrossRef]

67. Maughan-Brown, B.; Harrison, A.; Galarraga, O.; Kuo, C.; Smith, P.; Bekker, L.G.; Lurie, M.N. Factors affecting linkage to HIV care and ART initiation following referral for ART by a mobile health clinic in South Africa: Evidence from a multimethod study. *J. Behav. Med.* **2019**, *42*, 883–897. [CrossRef]

68. Monroe, A.K.; Lau, B.; Mugavero, M.J.; Mathews, W.C.; Mayer, K.H.; Napravnik, S.; Hutton, H.E.; Kim, H.S.; Jabour, S.; Moore, R.D.; et al. Heavy alcohol use is associated with worse retention in HIV care. *J. Acquir. Immune Defic. Syndr.* **2016**, *73*, 419–425. [CrossRef] [PubMed]

69. Cook, R.L.; Zhou, Z.; Kelso-Chichetto, N.E.; Janelle, J.; Morano, J.P.; Somboonwit, C. Alcohol consumption patterns and HIV viral suppression among persons receiving HIV care in Florida: An observational study. *Addict. Sci. Clin. Pract.* **2017**, *12*, 1–9. [CrossRef] [PubMed]

70. Fatch, R.; Bellows, B.; Bagenda, F.; Mulogo, E.; Weiser, S.; Hahn, J.A. Alcohol consumption as a barrier to prior HIV testing in a population-based study in rural Uganda. *AIDS Behav.* **2013**, *17*, 1713–1723. [CrossRef] [PubMed]

71. Kalichman, S.; Banas, E.; Kalichman, M.; Mathews, C. Stigmatisation of alcohol use among people receiving antiretroviral therapy for HIV infection, Cape Town, South Africa. *Glob. Public Health* **2020**, *15*, 1040–1049. [CrossRef]

72. Britton, M.K.; Porges, E.C.; Bryant, V.; Cohen, R.A. Neuroimaging and Cognitive Evidence for Combined HIV-Alcohol Effects on the Central Nervous System: A Review. *Alcohol. Clin. Exp. Res.* **2021**, *45*, 290–306. [CrossRef]

73. Heinz, A.J.; Fogler, K.A.; Newcomb, M.E.; Trafton, J.A.; Bonn-Miller, M.O. Problematic alcohol use among individuals with HIV: Relations with everyday memory functioning and HIV symptom severity. *AIDS Behav.* **2014**, *18*, 1302–1314. [CrossRef]

74. El-Krab, R.; Kalichman, S.C. Alcohol-antiretroviral therapy interactive toxicity beliefs and intentional medication nonadherence: Review of research with implications for interventions. *AIDS Behav.* **2021**, *25*, 1–14. [CrossRef]

75. Bagasra, O.; Kajdacsy-Balla, A.; Lischner, H.W.; Pomerantz, R.J. Alcohol intake increases human immunodeficiency virus type 1 replication in human peripheral blood mononuclear cells. *J. Infect. Dis.* **1993**, *167*, 789–797. [CrossRef]

76. Bagasra, O.; Bachman, S.E.; Jew, L.; Tawadros, R.; Cater, J.; Boden, G.; Ryan, I.; Pomerantz, R.J. Increased human immunodeficiency virus type 1 replication in human peripheral blood mononuclear cells induced by ethanol: Potential immunopathogenic mechanisms. *J. Infect. Dis.* **1996**, *173*, 550–558. [CrossRef] [PubMed]

77. Chen, H.; Zha, J.; Gowans, R.E.; Camargo, P.; Nishitani, J.; McQuirter, J.L.; Cole, S.W.; Zack, J.A.; Liu, X. Alcohol enhances HIV type 1 infection in normal human oral keratinocytes by up-regulating cell-surface CXCR4 coreceptor. *AIDS Res. Hum. Retrovir.* **2004**, *20*, 513–519. [CrossRef]

78. Liu, X.; Zha, J.; Nishitani, J.; Chen, H.; Zack, J.A. HIV-1 infection in peripheral blood lymphocytes (PBLs) exposed to alcohol. *Virology* **2003**, *307*, 37–44. [CrossRef]

79. Crum, R.M.; Galai, N.; Cohn, S.; Celentano, D.D.; Vlahov, D. Alcohol use and T-lymphocyte subsets among injection drug users with HIV-1 infection: A prospective analysis. *Alcohol. Clin. Exp. Res.* **1996**, *20*, 364–371. [CrossRef] [PubMed]

80. Baum, M.K.; Rafie, C.; Lai, S.; Sales, S.; Page, J.B.; Campa, A. Alcohol use accelerates HIV disease progression. *AIDS Res. Hum. Retrovir.* **2010**, *26*, 511–518. [CrossRef]

81. Samet, J.H.; Cheng, D.M.; Libman, H.; Nunes, D.P.; Alperen, J.K.; Saitz, R. Alcohol consumption and HIV disease progression. *J. Acquir. Immune Defic. Syndr.* **2007**, *46*, 194–199. [CrossRef] [PubMed]

82. Asiimwe, S.B.; Fatch, R.; Patts, G.; Winter, M.; Lloyd-Travaglini, C.; Emenyonu, N.; Muyindike, W.; Kekibiina, A.; Blokhina, E.; Gnatienko, N. Alcohol types and HIV disease progression among HIV-infected drinkers not yet on antiretroviral therapy in Russia and Uganda. *AIDS Behav.* **2017**, *21*, 204–215. [CrossRef]

83. Miguez, M.J.; Shor-Posner, G.; Morales, G.; Rodriguez, A.; Burbano, X. HIV treatment in drug abusers: Impact of alcohol use. *Addict. Biol.* **2003**, *8*, 33–37. [CrossRef] [PubMed]

84. Samet, J.H.; Horton, N.J.; Traphagen, E.T.; Lyon, S.M.; Freedberg, K.A. Alcohol consumption and HIV disease progression: Are they related? *Alcohol. Clin. Exp. Res.* **2003**, *27*, 862–867. [CrossRef]
85. McDowell, J.A.; Chittick, G.E.; Stevens, C.P.; Edwards, K.D.; Stein, D.S. Pharmacokinetic interaction of abacavir (1592U89) and ethanol in human immunodeficiency virus-infected adults. *Antimicrob. Agents Chemother.* **2000**, *44*, 1686–1690. [CrossRef]
86. Lieber, C.S.; DeCarli, L.M. Hepatic microsomal ethanol-oxidizing system. In vitro characteristics and adaptive properties in vivo. *J. Biol. Chem.* **1970**, *245*, 2505–2512. [CrossRef]
87. Neuman, M.G.; Monteiro, M.; Rehm, J. Drug interactions between psychoactive substances and antiretroviral therapy in individuals infected with human immunodeficiency and hepatitis viruses. *Subst. Use Misuse* **2006**, *41*, 1395–1463. [CrossRef]
88. Neuman, M.G.; Schneider, M.; Nanau, R.M.; Parry, C. HIV-antiretroviral therapy induced liver, gastrointestinal, and pancreatic injury. *Int. J. Hepatol.* **2012**, *2012*, 760706. [CrossRef]
89. Hahn, J.A.; Samet, J.H. Alcohol and HIV disease progression: Weighing the evidence. *Curr. HIV/AIDS Rep.* **2010**, *7*, 226–233. [CrossRef]
90. Scott-Sheldon, L.A.J.; Carey, K.B.; Johnson, B.T.; Carey, M.P. Behavioral interventions targeting alcohol use among people living with HIV/AIDS: A systematic review and meta-analysis. *AIDS Behav.* **2017**, *21*, 126–143. [CrossRef]
91. Farhadian, N.; Moradi, S.; Zamanian, M.H.; Farnia, V.; Rezaeian, S.; Farhadian, M.; Shahlaei, M. Effectiveness of naltrexone treatment for alcohol use disorders in HIV: A systematic review. *Subst. Abuse Treat. Prev. Policy* **2020**, *15*, 24. [CrossRef] [PubMed]
92. Collins, R.L.; Taylor, S.L.; Elliott, M.N.; Ringel, J.S.; Kanouse, D.E.; Beckman, R. Off-premise alcohol sales policies, drinking, and sexual risk among people living with HIV. *Am. J. Public Health* **2010**, *100*, 1890–1892. [CrossRef] [PubMed]
93. Cohen, D.A.; Wu, S.Y.; Farley, T.A. Structural interventions to prevent HIV/sexually transmitted disease: Are they cost-effective for women in the southern United States? *Sex. Transm. Dis.* **2006**, *33*, S46–S49. [CrossRef] [PubMed]
94. Shuper, P.A. The role of alcohol-related behavioral research in the design of HIV secondary prevention interventions in the era of antiretroviral therapy: Targeted research priorities moving forward. *AIDS Behav.* **2021**, *25*, 1–16. [CrossRef]
95. World Health Organization. *Global Tuberculosis Report*; World Health Organization: Geneva, Switzerland, 2020.
96. GBD 2016 Lower Respiratory Infections Collaborators. Estimates of the global, regional, and national morbidity, mortality, and aetiologies of lower respiratory infections in 195 countries, 1990–2016: A systematic analysis for the Global Burden of Disease Study 2016. *Lancet Infect. Dis.* **2018**, *18*, 1191–1210. [CrossRef]
97. Peltzer, K.; Davids, A.; Njuho, P. Alcohol use and problem drinking in South Africa: Findings from a national population-based survey. *Afr. J. Psychiatry* **2011**, *14*, 30–37. [CrossRef]
98. Imtiaz, S.; Shield, K.D.; Roerecke, M.; Samokhvalov, A.V.; Lönnroth, K.; Rehm, J. Alcohol consumption as a risk factor for tuberculosis: Meta-analyses and burden of disease. *Eur. Respir. J.* **2017**, *50*, 1700216. [CrossRef]
99. Francisco, J.; Oliveira, O.; Felgueiras, Ó.; Gaio, A.R.; Duarte, R. How much is too much alcohol in tuberculosis? *Eur. Respir. J.* **2017**, *49*, 1601468. [CrossRef] [PubMed]
100. Cords, O.; Martinez, L.; Warren, J.L.; O'Marr, J.M.; Walter, K.S.; Cohen, T.; Zheng, J.; Ko, A.I.; Croda, J.; Andrews, J.R. Incidence and prevalence of tuberculosis in incarcerated populations: A systematic review and meta-analysis. *Lancet Public Health* **2021**, *6*, e300–e308. [CrossRef]
101. Friedman, L.N.; Sullivan, G.M.; Bevilaqua, R.P.; Loscos, R. Tuberculosis screening in alcoholics and drug addicts. *Am. Rev. Respir. Dis.* **1987**, *136*, 1188–1192. [CrossRef] [PubMed]
102. Ragan, E.J.; Kleinman, M.B.; Sweigart, B.; Gnatienko, N.; Parry, C.D.; Horsburgh, C.R.; LaValley, M.P.; Myers, B.; Jacobson, K.R. The impact of alcohol use on tuberculosis treatment outcomes: A systematic review and meta-analysis. *Int. J. Tuberc. Lung. Dis.* **2020**, *24*, 73–82. [CrossRef] [PubMed]
103. Dos Santos, D.T.; Arroyo, L.H.; Alves, Y.M.; Alves, L.S.; Berra, T.Z.; Crispim, J.A.; Alves, J.D.; Ramos, D.A.C.; Alonso, J.B.; de Assis, I.S.; et al. Survival time among patients who were diagnosed with tuberculosis, the precocious deaths and associated factors in southern Brazil. *Trop. Med. Health* **2021**, *49*, 31. [CrossRef] [PubMed]
104. Workie, M.G.; Aycheh, M.W.; Birhanu, M.Y.; Tsegaye, T.B. Treatment interruption among drug-susceptible pulmonary tuberculosis patients in Southern Ethiopia. *Patient Prefer. Adherence* **2021**, *15*, 1143–1151. [CrossRef]
105. Arora, U.; Garg, P.; Agarwal, S.; Nischal, N.; Wig, N. Complexities in the treatment of coinfection with HIV, hepatitis B, hepatitis C, and tuberculosis. *Lancet Infect. Dis.* **2021**, *21*, e182. [CrossRef]
106. Thomas, B.E.; Thiruvengadam, K.; Kadam, D.; Ovung, S.; Sivakumar, S.; Bala Yogendra Shivakumar, S.V.; Paradkar, M.; Gupte, N.; Suryavanshi, N. Smoking, alcohol use disorder and tuberculosis treatment outcomes: A dual co-morbidity burden that cannot be ignored. *PLoS ONE* **2019**, *14*, e0220507. [CrossRef]
107. Kamarulzaman, A.; Reid, S.E.; Schwitters, A.; Wiessing, L.; El-Bassel, N.; Dolan, K.; Moazen, B.; Wirtz, A.L.; Verster, A.; Altice, F.L. Prevention of transmission of HIV, hepatitis B virus, hepatitis C virus, and tuberculosis in prisoners. *Lancet* **2016**, *388*, 1115–1126. [CrossRef]
108. Zelnick, J.R.; Daftary, A.; Hwang, C.; Labar, A.S.; Boodhram, R.; Maharaj, B.; Wolf, A.K.; Mondal, S.; Amico, K.R.; Orrell, C.; et al. Electronic dose monitoring identifies a high-risk subpopulation in the treatment of drug-resistant tuberculosis and HIV. *Clin. Infect. Dis.* **2020**. [CrossRef] [PubMed]
109. Myers, B.; Bouton, T.C.; Ragan, E.J.; White, L.F.; McIlleron, H.; Theron, D.; Parry, C.D.H.; Horsburgh, C.R.; Warren, R.M.; Jacobson, K.R. Impact of alcohol consumption on tuberculosis treatment outcomes: A prospective longitudinal cohort study protocol. *BMC Infect. Dis.* **2018**, *18*, 488. [CrossRef] [PubMed]

110. Louwagie, G.M.; Morojele, N.; Siddiqi, K.; Mdege, N.D.; Tumbo, J.; Omole, O.; Pitso, L.; Bachmann, M.O.; Ayo-Yusuf, O.A. Addressing tobacco smoking and drinking to improve TB treatment outcomes, in South Africa: A feasibility study of the ProLife program. *Transl. Behav. Med.* **2020**, *10*, 1491–1503. [CrossRef] [PubMed]

111. Moriarty, A.S.; Louwagie, G.M.; Mdege, N.D.; Morojele, N.; Tumbo, J.; Omole, O.B.; Bachmann, M.O.; Kanaan, M.; Turner, A.; Parrott, S.; et al. Improving TB outcomes by modifying life-style behaviours through a brief motivational intervention followed by short text messages (ProLife): Study protocol for a randomised controlled trial. *Trials* **2019**, *20*, 457. [CrossRef]

112. Myers, B.; Parry, C.D.H.; Morojele, N.K.; Nkosi, S.; Shuper, P.A.; Kekwaletswe, C.T.; Sorsdahl, K.R. "Moving forward with life": Acceptability of a brief alcohol reduction intervention for people receiving antiretroviral therapy in South Africa. *Int. J. Environ. Res. Public Health* **2020**, *17*, 5706. [CrossRef]

113. Magidson, J.F.; Satinsky, E.N.; Luberto, C.M.; Myers, B.; Funes, C.J.; Vanderkruik, R.; Andersen, L.S. "Cooling of the mind": Assessing the relevance of mindfulness training among people living with HIV using alcohol and other substances in South Africa. *Soc. Sci. Med.* **2020**, *266*, 113424. [CrossRef]

114. Greenfield, S.F.; Shields, A.; Connery, H.S.; Livchits, V.; Yanov, S.A.; Lastimoso, C.S.; Strelis, A.K.; Mishustin, S.P.; Fitzmaurice, G.; Mathew, T.A.; et al. Integrated management of physician-delivered alcohol care for tuberculosis patients: Design and implementation. *Alcohol. Clin. Exp. Res.* **2010**, *34*, 317–330. [CrossRef] [PubMed]

115. Calligaro, G.L.; de Wit, Z.; Cirota, J.; Orrell, C.; Myers, B.; Decker, S.; Stein, D.J.; Sorsdahl, K.; Dawson, R. Brief psychotherapy administered by non-specialised health workers to address risky substance use in patients with multidrug-resistant tuberculosis: A feasibility and acceptability study. *Pilot Feasibility Stud.* **2021**, *7*, 28. [CrossRef]

116. Peltzer, K.; Naidoo, P.; Louw, J.; Matseke, G.; Zuma, K.; Mchunu, G.; Tutshana, B.; Mabaso, M. Screening and brief interventions for hazardous and harmful alcohol use among patients with active tuberculosis attending primary public care clinics in South Africa: Results from a cluster randomized controlled trial. *BMC Public Health* **2013**, *13*, 699. [CrossRef]

117. Aiwale, A.S.; Patel, U.A.; Barvaliya, M.J.; Jha, P.R.; Tripathi, C. Isoniazid induced convulsions at therapeutic dose in an alcoholic and smoker patient. *Curr. Drug Saf.* **2015**, *10*, 94–95. [CrossRef]

118. Cross, F.S.; Long, M.W.; Banner, A.S.; Snider, D.E., Jr. Rifampin-isoniazid therapy of alcoholic and nonalcoholic tuberculous patients in a U.S. Public Health Service Cooperative Therapy Trial. *Am. Rev. Respir. Dis.* **1980**, *122*, 349–353. [CrossRef]

119. Karamanakos, P.N.; Pappas, P.; Boumba, V.; Vougiouklakis, T.; Marselos, M. The alcohol intolerance produced by isoniazid is not due to a disulfiram-like reaction despite aldehyde dehydrogenase inhibition. *Pharmacology* **2016**, *98*, 267–271. [CrossRef] [PubMed]

120. Pettit, A.C.; Bethel, J.; Hirsch-Moverman, Y.; Colson, P.W.; Sterling, T.R. Female sex and discontinuation of isoniazid due to adverse effects during the treatment of latent tuberculosis. *J. Infect.* **2013**, *67*, 424–432. [CrossRef]

121. Freiman, J.M.; Fatch, R.; Cheng, D.; Emenyonu, N.; Ngabirano, C.; Geadas, C.; Adong, J.; Muyindike, W.R.; Linas, B.P.; Jacobson, K.R.; et al. Prevalence of elevated liver transaminases and their relationship with alcohol use in people living with HIV on anti-retroviral therapy in Uganda. *PLoS ONE* **2021**, *16*, e0250368. [CrossRef] [PubMed]

122. Freiman, J.M.; Jacobson, K.R.; Muyindike, W.R.; Horsburgh, C.R.; Ellner, J.J.; Hahn, J.A.; Linas, B.P. Isoniazid preventive therapy for people with HIV who are heavy alcohol drinkers in high TB/HIV-burden countries: A risk-benefit analysis. *J. Acquir. Immune Defic. Syndr.* **2018**, *77*, 405–412. [CrossRef]

123. McClintock, A.H.; Eastment, M.; McKinney, C.M.; Pitney, C.L.; Narita, M.; Park, D.R.; Dhanireddy, S.; Molnar, A. Treatment completion for latent tuberculosis infection: A retrospective cohort study comparing 9 months of isoniazid, 4 months of rifampin and 3 months of isoniazid and rifapentine. *BMC Infect. Dis.* **2017**, *17*, 146. [CrossRef]

124. Mukherjee, T.I.; Hirsch-Moverman, Y.; Saito, S.; Gadisa, T.; Melaku, Z.; Howard, A.A. Determinants of alcohol use among people living with HIV initiating isoniazid preventive therapy in Ethiopia. *Drug Alcohol. Depend.* **2019**, *204*, 107465. [CrossRef] [PubMed]

125. Lodi, S.; Emenyonu, N.I.; Marson, K.; Kwarisiima, D.; Fatch, R.; McDonell, M.G.; Cheng, D.M.; Thirumurthy, H.; Gandhi, M.; Camlin, C.S.; et al. The Drinkers' Intervention to Prevent Tuberculosis (DIPT) trial among heavy drinkers living with HIV in Uganda: Study protocol of a 2 × 2 factorial trial. *Trials* **2021**, *22*, 355. [CrossRef] [PubMed]

126. Szabo, G.; Saha, B. Alcohol's effect on host defense. *Alcohol Res.* **2015**, *37*, 159–170. [PubMed]

127. Yeligar, S.M.; Chen, M.M.; Kovacs, E.J.; Sisson, J.H.; Burnham, E.L.; Brown, L.A. Alcohol and lung injury and immunity. *Alcohol* **2016**, *55*, 51–59. [CrossRef] [PubMed]

128. Pasala, S.; Barr, T.; Messaoudi, I. Impact of alcohol abuse on the adaptive immune system. *Alcohol Res.* **2015**, *37*, 185–197. [PubMed]

129. Haiqing, C.; Chen, L.; Yin, C.; Liao, Y.; Meng, X.; Lu, C.; Tang, S.; Li, X.; Wang, X. The effect of micro-nutrients on malnutrition, immunity and therapeutic effect in patients with pulmonary tuberculosis: A systematic review and meta-analysis of randomised controlled trials. *Tuberculosis* **2020**, *125*, 101994. [CrossRef]

130. Sinha, P.; Lönnroth, K.; Bhargava, A.; Heysell, S.K.; Sarkar, S.; Salgame, P.; Rudgard, W.; Boccia, D.; Van Aartsen, D.; Hochberg, N.S. Food for thought: Addressing undernutrition to end tuberculosis. *Lancet Infect. Dis.* **2021**. [CrossRef]

131. Enoh, J.E.; Cho, F.N.; Manfo, F.P.; Ako, S.E.; Akum, E.A. Abnormal levels of liver enzymes and hepatotoxicity in HIV-positive, TB, and HIV/TB-coinfected patients on treatment in Fako division, southwest region of Cameroon. *Biomed. Res. Int.* **2020**, *2020*, 9631731. [CrossRef]

132. Melikyan, N.; Huerga, H.; Atshemyan, H.; Kirakosyan, O.; Sargsyants, N.; Aydinyan, T.; Saribekyan, N.; Khachatryan, N.; Oganezova, I.; Falcao, J.; et al. Concomitant treatment of chronic hepatitis C with direct-acting antivirals and multidrug-resistant tuberculosis is effective and safe. *Open Forum. Infect. Dis.* **2021**, *8*, ofaa653. [CrossRef]

133. Yew, W.W.; Leung, C.C. Antituberculosis drugs and hepatotoxicity. *Respirology (Carlton Vic.)* **2006**, *11*, 699–707. [CrossRef]

134. Alghamdi, W.A.; Al-Shaer, M.H.; Kipiani, M.; Barbakadze, K.; Mikiashvili, L.; Kempker, R.R.; Peloquin, C.A. Pharmacokinetics of bedaquiline, delamanid and clofazimine in patients with multidrug-resistant tuberculosis. *J. Antimicrob. Chemother.* **2021**, *76*, 1019–1024. [CrossRef] [PubMed]

135. Torres, A.; Cilloniz, C.; Niederman, M.S.; Menéndez, R.; Chalmers, J.D.; Wunderink, R.G.; van der Poll, T. Pneumonia. *Nat. Rev. Dis. Primers* **2021**, *7*, 25. [CrossRef]

136. Global Health Data Exchange (GHDx). GBD Results Tool for the Global Burden of Disease 2019 Study. Available online: http://ghdx.healthdata.org/gbd-results-tool (accessed on 12 April 2020).

137. Shield, K.; Manthey, J.; Rylett, M.; Probst, C.; Wettlaufer, A.; Parry, C.D.H.; Rehm, J. National, regional, and global burdens of disease from 2000 to 2016 attributable to alcohol use: A comparative risk assessment study. *Lancet Public Health* **2020**, *5*, e51–e61. [CrossRef]

138. Torres, A.; Peetermans, W.E.; Viegi, G.; Blasi, F. Risk factors for community-acquired pneumonia in adults in Europe: A literature review. *Thorax* **2013**, *68*, 1057–1065. [CrossRef] [PubMed]

139. Shield, K.D.; Rehm, J. Societal development and the alcohol—Attributable burden of disease. *Addiction* **2021**, *116*, 2326–2338. [CrossRef] [PubMed]

140. Probst, C.; Roerecke, M.; Behrendt, S.; Rehm, J. Socioeconomic differences in alcohol-attributable mortality compared with all-cause mortality: A systematic review and meta-analysis. *Int. J. Epidemiol.* **2014**, *43*, 1314–1327. [CrossRef] [PubMed]

141. Probst, C.; Kilian, C.; Sanchez, S.; Lange, S.; Rehm, J. The role of alcohol use and drinking patterns in socioeconomic inequalities in mortality: A systematic review. *Lancet Public Health* **2020**, *5*, e324–e332. [CrossRef]

142. Babor, T.F.; Casswell, S.; Graham, K.; Huckle, T.; Livingston, M.; Österberg, E.; Rehm, J.; Room, R.; Rossow, I.; Sornpaisarn, B. *Alcohol: No Ordinary Commodity. Research and Public Policy*, 3rd ed.; Oxford University Press: Oxford, UK, 2021.

143. Bhatty, M.; Pruett, S.B.; Swiatlo, E.; Nanduri, B. Alcohol abuse and Streptococcus pneumoniae infections: Consideration of virulence factors and impaired immune responses. *Alcohol* **2011**, *45*, 523–539. [CrossRef]

144. Mehta, A.J.; Guidot, D.M. Alcohol and the lung. *Alcohol Res.* **2017**, *38*, 243–254.

145. Simet, S.M.; Sisson, J.H. Alcohol's effects on lung health and immunity. *Alcohol Res. Curr. Rev.* **2015**, *37*, 199.

146. Gamble, L.; Mason, C.; Nelson, S. The effects of alcohol on immunity and bacterial infection in the lung. *Med. Mal. Infect.* **2006**, *36*, 72–77. [CrossRef]

147. Heermans, E.H. Booze and blood: The effects of acute and chronic alcohol abuse on the hematopoietic system. *Clin. Lab. Sci.* **1998**, *11*, 229–232.

148. Nelson, S.; Zhang, P.; Bagby, G.J.; Happel, K.I.; Raasch, C.E. Alcohol abuse, immunosuppression, and pulmonary infection. *Curr. Drug Abuse Rev.* **2008**, *1*, 56–67. [CrossRef]

149. Szabo, G.; Mandrekar, P.; Catalano, D. Inhibition of superantigen-induced T cell proliferation and monocyte IL-1β, TNF-α, and IL-6 production by acute ethanol treatment. *J. Leukoc. Biol.* **1995**, *58*, 342–350. [CrossRef]

150. Samokhvalov, A.; Irving, H.; Rehm, J. Alcohol consumption as a risk factor for pneumonia: A systematic review and meta-analysis. *Epidemiol. Infect.* **2010**, *138*, 1789–1795. [CrossRef]

151. Simou, E.; Britton, J.; Leonardi-Bee, J. Alcohol and the risk of pneumonia: A systematic review and meta-analysis. *BMJ Open* **2018**, *8*, e022344. [CrossRef]

152. Rehm, J.; Anderson, P.; Gual, A.; Kraus, L.; Marmet, S.; Nutt, D.; Room, R.; Samokhvalov, A.; Scafato, E.; Shield, K. The tangible common denominator of substance use disorders: A reply to commentaries to Rehm et al. (2013a). *Alcohol Alcohol.* **2014**, *49*, 118–122. [CrossRef] [PubMed]

153. Schwarzinger, M.; Thiébaut, S.P.; Baillot, S.; Mallet, V.; Rehm, J. Alcohol use disorders and associated chronic disease: A national retrospective cohort study from France. *BMC Public Health* **2018**, *18*, 1–9. [CrossRef] [PubMed]

154. Fernández-Solá, J.; Junqué, A.; Estruch, R.; Monforte, R.; Torres, A.; Urbano-Márquez, A. High alcohol intake as a risk and prognostic factor for community-acquired pneumonia. *Arch. Intern. Med.* **1995**, *155*, 1649–1654. [CrossRef] [PubMed]

155. Koivula, I.; Sten, M.; Makela, P.H. Risk factors for pneumonia in the elderly. *Am. J. Med.* **1994**, *96*, 313–320. [CrossRef]

156. Luján, M.; Gallego, M.; Belmonte, Y.; Fontanals, D.; Valles, J.; Lisboa, T.; Rello, J. Influence of pneumococcal serotype group on outcome in adults with bacteraemic pneumonia. *Eur. Respir. J.* **2010**, *36*, 1073–1079. [CrossRef] [PubMed]

157. Trevejo-Nunez, G.; Kolls, J.K.; De Wit, M. Alcohol use as a risk factor in infections and healing: A clinician's perspective. *Alcohol Res.* **2015**, *37*, 177.

158. Gupta, N.M.; Deshpande, A.; Rothberg, M.B. Pneumonia and alcohol use disorder: Implications for treatment. *Cleve Clin. J. Med.* **2020**, *87*, 493–500. [CrossRef] [PubMed]

159. WHO Director General. WHO Director-General's Opening Remarks at the Media Briefing on COVID-19—11 March 2020. Available online: https://www.who.int/director-general/speeches/detail/who-director-general-s-opening-remarks-at-the-media-briefing-on-covid-19---11-march-2020 (accessed on 21 July 2021).

160. Worldometer. COVID-19 Coronavirus Pandemic. Available online: https://www.worldometers.info/coronavirus/ (accessed on 10 July 2021).

161. Islam, N.; Shkolnikov, V.M.; Acosta, R.J.; Klimkin, I.; Kawachi, I.; Irizarry, R.A.; Alicandro, G.; Khunti, K.; Yates, T.; Jdanov, D.A.; et al. Excess deaths associated with covid-19 pandemic in 2020: Age and sex disaggregated time series analysis in 29 high income countries. *BMJ* **2021**, *373*, n1137. [CrossRef]

162. Althobaiti, Y.S.; Alzahrani, M.A.; Alsharif, N.A.; Alrobaie, N.S.; Alsaab, H.O.; Uddin, M.N. The possible relationship between the abuse of tobacco, opioid, or alcohol with COVID-19. *Healthcare* **2020**, *9*, 2. [CrossRef]

163. Kendler, K.S.; Ohlsson, H.; Karriker-Jaffe, K.J.; Sundquist, J.; Sundquist, K. Social and economic consequences of alcohol use disorder: A longitudinal cohort and co-relative analysis. *Psychol. Med.* **2017**, *47*, 925–935. [CrossRef] [PubMed]

164. Buu, A.; Mansour, M.; Wang, J.; Refior, S.K.; Fitzgerald, H.E.; Zucker, R.A. Alcoholism effects on social migration and neighborhood effects on alcoholism over the course of 12 years. *Alcohol. Clin. Exp. Res.* **2007**, *31*, 1545–1551. [CrossRef] [PubMed]

165. Samuels-Kalow, M.E.; Dorner, S.; Cash, R.E.; Dutta, S.; White, B.; Ciccolo, G.E.; Brown, D.F.M.; Camargo, C.A., Jr. Neighborhood disadvantage measures and COVID-19 cases in Boston, 2020. *Public Health Rep.* **2021**, *136*, 368–374. [CrossRef] [PubMed]

166. Gurrieri, L.; Fairbairn, C.E.; Sayette, M.A.; Bosch, N. Alcohol narrows physical distance between strangers. *Proc. Natl. Acad. Sci. USA* **2021**, *118*, e2101937118. [CrossRef] [PubMed]

167. Osuchowski, M.F.; Winkler, M.S.; Skirecki, T.; Cajander, S.; Shankar-Hari, M.; Lachmann, G.; Monneret, G.; Venet, F.; Bauer, M.; Brunkhorst, F.M.; et al. The COVID-19 puzzle: Deciphering pathophysiology and phenotypes of a new disease entity. *Lancet Respir. Med.* **2021**, *9*, 622–642. [CrossRef]

168. Bailey, K.L.; Samuelson, D.R.; Wyatt, T.A. Alcohol use disorder: A pre-existing condition for COVID-19? *Alcohol* **2021**, *90*, 11–17. [CrossRef]

169. Vasudeva, A.; Patel, T.K. Alcohol consumption: An important epidemiological factor in COVID-19? *J. Glob. Health* **2020**, *10*, 020335. [CrossRef] [PubMed]

170. Huang, W.; Zhou, H.; Hodgkinson, C.; Montero, A.; Goldman, D.; Chang, S.L. Network meta-analysis on the mechanisms underlying alcohol augmentation of COVID-19 pathologies. *Alcohol. Clin. Exp. Res.* **2021**, *45*, 675–688. [CrossRef]

171. Wendt, F.R.; De Lillo, A.; Pathak, G.A.; De Angelis, F.; Polimanti, R. Host genetic liability for severe COVID-19 overlaps with alcohol drinking behavior and diabetic outcomes and in over 1 million participants. *medRxiv* **2020**, *12*. [CrossRef]

172. Benzano, D.; Ornell, F.; Schuch, J.B.; Pechansky, F.; Sordi, A.O.; von Diemen, L.; Kessler, F.H.P. Clinical vulnerability for severity and mortality by COVID-19 among users of alcohol and other substances. *Psychiatry Res.* **2021**, *300*, 113915. [CrossRef] [PubMed]

173. Fan, X.; Liu, Z.; Poulsen, K.L.; Wu, X.; Miyata, T.; Dasarathy, S.; Rotroff, D.M.; Nagy, L.E. Alcohol consumption is associated with poor prognosis in obese patients with COVID-19: A Mendelian randomization study using UK Biobank. *Nutrients* **2021**, *13*, 1592. [CrossRef] [PubMed]

174. Kushner, T.; Cafardi, J. Chronic liver disease and COVID-19: Alcohol use disorder/alcohol-associated liver disease, nonalcoholic fatty liver disease/nonalcoholic steatohepatitis, autoimmune liver disease, and compensated cirrhosis. *Clin. Liver Dis.* **2020**, *15*, 195–199. [CrossRef] [PubMed]

175. Moon, A.M.; Curtis, B.; Mandrekar, P.; Singal, A.K.; Verna, E.C.; Fix, O.K. Alcohol-associated liver disease before and after COVID-19: An overview and call for ongoing investigation. *Hepatol. Commun.* **2021**, *5*, 1616–1621. [CrossRef]

176. Cava, E.; Neri, B.; Carbonelli, M.G.; Riso, S.; Carbone, S. Obesity pandemic during COVID-19 outbreak: Narrative review and future considerations. *Clin. Nutr.* **2021**, *40*, 1637–1643. [CrossRef]

177. Bilal, B.; Saleem, F.; Fatima, S.S. Alcohol consumption and obesity: The hidden scare with COVID-19 severity. *Med. Hypotheses* **2020**, *144*, 110272. [CrossRef]

178. Alberca, R.W.; Rigato, P.O.; Ramos YÁ, L.; Teixeira, F.M.E.; Branco, A.C.C.; Fernandes, I.G.; Pietrobon, A.J.; Duarte, A.; Aoki, V.; Orfali, R.L.; et al. Clinical characteristics and survival analysis in frequent alcohol consumers with COVID-19. *Front. Nutr.* **2021**, *8*, 689296. [CrossRef]

179. Saurabh, S.; Verma, M.K.; Gautam, V.; Kumar, N.; Jain, V.; Goel, A.D.; Gupta, M.K.; Sharma, P.P.; Bhardwaj, P.; Singh, K.; et al. Tobacco, alcohol use and other risk factors for developing symptomatic COVID-19 vs asymptomatic SARS-CoV-2 infection: A case-control study from western Rajasthan, India. *Trans. R. Soc. Trop. Med. Hyg.* **2021**, *115*, 820–831. [CrossRef]

180. Varela Rodríguez, C.; Arias Horcajadas, F.; Martín-Arriscado Arroba, C.; Combarro Ripoll, C.; Juanes Gonzalez, A.; Esperesate Pajares, M.; Rodrigo Holgado, I.; Cadenas Manceñido, Á.; Sánchez Rodríguez, L.; Baselga Penalva, B.; et al. COVID-19-related neuropsychiatric symptoms in patients with alcohol abuse conditions during the SARS-CoV-2 pandemic: A retrospective cohort study using real world data from electronic health records of a tertiary hospital. *Front. Neurol.* **2021**, *12*, 630566. [CrossRef] [PubMed]

181. Wang, Q.Q.; Kaelber, D.C.; Xu, R.; Volkow, N.D. COVID-19 risk and outcomes in patients with substance use disorders: Analyses from electronic health records in the United States. *Mol. Psychiatry* **2021**, *26*, 30–39. [CrossRef] [PubMed]

182. Dai, M.; Tao, L.; Chen, Z.; Tian, Z.; Guo, X.; Allen-Gipson, D.S.; Tan, R.; Li, R.; Chai, L.; Ai, F.; et al. Influence of cigarettes and alcohol on the severity and death of COVID-19: A multicenter retrospective study in Wuhan, China. *Front. Physiol.* **2020**, *11*, 588553. [CrossRef]

183. Lv, L.; Zhou, Y.W.; Yao, R. Alcohol consumption and COVID-19 severity: A propensity score matched study in China. *Signa Vitae* **2021**, *17*, 112–120. [CrossRef]

184. Pro, G.; Gilbert, P.A.; Baldwin, J.A.; Brown, C.C.; Young, S.; Zaller, N. Multilevel modeling of county-level excessive alcohol use, rurality, and COVID-19 case fatality rates in the US. *PLoS ONE* **2021**, *16*, e0253466. [CrossRef] [PubMed]

185. Abbasi-Oshaghi, E.; Mirzaei, F.; Khodadadi, I. Alcohol misuse may increase the severity of COVID-19 infections. *Disaster. Med. Public Health Prep.* **2020**, *18*, 1–2. [CrossRef]

186. Saengow, U.; Assanangkornchai, S.; Casswell, S. Alcohol: A probable risk factor of COVID-19 severity. *Addiction* **2021**, *116*, 204–220. [CrossRef]

187. Testino, G. Are patients with alcohol use disorders at increased risk for COVID-19 infection? *Alcohol Alcohol.* **2020**, *55*, 344–346. [CrossRef]

188. Collin, S.M.; Wurie, F.; Muzyamba, M.C.; de Vries, G.; Lonnroth, K.; Migliori, G.B.; Abubakar, I.; Anderson, S.R.; Zenner, D. Effectiveness of interventions for reducing TB incidence in countries with low TB incidence: A systematic review of reviews. *Eur. Respir. Rev.* **2019**, *28*, 180107. [CrossRef] [PubMed]

189. Almirall, J.; Bolibar, I.; Serra-Prat, M. Risk factors for community-acquired pneumonia in adults: Recommendations for its prevention. *Community Acquir. Infect.* **2015**, *2*, 32. [CrossRef]

190. Pletz, M.W.; Rohde, G.G.; Welte, T.; Kolditz, M.; Ott, S. Advances in the prevention, management, and treatment of community-acquired pneumonia. *F1000Res* **2016**, *5*. [CrossRef]

191. Rivero-Calle, I.; Cebey-Lopez, M.; Pardo-Seco, J.; Yuste, J.; Redondo, E.; Vargas, D.A.; Mascaros, E.; Diaz-Maroto, J.L.; Linares-Rufo, M.; Jimeno, I.; et al. Lifestyle and comorbid conditions as risk factors for community-acquired pneumonia in outpatient adults (NEUMO-ES-RISK project). *BMJ Open Respir. Res.* **2019**, *6*, e000359. [CrossRef]

192. Carrasco, M.A.; Esser, M.B.; Sparks, A.; Kaufman, M.R. HIV-alcohol risk reduction interventions in sub-Saharan Africa: A systematic review of the literature and recommendations for a way forward. *AIDS Behav.* **2016**, *20*, 484–503. [CrossRef] [PubMed]

193. Morojele, N.K.; Ranchod, C. Review of interventions to reduce alcohol use-related sexual risk behaviour in Africa. *Afr. J. Drug Alcohol. Stud.* **2011**, *10*, 95–106.

194. Fritz, K.; McFarland, W.; Wyrod, R.; Chasakara, C.; Makumbe, K.; Chirowodza, A.; Mashoko, C.; Kellogg, T.; Woelk, G. Evaluation of a peer network-based sexual risk reduction intervention for men in beer halls in Zimbabwe: Results from a randomized controlled trial. *AIDS Behav.* **2011**, *15*, 1732–1744. [CrossRef]

195. Kalichman, S.C.; Simbayi, L.C.; Cain, D.; Carey, K.B.; Carey, M.P.; Eaton, L.; Harel, O.; Mehlomakhulu, V.; Mwaba, K. Randomized community-level HIV prevention intervention trial for men who drink in South African alcohol-serving venues. *Eur. J. Public Health* **2014**, *24*, 833–839. [CrossRef] [PubMed]

196. Morojele, N.K.; Kitleli, N.; Ngako, K.; Kekwaletswe, C.T.; Nkosi, S.; Fritz, K.; Parry, C.D. Feasibility and acceptability of a bar-based sexual risk reduction intervention for bar patrons in Tshwane, South Africa. *SAHARA J.* **2014**, *11*, 1–9. [CrossRef] [PubMed]

197. Raviglione, M.; Poznyak, V. Targeting harmful use of alcohol for prevention and treatment of tuberculosis: A call for action. *Eur. Respir. J.* **2017**, *50*. [CrossRef] [PubMed]

198. Shin, S.; Livchits, V.; Connery, H.S.; Shields, A.; Yanov, S.; Yanova, G.; Fitzmaurice, G.M.; Nelson, A.K.; Greenfield, S.F.; Tomsk Tuberculosis Alcohol Working Group. Effectiveness of alcohol treatment interventions integrated into routine tuberculosis care in Tomsk, Russia. *Addiction* **2013**, *108*, 1387–1396. [CrossRef] [PubMed]

199. Mehta, A.J. Alcoholism and critical illness: A review. *World J. Crit. Care Med.* **2016**, *5*, 27–35. [CrossRef]

200. Brown, J.L.; DeMartini, K.S.; Sales, J.M.; Swartzendruber, A.L.; DiClemente, R.J. Interventions to reduce alcohol use among HIV-infected individuals: A review and critique of the literature. *Curr. HIV/AIDS Rep.* **2013**, *10*, 356–370. [CrossRef]

201. Madhombiro, M.; Musekiwa, A.; January, J.; Chingono, A.; Abas, M.; Seedat, S. Psychological interventions for alcohol use disorders in people living with HIV/AIDS: A systematic review. *Syst. Rev.* **2019**, *8*, 244. [CrossRef]

202. Samet, J.H.; Walley, A.Y. Interventions targeting HIV-infected risky drinkers: Drops in the bottle. *Alcohol Res. Health* **2010**, *33*, 267–279. [PubMed]

203. Regenauer, K.S.; Myers, B.; Batchelder, A.W.; Magidson, J.F. "That person stopped being human": Intersecting HIV and substance use stigma among patients and providers in South Africa. *Drug Alcohol. Depend.* **2020**, *216*, 108322. [CrossRef] [PubMed]

204. Kalichman, S.C.; Katner, H.; Hill, M.; Kalichman, M.O.; Hernandez, D. Alcohol-related intentional antiretroviral nonadherence among people living with HIV: Test of an Interactive Toxicity Beliefs Process Model. *J. Int. Assoc. Provid. AIDS Care* **2019**, *18*, 2325958219826612. [CrossRef]

205. World Health Organization. *Tackling NCDs: 'Best Buys' and Other Recommended Interventions for the Prevention and Control of Noncommunicable Diseases*; World Health Organization: Geneva, Switzerland, 2017.

206. Ferreira-Borges, C.; Esser, M.B.; Dias, S.; Babor, T.; Parry, C.D. Alcohol control policies in 46 African countries: Opportunities for improvement. *Alcohol Alcohol.* **2015**, *50*, 470–476. [CrossRef]

207. Gururaj, G.; Gautham, M.S.; Arvind, B.A. Alcohol consumption in India: A rising burden and a fractured response. *Drug Alcohol Rev.* **2021**, *40*, 368–384. [CrossRef] [PubMed]

208. Medina-Mora, M.E.; Monteiro, M.; Rafful, C.; Samano, I. Comprehensive analysis of alcohol policies in the Latin America and the Caribbean. *Drug Alcohol Rev.* **2021**, *40*, 385–401. [CrossRef]

209. Morojele, N.K.; Dumbili, E.W.; Obot, I.S.; Parry, C.D.H. Alcohol consumption, harms and policy developments in sub-Saharan Africa: The case for stronger national and regional responses. *Drug Alcohol Rev.* **2021**, *40*, 402–419. [CrossRef] [PubMed]

nutrients

MDPI

Review

Alcohol's Impact on the Gut and Liver

Keith Pohl [1,2,†], Prebashan Moodley [1,2,†] and Ashwin D. Dhanda [1,2,*]

1   South West Liver Unit, University Hospitals Plymouth NHS Trust, Plymouth PL6 8DH, UK; k.pohl@nhs.net (K.P.); p.moodley@nhs.net (P.M.)
2   Hepatology Research Group, Faculty of Health, University of Plymouth, Plymouth PL4 8AA, UK
*   Correspondence: ashwin.dhanda@plymouth.ac.uk; Tel.: +44-1752-432723
†   These authors contributed equally to this work.

Abstract: Alcohol is inextricably linked with the digestive system. It is absorbed through the gut and metabolised by hepatocytes within the liver. Excessive alcohol use results in alterations to the gut microbiome and gut epithelial integrity. It contributes to important micronutrient deficiencies including short-chain fatty acids and trace elements that can influence immune function and lead to liver damage. In some people, long-term alcohol misuse results in liver disease progressing from fatty liver to cirrhosis and hepatocellular carcinoma, and results in over half of all deaths from chronic liver disease, over half a million globally per year. In this review, we will describe the effect of alcohol on the gut, the gut microbiome and liver function and structure, with a specific focus on micronutrients and areas for future research.

Keywords: alcohol; gut; liver; cirrhosis; hepatocellular carcinoma; microbiome

Citation: Pohl, K.; Moodley, P.; Dhanda, A.D. Alcohol's Impact on the Gut and Liver. *Nutrients* **2021**, *13*, 3170. https://doi.org/10.3390/nu13093170

Academic Editor: Peter Anderson

Received: 23 August 2021
Accepted: 7 September 2021
Published: 11 September 2021

Publisher's Note: MDPI stays neutral with regard to jurisdictional claims in published maps and institutional affiliations.

1. Introduction

Alcohol (ethanol) is a small water-soluble molecule that enters the blood stream via the stomach and proximal small intestine and is then distributed throughout the body. It first enters the portal vein, which drains directly into the liver, where the greatest exposure to alcohol occurs. The liver eliminates the majority of alcohol (90%), while 2–5% is excreted unchanged in urine, sweat and breath [1].

Alcohol consumption is engrained in many cultures, making alcohol the most commonly used drug worldwide. However, it is not without risk. Globally, alcohol is the seventh leading cause of death and disability-associated life years (DALYs) lost, and it caused 2.8 million deaths in 2016 [2]. In that year, among adults less than 50 years old, alcohol was the leading cause of death and DALYs lost, responsible for 3.8% and 12.2% of female and male deaths, respectively [2]. Alcohol is causally implicated in over 200 conditions, including cancers of the digestive tract and liver [3]. However, a large proportion of the global burden is due to alcohol-related liver disease, accounting for 27% of all deaths from chronic liver disease, and alcohol-related hepatocellular carcinoma, together responsible for over half a million deaths annually [4,5].

Here, we review the effect of alcohol on the gut and liver, focusing on its interaction with micronutrients.

2. Alcohol and the Gut

The pathological effects of alcohol on the digestive system hinge in part on the gut–liver axis. This bi-directional relationship facilitated by the enterohepatic circulation involves the transportation of digestive and bacterial products from the gut to the liver, and the return of bile, antibodies and cytokines to the gut [6]. Alcohol ingestion in both chronic and 'binge' settings has been shown to alter this axis through the disruption of gut microbial composition, the metabolome and the gut epithelial barrier. These disturbances ultimately have a knock-on effect on nutrient absorption [7,8].

2.1. The Effect of Alcohol on Microbial Composition and Gut Barrier Function

The human gut microbiome describes a complex community of bacteria, viruses, fungi and archaea, which varies both with environmental factors (such as diet and drugs) and age, but less so with host genetics [9–12]. Disruption of the microbiome (dysbiosis) has been linked with a wide range of conditions including diabetes, obesity, cardiovascular disease, inflammatory bowel disease and liver cirrhosis; however, it is unclear whether this is a cause or effect relationship [13,14]. Several studies have investigated the effect of alcohol consumption in both animal and human models, and have consistently shown that alcohol consumption is linked with the development of dysbiosis [13,15]. In brief, alcohol has been shown to increase the relative abundance of *Proteobacteria*, *Enterobacteriacea* and *Streptococcus* and decrease the abundance of *Bacteroides*, *Akkermansia* and *Faecalibacterium* [15]. The aetiological mechanism of this dysbiosis is not fully understood; however, it is likely that this is multifaceted, including alcohol-induced oxidative stress (which is poorly tolerated by obligate anaerobes such as *Bacteroides*) and the downregulation of antibacterial peptides such as α-defensins by alcohol [16,17].

Alcohol-induced dysbiosis contributes to the development of both acute (e.g., alcoholic hepatitis) and chronic (e.g., alcohol-related cirrhosis) liver diseases through its pathological effect on gut integrity. The intestinal mucous barrier has an essential role in the immune function of the gut, and its disruption leads to these disease states. In this barrier, neighbouring enterocytes are bound together by the apical 'tight junction' proteins claudins, occludin and zona occludens, preventing the unwanted translocation of luminal contents such as pathogen-associated molecular particles (PAMPs) and bacterial endotoxins into the portal circulation [18]. Dysbiosis induced by alcohol consumption has been linked to the disruption of these tight junctions. As a consequence, the subsequent immune dysfunction and increase in circulating pro-inflammatory cytokines such as tumour necrosis factor (TNF)-α and interleukin (IL)-1β further disrupts the gut barrier [18,19].

2.2. The Effect of Alcohol on the Metabolome

Alcohol-related dysbiosis inevitably affects the gut metabolome, and dramatic alterations in short-chain fatty acids (SCFAs), amino acids and bile acids have been documented.

The role of SCFAs in the maintenance of tight junctions is becoming increasingly apparent. SCFAs are fatty acids with fewer than six carbon atoms, and are the product of the anaerobic fermentation of indigestible dietary fibres by the gut microbiota [20]. Analysis of the faecal metabolome in humans with alcohol use disorders revealed a reduction in SCFAs, which is likely to be due in part to dysbiosis that negatively affects SCFA-producing bacteria such as *Faecalibacterium* [15,21]. Several murine models have reliably shown that supplementation with SCFAs in either the form of a high fibre diet, probiotic or dietary modification enhances gut epithelial integrity and reduces liver injury in alcoholic models, and work in this area is ongoing [22–24].

The gut metabolome also plays an important part in the metabolism and absorption of essential and non-essential amino acids, which appear to be altered by alcohol ingestion. Less work has been conducted to investigate this effect. However, several studies have identified that alcohol consumption lowers the concentration of almost all amino acids in the gut lumen [12,14,25]. Both essential, dietary-obtained amino acids (e.g., lysine) and non-essential amino acids (e.g., glutamic acid) are affected. It is postulated that this is a result of a disturbed microbial–host co-metabolism as a result of dysbiosis [14]. Although luminal amino acid concentrations fall with alcohol consumption, serum levels of some, such as tyrosine and phenylalanine, rise, suggesting an altered metabolic and absorption profile of the dysbiotic microbiome [14,26]. This metabolic imbalance may play a role in the generation of increased levels of reactive oxygen species (ROS) and toxic intermediates.

Bile acids have been shown to be altered in both the serum and luminal contents of humans and rats consuming alcohol [12,14,27]. Primary (synthesised by the liver) and secondary (from bacterial metabolism) bile acids perform a variety of functions predominantly in the small bowel and have crucial roles in lipid absorption, cholesterol homeostasis as well as hormonal actions through their steroid structure. In a healthy entero-hepatic circulation, primary bile acids are conjugated with either taurine or glycine to form bile salts that are secreted into the intestinal lumen. The intestinal microbiota then metabolises these to secondary bile acids, removing the taurine/glycine groups before recycling them back to the liver. Alcohol consumption appears to disrupt this by increasing the proportion of secondary bile acids and the total concentration of bile acids, as well as increasing the proportion conjugated with glycine instead of taurine [12,14]. It is felt that this is caused by dysbiosis decreasing the bioavailability of taurine and an increased rate of entero-hepatic cycling [12,14]. The consequence of this disruption is not fully understood; however, it is likely that the glycine-conjugated acids that are more prevalent during alcohol consumption are relatively more toxic, and the increased synthesis of bile acids despite high luminal concentrations contributes to hepatic steatosis [12].

2.3. The Effect of Alcohol Consumption on Nutritional Status

Chronic alcohol ingestion reduces nutrient absorption and contributes to malnutrition [28]. Alterations in intestinal permeability, bile acid profiles and the microbiome all contribute to this, and in addition, the toxic metabolites and ROS released during alcohol metabolism cause structural damage to the intestine. In particular, chronic alcohol use has been shown to cause cell death, mucosal erosions and the loss of epithelium at the villi tips [29]. The consequences of this are variable deficiencies in vitamins A, B1 (thiamine), B2 (riboflavin), B6 (pyridoxine), C, D, E and K as well as folate, calcium, magnesium, phosphate, iron and the trace elements zinc and selenium [28,30]. It is important that all patients with chronic alcohol use disorders undergo a full nutritional assessment as these deficiencies vary between individuals, with iron as an example that can be either deficient or found in excess. Alongside the mechanisms described above, heavy alcohol users obtain up to 50 percent of their daily caloric intake from nutritionally deplete alcoholic drinks [8]. Furthermore, it should be noted that alongside the symptomatic effects of chronic alcohol misuse (e.g., vomiting, anorexia and abdominal pain), social factors in this group such as poverty and access to a nutritionally 'complete' diet may also contribute to malnutrition [31].

The effect of alcohol on the gut is summarised in Figure 1.

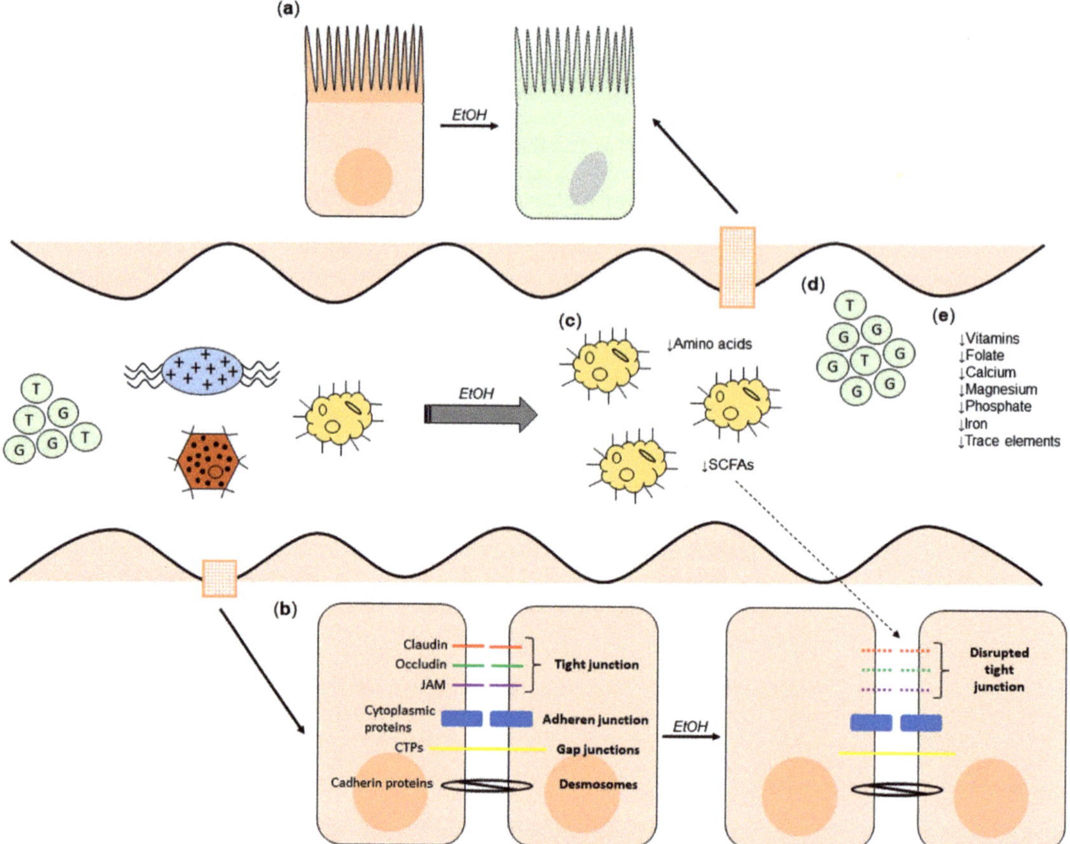

Figure 1. The effect of alcohol on the gut. (**a**) The histological effects of alcohol on the gut mucosa (cell death, mucosal erosions and loss of epithelium at villi tips). (**b**) Alcohol-induced disruption of tight junctions, exacerbated by reduced luminal SCFA concentrations. (**c**) Alcohol-induced dysbiosis leading to reduced SCFA and amino acid concentrations. (**d**) Increased concentration of secondary bile acids, and increased proportion conjugated with glycine. (**e**) Nutrient deficiencies as a consequence of (**a–d**). CTP: Connexin transmembrane proteins; JAM: Junctional adhesion molecule; EtOH: alcohol.

3. Alcohol and the Liver

3.1. Alcohol Metabolism

The metabolism of alcohol in the liver is key to understanding its role in the pathogenesis of alcohol-related liver disease. Alcohol is primarily metabolised in hepatocytes by alcohol dehydrogenase to acetaldehyde and then to acetate by aldehyde dehydrogenase. Acetate is converted to water and carbon dioxide mainly in peripheral tissue, which is easily excreted. A minority of alcohol is metabolised by the mitochondrial enzyme oxidation system (MEOS), through the action of the cytochrome P450 (CYP) enzyme CYP2E1, to acetaldehyde with the generation of ROS. A third minor pathway of alcohol metabolism to acetaldehyde is by the action of catalase and the conversion of H_2O_2 to H_2O.

It is the generation of acetaldehyde, a highly reactive protein, which contributes to liver damage. It binds to lipids, proteins and DNA to form potentially immunogenic adducts [32]. These adducts can generate an adaptive immune response leading to hepatocellular damage and inflammation [33]. Structural mitochondrial alteration can lead to functional impairment including decreased ATP generation, the production of ROS and decreased activity of acetaldehyde dehydrogenase. Acetaldehyde is also a key metabolite in the progression of liver fibrosis. It can promote the synthesis of collagen I in hepatic stellate cells (HSCs), and acetaldehyde adducts stimulate the release of inflammatory cytokines and chemokines [33].

The alcohol dehydrogenase pathway is efficient in metabolising alcohol in small quantities, but in chronic alcohol exposure, the pathway becomes saturated and there is significant induction of CYP2E1 [32]. The switch to the CYP pathway results in the generation of ROS, leading to oxidative stress. ROS bind to proteins, changing their structural and functional properties, and may act as neoantigens. ROS can also bind directly to DNA, causing damage, or lead to lipid peroxidation products such as 4-hydroxynonenal (4-HNE) and malondialdehyde (MDA) that generate highly carcinogenic DNA adducts (Figure 2) [34]. In addition, in chronic heavy alcohol ingestion, the antioxidant clearing system of the liver is impaired because of an acetaldehyde-mediated decrease in glutathione. The outcome of oxidative stress is the induction of hepatocyte apoptosis and necrosis [35].

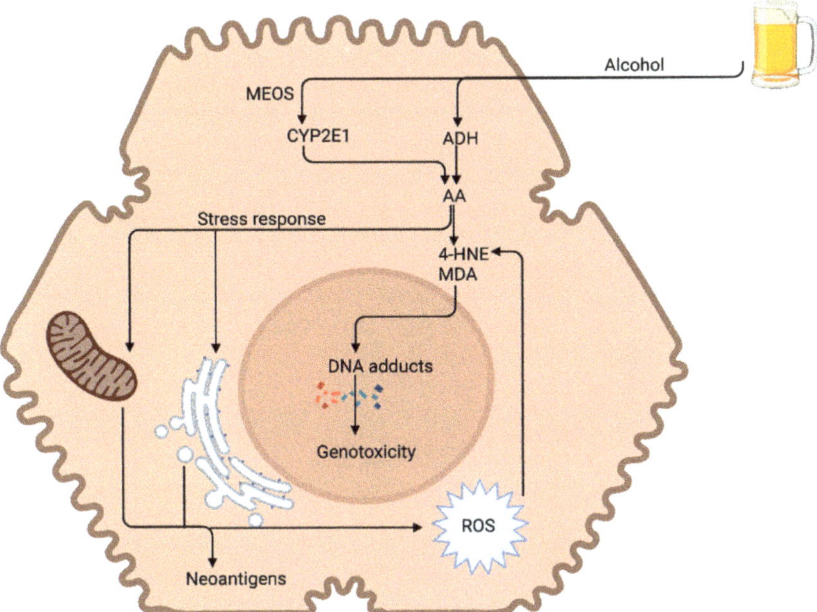

Figure 2. Alcohol-induced liver injury. Acetaldehyde (AA) is responsible for the majority of the toxic effects of alcohol on the liver. Acetaldehyde is extremely lipophilic, leading to the formation of acetaldehyde adducts—malondialdehyde (MDA) and 4-hydroxynonenal (4-HNE). This along with reactive oxygen species (ROS) leads to DNA damage and genotoxicity. Acetaldehyde also induces functional and structural alterations in various cell organelles (e.g., mitochondria and endoplasmic reticulum). MEOS: mitochondrial enzyme oxidation system; ADH: alcohol dehydrogenase. Image created at biorender.com (accessed on 20 August 2021).

3.2. Alcohol-Related Steatosis

Steatosis, characterised by the accumulation of fat (triglycerides, phospholipids and cholesterol esters) in hepatocytes, is the earliest response of the liver to chronic alcohol use and is almost universal in chronic heavy drinkers [36]. Although it is fully reversible upon a reduction in alcohol use, its presence is associated with the progression of alcohol-related liver disease, with a recent meta-analysis finding an annual progression rate to cirrhosis of 3% [37]. It is likely that hepatic steatosis increases the risk of liver inflammation (steatohepatitis), fibrosis and cirrhosis through greater lipid peroxidation and oxidative stress. However, progression only occurs in up to 20% and is not only influenced by the amount of alcohol but also other factors, including gender, co-existing liver disease, smoking and genetics [38].

Chronic alcohol ingestion leads to hepatic steatosis via increased hepatic lipogenesis and decreased hepatic lipolysis. Alcohol elevates the ratio of reduced NAD/oxidised NAD in hepatocytes, which interferes with mitochondrial beta oxidation of fatty acids, leading to their accumulation in hepatocytes [39]. Chronic alcohol use induces hepatic expression of sterol regulatory element binding protein-1c (SREBP1c), a transcription factor that stimulates the expression of lipogenic genes, resulting in increased fatty acid synthesis [40]. Alcohol also downregulates inhibitors of SREBP1c expression such as AMP-activated protein kinase (AMPK), Sirtuin-1, adiponectin and signal transducer and activator of transcription 3 (STAT3) [41]. Conversely, chronic alcohol use enhances adipose tissue breakdown and lipolysis, releasing free fatty acids, which are esterified in hepatocytes into triglycerides [42].

Alcohol inactivates peroxisome proliferator-activated-receptor (PPAR)-α, a nuclear hormone receptor that upregulates the expression of many genes involved in free fatty acid transport and oxidation. Acetaldehyde directly inhibits transcriptional activation activity and DNA binding of PPAR-α [43]. Alcohol also indirectly inhibits PPAR-α via CYP2E1-derived oxidative stress, adenosine, the downregulation of adiponectin and zinc deficiency (a common state in patients with alcohol-related liver disease) [30,44]. The inactivation of PPAR-α results in reduced hepatic lipolysis (Figure 3).

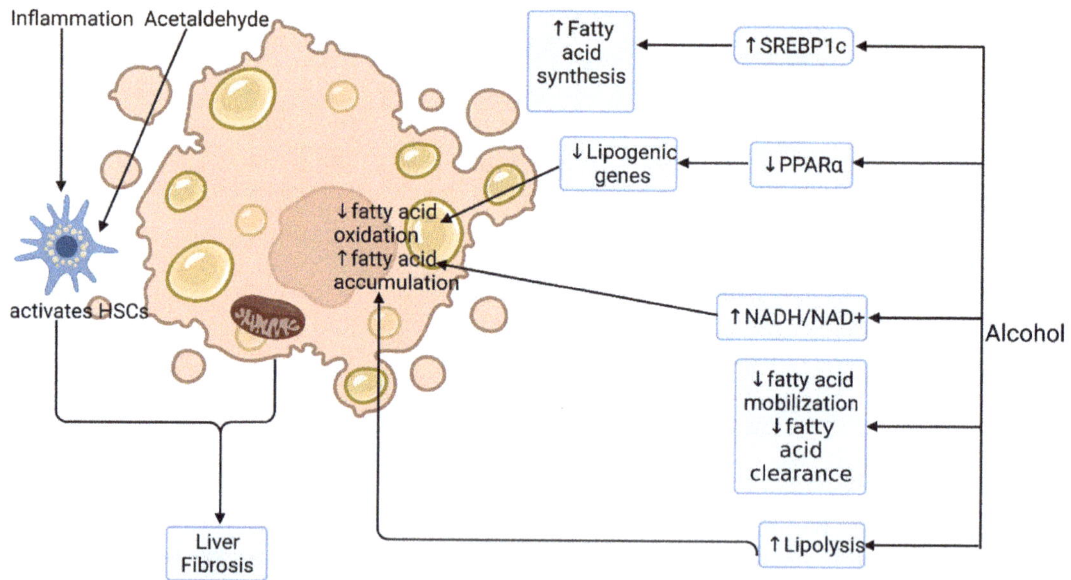

Figure 3. Alcohol-induced steatosis. Alcohol induces hepatic steatosis by multiple mechanisms. It alters the redox ratio in the cell (NADH/NAD+), thereby inhibiting fatty acid oxidation and promoting its accumulation. It increases transcription factor SREBP1c, which leads to increased fatty acid synthesis and deposition. Alcohol inactivates PPARα, a nuclear hormone receptor that regulates many of the genes involved in fatty acid transport and oxidation. Alcohol has a direct inhibitory effect on fatty acid clearance and mobilisation. ↑: increased; ↓: decreased; HSC: hepatic stellate cell. Image created at biorender.com (accessed on 20 August 2021).

3.3. Alcoholic Steatohepatitis

Hepatic inflammation strongly influences the development of fibrosis, cirrhosis and ultimately hepatocellular carcinoma. The alcohol-induced leaky gut (as described above) leads to the delivery of PAMPs to the liver. Together with damage-associated molecular patterns, released from damaged cells, PAMPs activate innate receptors (Toll-like receptors (TLRs) and NOD-like receptors (NLRs)) on monocytes, macrophages, Kupffer cells and hepatic parenchymal cells. Signalling through these receptors leads to increased transcription of pro-inflammatory transcription factors including nuclear-factor κB (NFκB) and the production of pro-inflammatory chemokines and cytokines (reviewed in detail in ref [45]). The net effect is the influx of monocytes, neutrophils and T cells, the release of soluble mediators that cause cell death and hepatic stellate cell (HSC) activation (Figure 4). In addition to an activated pro-inflammatory immune response to alcohol, patients with alcoholic hepatitis have evidence of immune dysfunction. The activation of monocytes by gut-derived PAMPs leads to T cell exhaustion with reduced numbers of anti-inflammatory IL-10 producing T cells and functionally impaired monocytes and neutrophils [46,47].

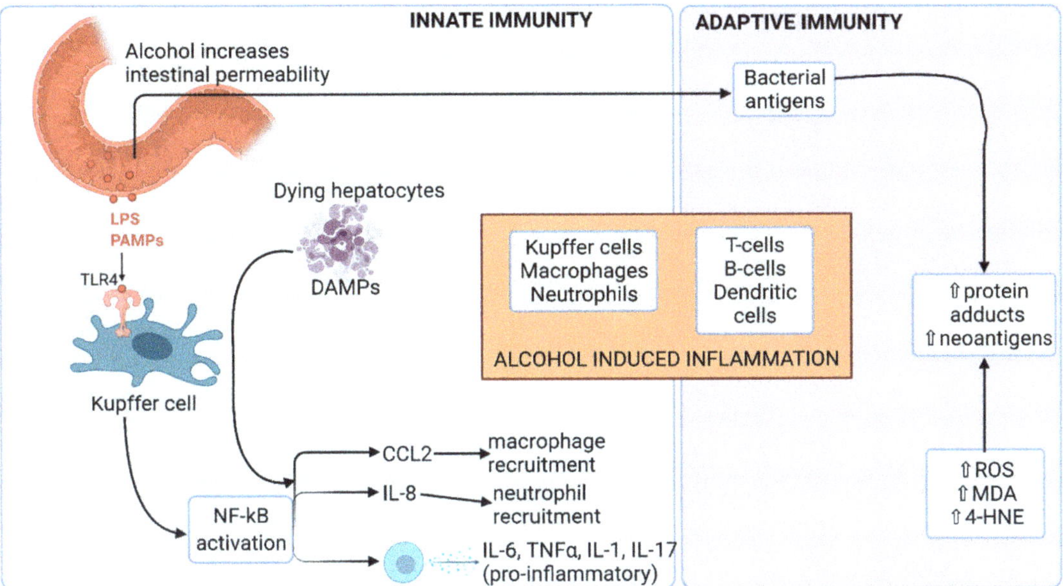

Figure 4. Alcohol-induced inflammation. Alcohol exerts its effects on both the innate and adaptive immunity. Alcohol not only induces enteric dysbiosis, but also increases intestinal permeability. Pathogen-associated molecular patterns (PAMPs) such as lipopolysaccharide (LPS) interact with TLR4 receptor on Kupffer cells and produce proinflammatory cytokines and chemokines via the NF-κB pathway, leading to liver inflammation. Acetaldehyde induces structural changes in various proteins and generates neoantigens, which elicit an adaptive immune response and contribute to liver inflammation. CCL2: C-C motif chemokine ligand 2; DAMPs: damage-associated molecular patterns; 4-HNE: 4-hydroxynonenal; IL: interleukin; MDA: malondialdehyde; NF-κB: nuclear factor kappa B; ROS: reactive oxygen species; TLR4: toll-like receptor 4; TNFα: tumor necrosis factor alpha; ↑: increased; ↓: decreased. Image created at biorender.com (accessed on 20 August 2021).

Hepatocyte cell death occurs through several mechanisms including apoptosis, pyroptosis, necrosis and necroptosis [48]. Apoptosis is induced by direct alcohol-mediated hepatotoxicity, the induction of oxidative stress, the inhibition of survival genes (*C-met*) and the induction of pro-apoptotic signalling molecules (TNF-α and Fas ligand) [49]. Necrosis, cell swelling and membrane rupture can also occur via a programmed pathway known as necroptosis, while pyroptosis is a programmed cell death dependent on caspase-1. The mode of cell death is likely to be influenced by the disease state, with apoptosis predominating in early alcohol-related liver disease but inflammasome activation driving pyroptosis and propagating liver injury in alcoholic hepatitis [50].

MicroRNAs (miRNAs) are small non-coding RNAs that have a role in the post-transcriptional regulation of their target genes. Two key miRNAs are differentially expressed in patients with alcohol-related liver disease. miRNA-155, a key regulator of inflammation, is increased in the liver and circulation in mouse models of alcohol-related liver disease [51]. Chronic alcohol consumption increases the expression of miRNA-155 in Kupffer cells, which contributes to increased LPS-triggered TNF production [52]. miR-181b-3p, a negative regulator of TLR4 signalling in Kupffer cells, is downregulated in patients with alcohol-related liver disease [53].

3.4. Alcohol-Induced Fibrosis and Cirrhosis

Fibrosis is the liver's wound healing response to a damaging stimulus, reversible on removal of the stimulus. In the presence of heavy long-term alcohol consumption, there is chronic inflammation and fibrogenesis causing the deposition of broad bands of fibrous tissue, distorting the liver architecture and altering hepatic blood flow, leading to portal hypertension and its associated complications. Extracellular matrix deposition by activated HSCs is the key event in the development and progression of liver fibrosis. Other cells (portal fibroblasts and myofibroblasts) contribute to a smaller extent [54]. HSCs are activated both by inflammatory cytokines and directly by alcohol and its metabolites and ROS. Activated HSCs perpetuate the inflammatory response by the secretion of chemokines and the expression of adhesion molecules that attract and stimulate circulating immune cells, which in turn activate quiescent HSCs [55].

3.5. Hepatocellular Carcinoma

Cirrhosis is a precancerous state increasing the risk of primary liver cancer, the commonest being hepatocellular carcinoma (HCC). Globally, around 30% of HCCs are due to alcohol. Alcohol itself is a carcinogen and in the context of HCC plays specific roles in its development through ROS-induced damage, inflammatory mechanisms and its reactive metabolite, acetaldehyde.

In heavy alcohol drinkers, increased activity of the CYP pathway generates ROS, leading to DNA damage that causes cell cycle arrest and apoptosis and disrupts gene function, increasing carcinogenesis and propagation [56]. The activation of inflammatory pathways in patients with alcohol-related liver disease is associated with increased cancer risk, although mechanisms have not been fully elucidated but are likely to involve pro-inflammatory cytokine promotion of ROS accumulation [57,58]. Cytokine generation is also associated with the upregulation of the angiogenesis and metastasis development [59]. Additionally, alcohol suppresses the anti-tumour response of CD8+ T cells [60].

Acetaldehyde is highly reactive and forms adducts with DNA and proteins that cause mitochondrial damage and disruption of DNA repair mechanisms. Increased levels of acetaldehyde found in people with genetic variations that confer altered activity of alcohol dehydrogenase and aldehyde dehydrogenase are associated with a higher risk of HCC in heavy drinkers [61].

4. Research Priorities and Future Perspectives

The human body is able to metabolise and eliminate small volumes of alcohol without long-term sequelae. However, excessive alcohol use leads to alterations in the gut microbiome, metabolome, epithelial integrity and immune signalling culminating in progressive liver disease. Our recent improved understanding of these changes has identified potential new therapies to delay or reverse liver disease. Here, we suggest areas in need of further research and potential strategies for therapy.

4.1. Microbiome

Alcohol's effect on the microbiome is highly heterogeneous with multiple genera up- or downregulated, making a single probiotic treatment challenging to identify. Probiotics marginally improve liver function tests in patients with liver disease [62], but few trials in patients with alcohol-related liver disease have been conducted and they have failed to demonstrate any clinical benefit in patients with cirrhosis [63]. However, combination therapy to restore gut dysbiosis may be a more successful strategy.

Faecal microbiota transplant has recently been piloted as a treatment for patients with alcohol use disorder [64]. Healthy donor stool transplant by nasogastric injection was associated with a partial improvement in bacterial diversity and reversal in dysbiosis, an improvement in gut SCFA production and reduced severity of alcohol use disorder [64]. Although not designed to treat or prevent cirrhosis, this therapy may prove beneficial in this patient population.

4.2. Short-Chain Fatty Acids

SCFAs are a product of the bacterial digestion of dietary fibre and are essential to maintain gut epithelial integrity. Strategies to increase intestinal SCFA may reduce gut permeability and exposure of the liver to gut-derived toxins, thus preventing the progression of liver disease. Studies of SCFA treatment have yet to be conducted in patients with alcohol-related liver disease. Delivery of the SCFA butyrate by enema reduced gut oxidative stress and inflammation in patients with inflammatory bowel disease [65]. Indirect methods to increase SCFAs such as by faecal microbiota transplant or augmenting SCFA-producing bacteria with specific probiotics (e.g., *Clostridium butyricum*) may also be beneficial. In a randomised trial, *C. butyricum* in combination with *Bifidobacterium infantis* reduced symptoms of minimal hepatic encephalopathy in patients with hepatitis B cirrhosis as well as measures of gut permeability [66]. However, supplementation with other probiotics (*Bifidobacterium*, *Lactobacillus* and *Lactococcus* genera) did not improve gut barrier function [67], suggesting the importance of targeting SCFA-producing species.

Interestingly, aerobic fitness also influences the microbiome and response to dietary fat or alcohol. An animal study demonstrated that high-aerobically fit rats fed a high fat diet were protected from steatosis [68]. Conversely, low-aerobically fit rats had reduced SCFA-producing bacteria and altered microbiome metabolism of carbohydrates and energy, leading to hepatic steatosis [69]. A combined study enhancing both aerobic fitness and SCFA-producing bacteria in patients with alcohol-related liver disease may yield beneficial results.

4.3. Dietary Manipulation

In a cohort study comparing the gut microbiome in patients with cirrhosis from the USA and Turkey, several dietary factors were associated with a reduced risk of progression mediated by increased microbial diversity [70]. Fermented milk, vegetables, cereals, coffee and tea were associated with greater diversity, while carbonated drinks, pork and poultry were associated with lower diversity. Increased intake of insoluble fibre may also increase the bacterial production of SCFAs. Trials have been conducted in patients with hepatic encephalopathy demonstrating the safety of a high protein and high fibre diet [71,72], but none has yet been conducted evaluating the effects of long-term clinical outcomes. Further carefully designed trials of dietary manipulation are required in patients with alcohol-related liver disease.

4.4. Micronutrient Supplementation

Patients with alcohol-related liver disease are deficient in micronutrients including zinc and selenium [30]. Both these trace elements are essential for cellular and immune function [73,74] and their deficiency is associated with more advanced disease [75] and mortality from alcoholic hepatitis [30]. Supplementation may improve clinical outcomes through the modulation of immune function, but existing studies are too small and heterogeneous to demonstrate improved survival [76]. An ongoing trial of zinc supplementation in patients with alcohol-related cirrhosis (NCT02072746) demonstrated an improvement in liver inflammation in an interim analysis [77].

Vitamin deficiencies, especially of vitamin C and D, in patients with alcohol-related liver disease are common. Vitamin supplementation may modulate the gut microbiome resulting in reduced dysbiosis, which may reduce gut permeability and liver inflammation; studies in healthy individuals have demonstrated beneficial shifts in microbial genera with vitamin C [78] and D [79]. Although studies of antioxidant and vitamin supplementation in patients with severe alcoholic hepatitis did not demonstrate a short-term survival benefit [80], trials of long-term vitamin supplementation in patients with less acute presentations of alcohol-related liver disease are lacking.

4.5. Immune Dysregulation

Patients with alcohol-related liver disease have simultaneous immune activation and exhaustion [19]. Interventions targeting inflammation broadly with corticosteroids [81] or specific pro-inflammatory cytokines such as TNF-α [82] have been investigated in severe forms of alcohol-related liver disease with alcoholic hepatitis. However, increased mortality from infective complications was noted, demonstrating the delicate balance of the immune response in these patients [83]. Strategies to target immune exhaustion may promote liver repair mechanisms at the same time as maintaining defence against pathogens. Checkpoint inhibitors such as anti-PD-1 monoclonal antibodies improve the host immune response and have been licenced for the treatment of cancers. Such treatment may reduce PAMP-induced CD8+ T cell exhaustion [47] and improve the healing response in alcohol-related liver disease.

4.6. Repurposing of Existing Therapies

Clinical trials of commonly used drugs such as metformin, pioglitazone and statins have shown benefit on liver biochemistry or histological changes in patients with non-alcoholic steatohepatitis [84–86]. Benefits are mainly mediated by a reduction in insulin resistance, a mechanism that is not a strong contributor to alcohol-related liver disease. However, these drugs also alter the gut microbiome [87]; metformin, for example, increases the abundance of SCFA-producing bacteria [88]. Further trials of these agents are required to evaluate their benefit in the setting of alcohol-related liver disease.

5. Conclusions

In conclusion, alcohol has wide-ranging effects on the gut and liver, altering the gut microbiome, barrier function and immune function resulting in liver inflammation, fibrosis and cirrhosis. Trials of new interventional strategies to target gut dysbiosis and immune dysfunction are now required.

Author Contributions: Conceptualization, A.D.D.; writing—original draft preparation, K.P. and P.M.; writing—review and editing, A.D.D. All authors have read and agreed to the published version of the manuscript.

Funding: This research received no external funding.

Institutional Review Board Statement: Not applicable.

Informed Consent Statement: Not applicable.

Conflicts of Interest: The authors declare no conflict of interest.

References

1. Paton, A. Alcohol in the body. *BMJ* **2005**, *330*, 85–87. [CrossRef]
2. GBD Alcohol Drug Use Collaborators. The global burden of disease attributable to alcohol and drug use in 195 countries and territories, 1990–2016: A systematic analysis for the Global Burden of Disease Study 2016. *Lancet Psychiatry* **2018**, *5*, 987–1012. [CrossRef]
3. World Health Organization. Alcohol Factsheet. Available online: https://www.who.int/en/news-room/fact-sheets/detail/alcohol (accessed on 20 August 2021).
4. GBD Causes of Death Collaborators. Global, regional, and national age-sex specific mortality for 264 causes of death, 1980–2016: A systematic analysis for the global burden of disease study 2016. *Lancet* **2017**, *390*, 1151–1210. [CrossRef]

5. Global Burden of Disease Liver Cancer Collaboration; Akinyemiju, T.; Abera, S.; Ahmed, M.; Alam, N.; Alemayohu, M.A.; Allen, C.; Al-Raddadi, R.; Alvis-Guzman, N.; Amoako, Y.; et al. The burden of primary liver cancer and underlying etiologies from 1990 to 2015 at the global, regional, and national level: Results from the global burden of disease study 2015. *JAMA Oncol.* **2017**, *3*, 1683–1691. [CrossRef]

6. Albillos, A.; de Gottardi, A.; Rescigno, M. The gut-liver axis in liver disease: Pathophysiological basis for therapy. *J. Hepatol.* **2020**, *72*, 558–577. [CrossRef] [PubMed]

7. Kamran, U.; Towey, J.; Khanna, A.; Chauhan, A.; Rajoriya, N.; Holt, A. Nutrition in alcohol-related liver disease: Physiopathology and management. *World J. Gastroenterol. WJG* **2020**, *26*, 2916–2930. [CrossRef] [PubMed]

8. Lieber, C.S. Relationships between nutrition, alcohol use, and liver disease. *Alcohol Res. Health* **2003**, *27*, 220–231. [PubMed]

9. Human Microbiome Project Consortium. Structure, function and diversity of the healthy human microbiome. *Nature* **2012**, *486*, 207–214. [CrossRef] [PubMed]

10. Rothschild, D.; Weissbrod, O.; Barkan, E.; Kurilshikov, A.; Korem, T.; Zeevi, D.; Costea, P.I.; Godneva, A.; Kalka, I.N.; Bar, N.; et al. Environment dominates over host genetics in shaping human gut microbiota. *Nature* **2018**, *555*, 210–215. [CrossRef]

11. David, L.A.; Maurice, C.F.; Carmody, R.N.; Gootenberg, D.B.; Button, J.E.; Wolfe, B.E.; Ling, A.V.; Devlin, A.S.; Varma, Y.; Fischbach, M.A.; et al. Diet rapidly and reproducibly alters the human gut microbiome. *Nature* **2014**, *505*, 559–563. [CrossRef] [PubMed]

12. Bajaj, J.S. Alcohol, liver disease and the gut microbiota. *Nat. Rev. Gastroenterol. Hepatol.* **2019**, *16*, 235–246. [CrossRef] [PubMed]

13. Engen, P.A.; Green, S.J.; Voigt, R.M.; Forsyth, C.B.; Keshavarzian, A. The gastrointestinal microbiome: Alcohol effects on the composition of intestinal microbiota. *Alcohol Res.* **2015**, *37*, 223–236. [PubMed]

14. Zhong, W.; Zhou, Z. Alterations of the gut microbiome and metabolome in alcoholic liver disease. *World J. Gastrointest. Pathophysiol.* **2014**, *5*, 514–522. [CrossRef] [PubMed]

15. Litwinowicz, K.; Choroszy, M.; Waszczuk, E. Changes in the composition of the human intestinal microbiome in alcohol use disorder: A systematic review. *Am. J. Drug Alcohol Abus.* **2020**, *46*, 4–12. [CrossRef]

16. Keshavarzian, A.; Farhadi, A.; Forsyth, C.B.; Rangan, J.; Jakate, S.; Shaikh, M.; Banan, A.; Fields, J.Z. Evidence that chronic alcohol exposure promotes intestinal oxidative stress, intestinal hyperpermeability and endotoxemia prior to development of alcoholic steatohepatitis in rats. *J. Hepatol.* **2009**, *50*, 538–547. [CrossRef]

17. Shukla, P.K.; Meena, A.S.; Rao, V.; Rao, R.G.; Balazs, L.; Rao, R. Human defensin-5 blocks ethanol and colitis-induced dysbiosis, tight junction disruption and inflammation in mouse intestine. *Sci. Rep.* **2018**, *8*, 16241. [CrossRef]

18. Starkel, P.; Leclercq, S.; de Timary, P.; Schnabl, B. Intestinal dysbiosis and permeability: The yin and yang in alcohol dependence and alcoholic liver disease. *Clin. Sci.* **2018**, *132*, 199–212. [CrossRef]

19. Dhanda, A.D.; Collins, P.L. Immune dysfunction in acute alcoholic hepatitis. *World J. Gastroenterol. WJG* **2015**, *21*, 11904–11913. [CrossRef]

20. Silva, Y.P.; Bernardi, A.; Frozza, R.L. The role of short-chain fatty acids from gut microbiota in gut-brain communication. *Front. Endocrinol.* **2020**, *11*, 25. [CrossRef]

21. Couch, R.D.; Dailey, A.; Zaidi, F.; Navarro, K.; Forsyth, C.B.; Mutlu, E.; Engen, P.A.; Keshavarzian, A. Alcohol induced alterations to the human fecal VOC metabolome. *PLoS ONE* **2015**, *10*, e0119362. [CrossRef]

22. Cresci, G.A.; Bush, K.; Nagy, L.E. Tributyrin supplementation protects mice from acute ethanol-induced gut injury. *Alcohol. Clin. Exp. Res.* **2014**, *38*, 1489–1501. [CrossRef]

23. Cresci, G.A.; Glueck, B.; McMullen, M.R.; Xin, W.; Allende, D.; Nagy, L.E. Prophylactic tributyrin treatment mitigates chronic-binge ethanol-induced intestinal barrier and liver injury. *J. Gastroenterol. Hepatol.* **2017**, *32*, 1587–1597. [CrossRef]

24. Jiang, X.W.; Li, Y.T.; Ye, J.Z.; Lv, L.X.; Yang, L.Y.; Bian, X.Y.; Wu, W.R.; Wu, J.J.; Shi, D.; Wang, Q.; et al. New strain of Pediococcus pentosaceus alleviates ethanol-induced liver injury by modulating the gut microbiota and short-chain fatty acid metabolism. *World J. Gastroenterol. WJG* **2020**, *26*, 6224–6240. [CrossRef] [PubMed]

25. Wei, X.; Jiang, S.; Zhao, X.; Li, H.; Lin, W.; Li, B.; Lu, J.; Sun, Y.; Yuan, J. Community-metabolome correlations of gut microbiota from child-turcotte-pugh of A and B patients. *Front. Microbiol.* **2016**, *7*, 1856. [CrossRef]

26. Voutilainen, T.; Karkkainen, O. Changes in the human metabolome associated with alcohol use: A review. *Alcohol Alcohol.* **2019**, *54*, 225–234. [CrossRef] [PubMed]

27. Kakiyama, G.; Hylemon, P.B.; Zhou, H.; Pandak, W.M.; Heuman, D.M.; Kang, D.J.; Takei, H.; Nittono, H.; Ridlon, J.M.; Fuchs, M.; et al. Colonic inflammation and secondary bile acids in alcoholic cirrhosis. *Am. J. Physiol. Gastrointest. Liver Physiol.* **2014**, *306*, G929–G937. [CrossRef]

28. Lieber, C.S. ALCOHOL: Its metabolism and interaction with nutrients. *Annu. Rev. Nutr.* **2000**, *20*, 395–430. [CrossRef]

29. Bishehsari, F.; Magno, E.; Swanson, G.; Desai, V.; Voigt, R.M.; Forsyth, C.B.; Keshavarzian, A. Alcohol and gut-derived inflammation. *Alcohol Res.* **2017**, *38*, 163–171. [PubMed]

30. Dhanda, A.; Atkinson, S.; Vergis, N.; Enki, D.; Fisher, A.; Clough, R.; Cramp, M.; Thursz, M. Trace element deficiency is highly prevalent and associated with infection and mortality in patients with alcoholic hepatitis. *Aliment. Pharmacol. Ther.* **2020**, *52*, 537–544. [CrossRef] [PubMed]

31. Katikireddi, S.V.; Whitley, E.; Lewsey, J.; Gray, L.; Leyland, A.H. Socioeconomic status as an effect modifier of alcohol consumption and harm: Analysis of linked cohort data. *Lancet Public Health* **2017**, *2*, e267. [CrossRef]

32. Cederbaum, A.I.; Lu, Y.; Wu, D. Role of oxidative stress in alcohol-induced liver injury. *Arch. Toxicol.* **2009**, *83*, 519–548. [CrossRef]

33. Mello, T.; Ceni, E.; Surrenti, C.; Galli, A. Alcohol induced hepatic fibrosis: Role of acetaldehyde. *Mol. Asp. Med.* **2008**, *29*, 17–21. [CrossRef] [PubMed]

34. Wang, Y.; Millonig, G.; Nair, J.; Patsenker, E.; Stickel, F.; Mueller, S.; Bartsch, H.; Seitz, H.K. Ethanol-induced cytochrome P4502E1 causes carcinogenic etheno-DNA lesions in alcoholic liver disease. *Hepatology* **2009**, *50*, 453–461. [CrossRef] [PubMed]

35. Tan, H.K.; Yates, E.; Lilly, K.; Dhanda, A.D. Oxidative stress in alcohol-related liver disease. *World J. Hepatol.* **2020**, *12*, 332–349. [CrossRef] [PubMed]

36. Lieber, C.S. Alcoholic fatty liver: Its pathogenesis and mechanism of progression to inflammation and fibrosis. *Alcohol* **2004**, *34*, 9–19. [CrossRef] [PubMed]

37. Parker, R.; Aithal, G.P.; Becker, U.; Gleeson, D.; Masson, S.; Wyatt, J.I.; Rowe, I.A.; WALDO Study Group. Natural history of histologically proven alcohol-related liver disease: A systematic review. *J. Hepatol.* **2019**, *71*, 586–593. [CrossRef] [PubMed]

38. Seitz, H.K.; Bataller, R.; Cortez-Pinto, H.; Gao, B.; Gual, A.; Lackner, C.; Mathurin, P.; Mueller, S.; Szabo, G.; Tsukamoto, H. Alcoholic liver disease. *Nat. Rev. Dis. Primers* **2018**, *4*, 16. [CrossRef]

39. Baraona, E.; Lieber, C.S. Effects of ethanol on lipid metabolism. *J. Lipid Res.* **1979**, *20*, 289–315. [CrossRef]

40. You, M.; Fischer, M.; Deeg, M.A.; Crabb, D.W. Ethanol induces fatty acid synthesis pathways by activation of sterol regulatory element-binding protein (SREBP). *J. Biol. Chem.* **2002**, *277*, 29342–29347. [CrossRef]

41. Purohit, V.; Gao, B.; Song, B.J. Molecular mechanisms of alcoholic fatty liver. *Alcohol. Clin. Exp. Res.* **2009**, *33*, 191–205. [CrossRef]

42. Wei, X.; Shi, X.; Zhong, W.; Zhao, Y.; Tang, Y.; Sun, W.; Yin, X.; Bogdanov, B.; Kim, S.; McClain, C.; et al. Chronic alcohol exposure disturbs lipid homeostasis at the adipose tissue-liver axis in mice: Analysis of triacylglycerols using high-resolution mass spectrometry in combination with in vivo metabolite deuterium labeling. *PLoS ONE* **2013**, *8*, e55382. [CrossRef] [PubMed]

43. Li, H.H.; Tyburski, J.B.; Wang, Y.W.; Strawn, S.; Moon, B.H.; Kallakury, B.V.; Gonzalez, F.J.; Fornace, A.J., Jr. Modulation of fatty acid and bile acid metabolism by peroxisome proliferator-activated receptor alpha protects against alcoholic liver disease. *Alcohol. Clin. Exp. Res.* **2014**, *38*, 1520–1531. [CrossRef] [PubMed]

44. Galli, A.; Pinaire, J.; Fischer, M.; Dorris, R.; Crabb, D.W. The transcriptional and DNA binding activity of peroxisome proliferator-activated receptor alpha is inhibited by ethanol metabolism. A novel mechanism for the development of ethanol-induced fatty liver. *J. Biol. Chem.* **2001**, *276*, 68–75. [CrossRef] [PubMed]

45. Gao, B.; Ahmad, M.F.; Nagy, L.E.; Tsukamoto, H. Inflammatory pathways in alcoholic steatohepatitis. *J. Hepatol.* **2019**, *70*, 249–259. [CrossRef] [PubMed]

46. Dhanda, A.D.; Williams, E.L.; Yates, E.; Lait, P.J.P.; Schewitz-Bowers, L.P.; Hegazy, D.; Cramp, M.E.; Collins, P.L.; Lee, R.W.J. Intermediate monocytes in acute alcoholic hepatitis are functionally activated and induce IL-17 expression in CD4(+) T cells. *J. Immunol.* **2019**, *203*, 3190–3198. [CrossRef] [PubMed]

47. Markwick, L.J.; Riva, A.; Ryan, J.M.; Cooksley, H.; Palma, E.; Tranah, T.H.; Manakkat Vijay, G.K.; Vergis, N.; Thursz, M.; Evans, A.; et al. Blockade of PD1 and TIM3 restores innate and adaptive immunity in patients with acute alcoholic hepatitis. *Gastroenterology* **2015**, *148*, 590–602. [CrossRef]

48. Wang, S.; Pacher, P.; De Lisle, R.C.; Huang, H.; Ding, W.X. A mechanistic review of cell death in alcohol-induced liver injury. *Alcohol. Clin. Exp. Res.* **2016**, *40*, 1215–1223. [CrossRef]

49. Feldstein, A.E.; Gores, G.J. Apoptosis in alcoholic and nonalcoholic steatohepatitis. *Front. Biosci. A J. Virtual Libr.* **2005**, *10*, 3093–3099. [CrossRef]

50. Gaul, S.; Leszczynska, A.; Alegre, F.; Kaufmann, B.; Johnson, C.D.; Adams, L.A.; Wree, A.; Damm, G.; Seehofer, D.; Calvente, C.J.; et al. Hepatocyte pyroptosis and release of inflammasome particles induce stellate cell activation and liver fibrosis. *J. Hepatol.* **2021**, *74*, 156–167. [CrossRef]

51. Bala, S.; Csak, T.; Saha, B.; Zatsiorsky, J.; Kodys, K.; Catalano, D.; Satishchandran, A.; Szabo, G. The pro-inflammatory effects of miR-155 promote liver fibrosis and alcohol-induced steatohepatitis. *J. Hepatol.* **2016**, *64*, 1378–1387. [CrossRef]

52. Bala, S.; Csak, T.; Kodys, K.; Catalano, D.; Ambade, A.; Furi, I.; Lowe, P.; Cho, Y.; Iracheta-Vellve, A.; Szabo, G. Alcohol-induced miR-155 and HDAC11 inhibit negative regulators of the TLR4 pathway and lead to increased LPS responsiveness of Kupffer cells in alcoholic liver disease. *J. Leukoc. Biol.* **2017**, *102*, 487–498. [CrossRef]

53. Saikia, P.; Bellos, D.; McMullen, M.R.; Pollard, K.A.; de la Motte, C.; Nagy, L.E. MicroRNA 181b-3p and its target importin alpha5 regulate toll-like receptor 4 signaling in Kupffer cells and liver injury in mice in response to ethanol. *Hepatology* **2017**, *66*, 602–615. [CrossRef] [PubMed]

54. Tsuchida, T.; Friedman, S.L. Mechanisms of hepatic stellate cell activation. *Nat. Rev. Gastroenterol. Hepatol.* **2017**, *14*, 397–411. [CrossRef] [PubMed]

55. Friedman, S.L. Hepatic stellate cells: Protean, multifunctional, and enigmatic cells of the liver. *Physiol. Rev.* **2008**, *88*, 125–172. [CrossRef] [PubMed]

56. Goldar, S.; Khaniani, M.S.; Derakhshan, S.M.; Baradaran, B. Molecular mechanisms of apoptosis and roles in cancer development and treatment. *Asian Pac. J. Cancer Prev.* **2015**, *16*, 2129–2144. [CrossRef] [PubMed]

57. Park, E.J.; Lee, J.H.; Yu, G.Y.; He, G.; Ali, S.R.; Holzer, R.G.; Osterreicher, C.H.; Takahashi, H.; Karin, M. Dietary and genetic obesity promote liver inflammation and tumorigenesis by enhancing IL-6 and TNF expression. *Cell* **2010**, *140*, 197–208. [CrossRef]

58. Chiba, T.; Marusawa, H.; Ushijima, T. Inflammation-associated cancer development in digestive organs: Mechanisms and roles for genetic and epigenetic modulation. *Gastroenterology* **2012**, *143*, 550–563. [CrossRef] [PubMed]

59. Hagymasi, K.; Blazovics, A.; Lengyel, G.; Kocsis, I.; Feher, J. Oxidative damage in alcoholic liver disease. *Eur. J. Gastroenterol. Hepatol.* **2001**, *13*, 49–53. [CrossRef] [PubMed]
60. Yan, G.; Wang, X.; Sun, C.; Zheng, X.; Wei, H.; Tian, Z.; Sun, R. Chronic alcohol consumption promotes diethylnitrosamine-induced hepatocarcinogenesis via immune disturbances. *Sci. Rep.* **2017**, *7*, 2567. [CrossRef]
61. Sakamoto, T.; Hara, M.; Higaki, Y.; Ichiba, M.; Horita, M.; Mizuta, T.; Eguchi, Y.; Yasutake, T.; Ozaki, I.; Yamamoto, K.; et al. Influence of alcohol consumption and gene polymorphisms of ADH2 and ALDH2 on hepatocellular carcinoma in a Japanese population. *Int. J. Cancer* **2006**, *118*, 1501–1507. [CrossRef]
62. Khalesi, S.; Johnson, D.W.; Campbell, K.; Williams, S.; Fenning, A.; Saluja, S.; Irwin, C. Effect of probiotics and synbiotics consumption on serum concentrations of liver function test enzymes: A systematic review and meta-analysis. *Eur. J. Nutr.* **2018**, *57*, 2037–2053. [CrossRef]
63. Li, F.; Duan, K.; Wang, C.; McClain, C.; Feng, W. Probiotics and alcoholic liver disease: Treatment and potential mechanisms. *Gastroenterol. Res. Pract.* **2016**, *2016*, 5491465. [CrossRef] [PubMed]
64. Bajaj, J.S.; Gavis, E.A.; Fagan, A.; Wade, J.B.; Thacker, L.R.; Fuchs, M.; Patel, S.; Davis, B.; Meador, J.; Puri, P.; et al. A randomized clinical trial of fecal microbiota transplant for alcohol use disorder. *Hepatology* **2021**, *73*, 1688–1700. [CrossRef] [PubMed]
65. Hamer, H.M.; Jonkers, D.M.; Vanhoutvin, S.A.; Troost, F.J.; Rijkers, G.; de Bruine, A.; Bast, A.; Venema, K.; Brummer, R.J. Effect of butyrate enemas on inflammation and antioxidant status in the colonic mucosa of patients with ulcerative colitis in remission. *Clin. Nutr.* **2010**, *29*, 738–744. [CrossRef] [PubMed]
66. Xia, X.; Chen, J.; Xia, J.; Wang, B.; Liu, H.; Yang, L.; Wang, Y.; Ling, Z. Role of probiotics in the treatment of minimal hepatic encephalopathy in patients with HBV-induced liver cirrhosis. *J. Int. Med. Res.* **2018**, *46*, 3596–3604. [CrossRef] [PubMed]
67. Horvath, A.; Leber, B.; Schmerboeck, B.; Tawdrous, M.; Zettel, G.; Hartl, A.; Madl, T.; Stryeck, S.; Fuchs, D.; Lemesch, S.; et al. Randomised clinical trial: The effects of a multispecies probiotic vs. placebo on innate immune function, bacterial translocation and gut permeability in patients with cirrhosis. *Aliment. Pharmacol. Ther.* **2016**, *44*, 926–935. [CrossRef]
68. Morris, E.M.; Jackman, M.R.; Johnson, G.C.; Liu, T.W.; Lopez, J.L.; Kearney, M.L.; Fletcher, J.A.; Meers, G.M.; Koch, L.G.; Britton, S.L.; et al. Intrinsic aerobic capacity impacts susceptibility to acute high-fat diet-induced hepatic steatosis. American journal of physiology. *Endocrinol. Metab.* **2014**, *307*, E355–E364. [CrossRef]
69. Panasevich, M.R.; Morris, E.M.; Chintapalli, S.V.; Wankhade, U.D.; Shankar, K.; Britton, S.L.; Koch, L.G.; Thyfault, J.P.; Rector, R.S. Gut microbiota are linked to increased susceptibility to hepatic steatosis in low-aerobic-capacity rats fed an acute high-fat diet. *Am. J. Physiol. Gastrointest. Liver Physiol.* **2016**, *311*, G166–G179. [CrossRef] [PubMed]
70. Bajaj, J.S.; Idilman, R.; Mabudian, L.; Hood, M.; Fagan, A.; Turan, D.; White, M.B.; Karakaya, F.; Wang, J.; Atalay, R.; et al. Diet affects gut microbiota and modulates hospitalization risk differentially in an international cirrhosis cohort. *Hepatology* **2018**, *68*, 234–247. [CrossRef] [PubMed]
71. Ruiz-Margain, A.; Macias-Rodriguez, R.U.; Rios-Torres, S.L.; Roman-Calleja, B.M.; Mendez-Guerrero, O.; Rodriguez-Cordova, P.; Torre, A. Effect of a high-protein, high-fiber diet plus supplementation with branched-chain amino acids on the nutritional status of patients with cirrhosis. *Rev. Gastroenterol. Mex.* **2018**, *83*, 9–15. [CrossRef]
72. Liu, Q.; Duan, Z.P.; Ha, D.K.; Bengmark, S.; Kurtovic, J.; Riordan, S.M. Synbiotic modulation of gut flora: Effect on minimal hepatic encephalopathy in patients with cirrhosis. *Hepatology* **2004**, *39*, 1441–1449. [CrossRef]
73. Burk, R.F.; Hill, K.E.; Motley, A.K.; Byrne, D.W.; Norsworthy, B.K. Selenium deficiency occurs in some patients with moderate-to-severe cirrhosis and can be corrected by administration of selenate but not selenomethionine: A randomized controlled trial. *Am. J. Clin. Nutr.* **2015**, *102*, 1126–1133. [CrossRef] [PubMed]
74. Maret, W. Zinc and human disease. *Met. Ions Life Sci.* **2013**, *13*, 389–414. [CrossRef]
75. Nangliya, V.; Sharma, A.; Yadav, D.; Sunder, S.; Nijhawan, S.; Mishra, S. Study of trace elements in liver cirrhosis patients and their role in prognosis of disease. *Biol. Trace Elem. Res.* **2015**, *165*, 35–40. [CrossRef]
76. Tan, H.K.; Streeter, A.; Cramp, M.E.; Dhanda, A.D. Effect of zinc treatment on clinical outcomes in patients with liver cirrhosis: A systematic review and meta-analysis. *World J. Hepatol.* **2020**, *12*, 389–398. [CrossRef] [PubMed]
77. Mohammad, M.; Song, M.; Falkner, K.; McClain, C.; Cave, M. Zinc sulfate for alcoholic cirrhosis (ZAC) clinical trial—Interim analysis of liver injury/inflammation biomarkers. *Hepatology* **2014**, *60*, 794A.
78. Otten, A.T.; Bourgonje, A.R.; Peters, V.; Alizadeh, B.Z.; Dijkstra, G.; Harmsen, H.J.M. Vitamin C supplementation in healthy individuals leads to shift of bacterial populations in the gut—A pilot study. *Antioxidants* **2021**, *10*, 1278. [CrossRef] [PubMed]
79. Singh, P.; Rawat, A.; Alwakeel, M.; Sharif, E.; Al Khodor, S. The potential role of vitamin D supplementation as a gut microbiota modifier in healthy individuals. *Sci. Rep.* **2020**, *10*, 21641. [CrossRef]
80. Phillips, M.; Curtis, H.; Portmann, B.; Donaldson, N.; Bomford, A.; O'Grady, J. Antioxidants versus corticosteroids in the treatment of severe alcoholic hepatitis–a randomised clinical trial. *J. Hepatol.* **2006**, *44*, 784–790. [CrossRef]
81. Thursz, M.R.; Richardson, P.; Allison, M.; Austin, A.; Bowers, M.; Day, C.P.; Downs, N.; Gleeson, D.; MacGilchrist, A.; Grant, A.; et al. Prednisolone or pentoxifylline for alcoholic hepatitis. *N. Engl. J. Med.* **2015**, *372*, 1619–1628. [CrossRef]
82. Naveau, S.; Chollet-Martin, S.; Dharancy, S.; Mathurin, P.; Jouet, P.; Piquet, M.A.; Davion, T.; Oberti, F.; Broet, P.; Emilie, D.; et al. A double-blind randomized controlled trial of infliximab associated with prednisolone in acute alcoholic hepatitis. *Hepatology* **2004**, *39*, 1390–1397. [CrossRef] [PubMed]

83. Vergis, N.; Atkinson, S.R.; Knapp, S.; Maurice, J.; Allison, M.; Austin, A.; Forrest, E.H.; Masson, S.; McCune, A.; Patch, D.; et al. In patients with severe alcoholic hepatitis, prednisolone increases susceptibility to infection and infection-related mortality, and is associated with high circulating levels of bacterial DNA. *Gastroenterology* **2017**, *152*, 1068–1077.e1064. [CrossRef] [PubMed]
84. Kumar, J.; Memon, R.S.; Shahid, I.; Rizwan, T.; Zaman, M.; Menezes, R.G.; Kumar, S.; Siddiqi, T.J.; Usman, M.S. Antidiabetic drugs and non-alcoholic fatty liver disease: A systematic review, meta-analysis and evidence map. *Dig. Liver Dis. Off. J. Ital. Soc. Gastroenterol. Ital. Assoc. Study Liver* **2021**, *53*, 44–51. [CrossRef]
85. Athyros, V.G.; Alexandrides, T.K.; Bilianou, H.; Cholongitas, E.; Doumas, M.; Ganotakis, E.S.; Goudevenos, J.; Elisaf, M.S.; Germanidis, G.; Giouleme, O.; et al. The use of statins alone, or in combination with pioglitazone and other drugs, for the treatment of non-alcoholic fatty liver disease/non-alcoholic steatohepatitis and related cardiovascular risk. An Expert Panel Statement. *Metab. Clin. Exp.* **2017**, *71*, 17–32. [CrossRef] [PubMed]
86. Jalali, M.; Rahimlou, M.; Mahmoodi, M.; Moosavian, S.P.; Symonds, M.E.; Jalali, R.; Zare, M.; Imanieh, M.H.; Stasi, C. The effects of metformin administration on liver enzymes and body composition in non-diabetic patients with non-alcoholic fatty liver disease and/or non-alcoholic steatohepatitis: An up-to date systematic review and meta-analysis of randomized controlled trials. *Pharmacol. Res.* **2020**, *159*, 104799. [CrossRef]
87. Weersma, R.K.; Zhernakova, A.; Fu, J. Interaction between drugs and the gut microbiome. *Gut* **2020**, *69*, 1510–1519. [CrossRef]
88. Forslund, K.; Hildebrand, F.; Nielsen, T.; Falony, G.; Le Chatelier, E.; Sunagawa, S.; Prifti, E.; Vieira-Silva, S.; Gudmundsdottir, V.; Pedersen, H.K.; et al. Disentangling type 2 diabetes and metformin treatment signatures in the human gut microbiota. *Nature* **2015**, *528*, 262–266. [CrossRef]

Review

Alcohol and the Risk of Injury

Tanya Chikritzhs [1,*] and Michael Livingston [1,2]

[1] National Drug Research Institute, Faculty of Health Sciences, Curtin University, Perth, WA 6008, Australia; michael.livingston@curtin.edu.au
[2] Centre for Alcohol Policy Research, La Trobe University, Bundoora, Melbourne, VIC 3086, Australia
* Correspondence: t.n.chikritzhs@curtin.edu.au

Abstract: Globally, almost four and a half million people died from injury in 2019. Alcohol's contribution to injury-related premature loss of life, disability and ill-health is pervasive, touching individuals, families and societies throughout the world. We conducted a review of research evidence for alcohol's causal role in injury by focusing on previously published systematic reviews, meta-analyses and where indicated, key studies. The review summarises evidence for pharmacological and physiological effects that support postulated causal pathways, highlights findings and knowledge gaps relevant to specific forms of injury (i.e., violence, suicide and self-harm, road injury, falls, burns, workplace injuries) and lays out options for evidence-based prevention.

Keywords: alcohol; injury; review; risk; mortality; morbidity; policy; intervention

1. Introduction

Globally, almost four and a half million people died from injury in 2019 [1], with 7% of these deaths directly attributable to alcohol. Alcohol's role in injury-related premature loss of life, disability and ill-health is pervasive, touching individuals, families and societies the world over. Alcohol use, particularly intoxication, plays a major role in a wide range of injuries, some of which are readily recognisable as alcohol-related (e.g., road injuries, violent assault) and others which are less so (e.g., falls, drownings, injuries in the workplace).

Alcohol-attributable injury accounts for around one-tenth of the total impact of alcohol on health (9.9% and 12.6% in low- and high-income countries, respectively) [1]. Males (90%) and young people aged 15–39 years (40%) dominate alcohol-attributable injury deaths [1]. Impacts on health systems are considerable, with alcohol contributing to between 5% and 40% of all emergency department (ED) injury presentations across 27 countries [2]. This comes with significant costs. In 2014, injuries caused by alcohol in the USA were an estimated 8% of all injury-related ED presentations at a cost of nearly USD 9 billion, and when in-patient admissions were added, costs almost tripled (USD 26 billion) [3]. Canadian estimates of hospitalisation and day surgery costs for alcohol-attributable injuries in 2017 were just under CAD 1 billion [4]. Large economic impacts are not limited to high-income countries. In Sri Lanka, alcohol-attributable injury costs exceeded an estimated USD 380 million, nearly half of the total costs of alcohol in that country in 2015 [5]. In Latin America, where around 30% of road fatalities are attributable to alcohol [6], the burden of road crashes overall was between 1.5% and 3.9% of gross domestic product in 2013 compared to about 2% in the USA [7].

Alcohol-related injuries thus represent a significant economic burden in many societies globally and require substantial resources from overstretched health systems to manage. Importantly though, alcohol-related injuries are preventable and there are clear examples of effective interventions to reduce them. For instance, deaths due to drink-driving declined rapidly in many high-income countries during the 1980s [8,9], driven by tougher drink-driving laws and enforcement alongside broader alcohol policy shifts such as increases in the legal minimum drinking age. However, from a global perspective, rates of overall

Citation: Chikritzhs, T.; Livingston, M. Alcohol and the Risk of Injury. *Nutrients* **2021**, *13*, 2777. https://doi.org/10.3390/nu13082777

Academic Editor: Peter Anderson

Received: 12 July 2021
Accepted: 10 August 2021
Published: 13 August 2021

alcohol-related injury have remained largely stable over time [1] and are likely to increase in coming years as alcohol consumption in lower- and middle-income countries increases [10].

This review will summarise research evidence linking alcohol to physical injuries, including potential causal mechanisms involved. The review also highlights key research findings specific to various forms of injury and lays out options available for prevention. Given the breadth of information we aim to cover here, we take a narrative review approach, relying largely on previously published systematic reviews, meta-analyses and focusing on key studies where appropriate.

2. Alcohol as a Cause of Physical Injury

Research evidence for alcohol as a cause of injury has clearly emerged for many types of injury, across multiple settings, and using a wide range of study designs. Key forms of research evidence include: laboratory experiments conducted under controlled conditions [11], real-world emergency department studies [12], driving simulation studies [13], studies linking population level drinking and injury rates [14] and retrospective time-series studies showing that alcohol policy changes and interventions can influence population rates of injury [15].

In one of the most comprehensive reviews of individual-level data, Taylor et al. [16] demonstrated strong dose–response relationships between amount of alcohol consumed in the past 3 h and odds of both motor vehicle and non-motor vehicle injury. Their meta-analysis estimated that even relatively moderate consumption levels (24 g of pure alcohol) roughly doubled the odds of injury, but that risks increased sharply at higher levels of consumption, such that someone who had consumed 120 g of alcohol had a more than 50 times higher risk of a motor vehicle injury than a non-drinker. Other reviews have shown that these effects are broadly consistent across different study designs and alcohol recall periods, suggesting robust relationships [17,18].

At the population level, time-series analyses have shown that changes in per-capita alcohol consumption are associated with changes in mortality rates related to road injuries [19], suicide [20] and homicide [21,22]. These studies (see [14]) clearly showed that the amount of alcohol consumed in a given society is a key driver of injury rates, although there is significant variation cross-nationally, reflecting variation in drinking patterns and prevention policies at the country level. These established relationships between alcohol use and physical injury have underpinned regulation and public policy in many countries. Some applications, such as legal blood alcohol limits for driving [23], reach back many decades, while others, such as integration into national drinking guidelines [24], are relatively recent.

Plausible Causal Pathways: Pharmacological and Physiological Actions of Alcohol on the Human Brain and Central Nervous System

Alcohol is a known neurotoxin and central nervous system depressant. Even at low to moderate levels, alcohol has been observed to impair balance, visual focus, reaction time, judgment and to change behaviour (e.g., [25]). At high enough doses, intoxication can result in loss of consciousness, coma, respiratory failure (i.e., due to airway obstruction), aspiration pneumonia and ultimately, death [26].

Regarding plausible causal pathways that explain links between alcohol and injury, experimental studies offer the firmest evidence by virtue of their ability to randomly assign participants to placebo and exposure groups, subjectively measure functional biomarkers and control alcohol dosage. Though not immune, experimental studies are also best equipped to separate out pharmacological/physiological effects from 'expectancy' effects, i.e., personal beliefs about how alcohol affects behaviour, such as physical aggression, which can vary widely among individuals and cultures [27,28].

Laboratory studies which test human performance on various tasks designed to detect alcohol effects on specific brain systems have identified substantial impairments across multiple measures of cognitive (e.g., information processing) and psychomotor functions (e.g., eye-brain-hand-foot coordination) that directly bear on all forms of injury risk. An extensive review of more than 200 controlled experimental studies on alcohol's acute effects on the brain and central nervous system found impairments for visuo-motor control, divided attention, focused attention, reaction time, response inhibition and working memory. Effects were highly consistent at blood/breath alcohol concentrations (BACs) of 0.05% and higher, and some effects were found even at lower levels [29].

Extending what has been learnt from standard laboratory experiments, in vivo neuroimaging studies can detect alcohol's pharmacological effect on the human brain and bring potential mechanisms for alcohol-caused injury into clearer focus. Reviews of neuroimaging studies consistently support findings of diminished cognitive and psychomotor functions identified by laboratory experiments [28]. Several reviews, incorporating studies with a wide range of designs, show that beginning at low levels, acute alcohol intake reduces overall brain glucose metabolism (a proxy for neuronal activity) and increases metabolism of acetate (a product of acetaldehyde oxidation) in a dose–response manner. Reduced glucose metabolism is most concentrated in the cerebellum (implicated in motor impairment), while limbic regions (implicated in reward-seeking behaviour and addiction) show increased metabolism [28,30,31]. These studies also show that brain centres most affected by alcohol (i.e., cerebellum, hippocampus, occipital cortex, striatum, amygdala) are regions where balance, movement coordination, attention focus, self-control, processing of emotional stimuli (e.g., threat detection), motivation and reward-seeking, spatial learning and memory are believed to occur.

There is strong concurrence, therefore, between reviews of experimental laboratory studies demonstrating cognitive and performance deficits and neuroimaging studies demonstrating pharmacological and physiological actions of alcohol on the brain that strongly implicate causal pathways to injury risk. It is crucial, however, to bear in mind that the relationship between alcohol and injury is by no means inexorable. Outside of the laboratory, observational studies confirm every-day experience that not all alcohol use, or even intoxication, necessarily results in injury. Risk of injury from alcohol can be influenced by individual differences and expectancies about appropriate or permissible behaviours [27,32,33] as can social and cultural norms (e.g., community acceptance or rejection of drinking and driving). External factors such as setting (e.g., home, pub, park), price and physical availability of alcohol also have major impacts on alcohol-caused injuries at a population level [34,35].

3. Specific Injury Types

The injury literature often distinguishes between injuries arising from intentional behaviours and those that are most often unintentional or accidental. Interpersonal violence, self-harm and suicide are all considered intentional injuries as they arise from purposeful actions directed towards oneself or others. Unintentional injuries on the other hand include road injuries, other transport injuries, falls, drownings, burns, poisonings, workplace injuries and other 'accidents' (e.g., freezing), and are often further categorised into transport and non-transport. Categorizing injuries according to intention is commonplace in the literature and assumed to have utility for injury management and prevention. For instance, prevention approaches to intentional injuries often focus on characteristics of individuals and their behaviours, while unintentional prevention initiatives are more often concerned with how people, objects and environments interact [36]. It is nevertheless worth noting that some have questioned whether categorisation limits collaboration and advancement of prevention efforts, as the underlying motivations of individuals are not always clear-cut (e.g., some burns are intentional, as are some road injuries), and groups share many similar characteristics, including effective prevention approaches (see Section 4 on policy and interventions below) [37].

In terms of morbidity and mortality, alcohol-attributable intentional and unintentional injuries each account for roughly 50% of total numbers of Disability Adjusted Life Years (DALYs), Deaths and Years of Life Lost (YLL), with transport (i.e., mostly road injury) making up the majority of unintentional injuries [1]. Figure 1 presents the estimated global impact of alcohol-related injury in terms of Disability-Adjusted Life Years (DALYs), breaking down the impact into the broad categories of injury.

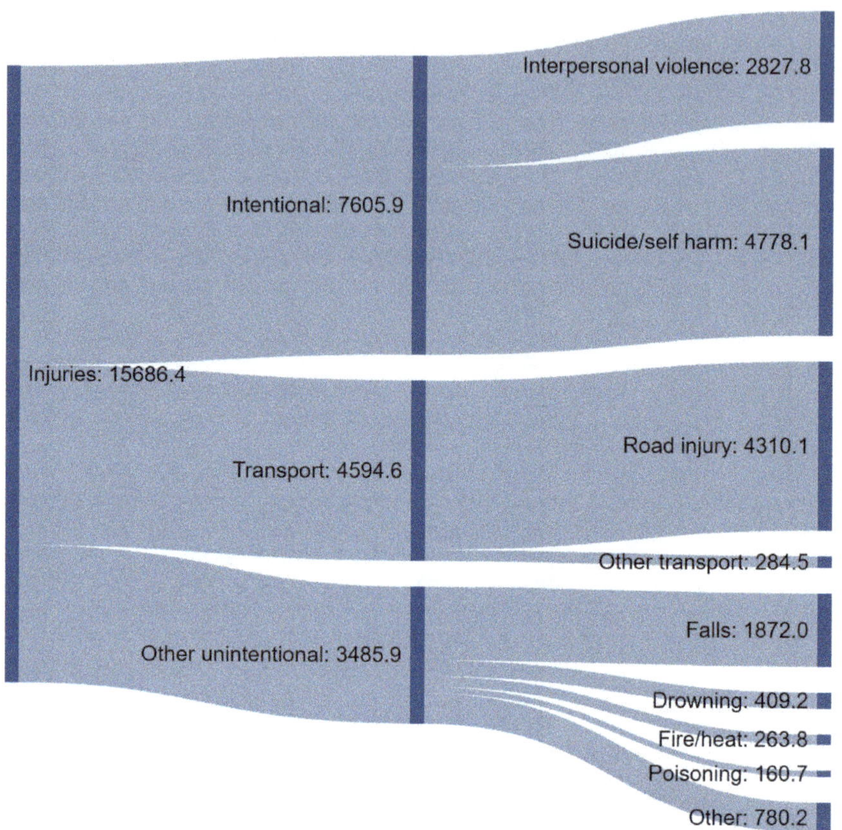

Figure 1. Global Disability Adjusted Life Years (DALYs '000) lost due to alcohol-attributable injuries, 2019. Data source: Institute for Health Metrics and Evaluation [1].

3.1. Interpersonal Violence

Interpersonal violence (IV) refers to intentional use of physical force (sexual or non-sexual) by an individual or small group of individuals against another individual or small group, and excludes larger scale conflict-related violence (e.g., warfare, rioting) [38]. In addition to sexual and/or non-sexual physical aspects, IV may also involve deprivation, neglect and psychological aspects. In the past decade or so, there has been increased focus on IV as a global problem posing major challenges to sustainable development goals including poverty, health and wellbeing, human rights and gender equality, particularly for women and girls (e.g., EU and UN Spotlight Initiative [39]).

Findings from many meta-analyses, some of which focused on laboratory studies (e.g., [40]), others on community-based studies (e.g., [41]), and even a recent meta-meta-analysis of 18 reviews covering multiple designs, settings and definitions of violence [42], strongly support alcohol use, especially by males, as a causal factor in IV. Although at times controversial, there is also robust evidence supporting the conclusion that alcohol use by victims at the time of the offence increases the risk of IV [42,43]. The role of alcohol use by females and IV has been less well-studied than for males, however, female alcohol use has also been identified as a risk factor for both perpetration and victimisation. Moreover, alcohol use is more strongly linked to victimisation among women than victimisation among men in intimate partner violence (IPV) [44,45].

Alcohol use by parents and caregivers, particularly at harmful or hazardous levels, also increases the risk of child physical injury, such as burns, fractures and, occasionally, death, arising from maltreatment [46]. Heavy alcohol use by either victim or perpetrator is also a risk factor for the physical abuse of older people by offspring, partners or other relatives in a caregiver role, as well as professional caregivers to a lesser extent [47].

Global burden of disease estimates indicate that as a proportion of total alcohol-attributable injuries, IV accounts for about 16% of deaths and 18% of DALYs [1]. However, IV is highly prone to under-reporting in official statistics, often as a result of victims avoiding authorities or official agencies for fear of further victimisation. In addition, effects of IV can extend well beyond immediate physical consequences that might compel a victim to seek treatment (e.g., emergency department attendance) or formal assistance (e.g., police, social services). Experience of trauma for instance, especially at a young age, increases risks of developing mental health problems, reproductive and sexual health problems, substance misuse, chronic illness (e.g., cardiovascular disease, diabetes, cancer) and of living in poverty later in life, very little of which is readily quantifiable [48].

3.2. Suicide and Self-Harm

Suicide and self-harm are second only to road injuries in terms of injury-related burden of disease contribution [49] and a major cause of death for young people [50]. The GBD estimates that around 15% of all suicide deaths are attributable to alcohol, meaning over 100,000 people die each year from alcohol-related self-harm. A series of systematic reviews have found strong and consistent evidence that alcohol and self-harm are strongly linked. This includes individual-level studies showing that people with alcohol use disorders (AUDs) are at increased risk of suicidal ideation, self-harm and completed suicide [51], studies showing that heavy drinking in-the-event increases suicide risk [52] and aggregate studies showing population-level links between alcohol consumption and suicide rates [20]. Evidence also points to violent methods of suicide, such as by firearm or hanging, involving heavier drinking in-the-event compared to poisoning (e.g., [53,54]).

Alcohol's role in self-harm appears to be mediated through cultural factors—for example, there is some evidence that alcohol is more strongly associated with self-harm for men and for cultures where intoxication-oriented drinking is more common [20]. Further, while the evidence is reasonably clear that alcohol contributes to suicide and self-harm, there remains uncertainty about the magnitude of the causal relationship, with at least some potential for alcohol use disorder, intoxication and self-harm to have common underlying drivers [20].

3.3. Road Injuries

Road injuries typically include fatal and non-fatal injuries that occur on public roads as a result of accidents involving one or more motor vehicles (e.g., cars, motorcycles, trucks), pedestrians or cyclists. Road injuries are currently ranked 7th for their contribution to total global DALYs (2.9%) for all ages [55]. Beginning in about 1980, many high-income countries reported substantial reductions in road injury rates. Rapid declines in national road tolls continued for about 15 years, largely as a result of concerted prevention efforts including legislation and enforcement of maximum legal blood alcohol concentration levels for driving [9]. After this time, improvements in high-income countries slowed considerably (e.g., [56]), although downward trends continued on a global scale with age-standardised DALYs for road injury declining for all ages (31%), 10–24 year olds (33.6%) and 25–49 year olds (22.5%) [55] between 1990 and 2019. Even so, worldwide, road injury remains the leading cause of death and disability for 10–24 year olds (6.6%) and 25–49 year olds (5.1%) [55].

Global statistics obscure large differences in road injury rates across nations and regions, particularly between high-income and low to low/middle-income nations. In the

African region for instance (where under-reporting is widespread) [57], road fatality rates lead the world, are at least double that for the European region [58] and have shown only marginal improvement over recent decades [57]. This suggests that a great deal of road injury prevention work remains to be carried out, particularly in Africa and South-East Asia where road safety laws, including for drink-driving, rarely meet best-practice standards [58].

The causal, dose-dependent role that alcohol plays in fatal and non-fatal road crashes has been well-established over decades of extensive observational, laboratory and driving simulation research [13,16,59]. Alcohol has been shown to impair driving performance at blood alcohol concentrations as low as 0.02% [60,61] and well before the driver or observers are able to detect signs of intoxication [62]. With a few notable exceptions (e.g., Sweden [63]), most countries maintain maximum legal blood alcohol concentration levels for non-probationary drivers (e.g., 0.05%, 0.08% [58]) that are higher than levels now known to significantly increase crash risk.

Alcohol-impaired drivers increase road injury risk to themselves and others, including passengers, pedestrians and other drivers [58,64]. Besides motor vehicle operators, alcohol-positive pedestrians [65] and cyclists [66,67] are also at increased risk of road injury. Drink-for-drink, young and inexperienced drivers are at much greater risk of serious road injury than their more experienced counterparts [56,61]. Among drivers with alcohol use disorder, road crash risk is at least twice that for non-dependent drivers (e.g., [68]).

Estimates of country-specific alcohol-attributable fractions for road fatalities vary considerably, however, averages for broad regions range from about 2% in the Eastern Mediterranean where alcohol consumption is largely prohibited, to almost 38% in Europe where alcohol is widely available [6]. Among all alcohol-attributable injuries, road injuries account for over one quarter (27.5%) of total DALYs and they also account for more than half of all unintentional injuries (see Figure 1).

3.4. Falls

Falls represent a major contributor to morbidity and mortality, ranking 21st across all ages (second only to road injuries in terms of injury) and 8th for people aged 75 years and older as a cause of age-standardised DALYs [55]. Not surprisingly, for those aged over 70, falls are the most common cause of an injury-related death [49].

Systematic reviews of studies that examine usual drinking practices and fall risk generally produce mixed results [69,70], however the majority fail to account for patterns of drinking. In a pooled analysis of case-control studies from emergency departments in 28 countries, Cherpitel et al., showed that both frequent and episodic heavy drinking were strong predictors of alcohol-involved falls [71]. Studies that have examined drinking in-the-event are even more compelling. In a meta-analysis of five studies that used acute measures of drinking, Taylor et al. [16] found a clear dose–response relationship, with odds of a fall-related injury increasing by 1.15 for each 10 g of alcohol consumed. This is supported by other reviews [70] and pooled analyses of ED studies [12,72].

Recent work has identified concerns about interactions between alcohol consumption and use of medication among older populations, where falls represent a disproportionately large cause of morbidity [73]. This points towards one possible specific intervention to reduce the burden of falls—better assessment and management of alcohol consumption risks by primary healthcare workers when administering/prescribing medications, especially those related to the central nervous system [73].

Studies have repeatedly shown that accidental fall injuries result in substantial costs [74]—for example, a US study estimated the annual cost of fall injuries at more than USD 80 billion [75]. National populations throughout the world are ageing [76] and, although trends vary across countries, some studies have reported increasing levels of risky drinking among older age groups [77–80]. Alcohol-related falls are therefore likely to present an increasing social burden for many countries in coming decades.

3.5. Drowning

A recent systematic review [81] found that around half of all drowning deaths and more than one-third of all drowning-related injuries involved alcohol, but that the prevalence of alcohol involvement varied markedly between studies. The literature relies heavily on post-mortem assessments of BAC or relatively crude cross-sectional surveys, meaning the strength of the causal evidence is relatively low, but the best estimates for proportion of all drownings causally attributable to alcohol range from 10% to 30% [82], with people who have blood alcohol concentrations of 0.10% or higher increasing their risk of drowning ten-fold [82].

3.6. Injuries from Excessive Heat and Cold

Alcohol intoxication raises the risk of sustaining serious injuries from excessive heat, such as burns from household fires, and from excessive cold, such as hypothermia or death from freezing when drinking outdoors in cold weather. Systematic reviews consistently identify alcohol intoxication as a key risk factor for residential fire mortality [83,84], with around half of all house fire fatalities tested returning positive BACs [85]. In a robust US case-control study, presence of an intoxicated person in the household was the single strongest predictor of fire leading to fatality [86]. Alcohol intoxication delays escape and increases risk of fire ignition, particularly in conjunction with smoking (e.g., falling asleep while drinking and smoking) [87]. Acute and chronic heavy alcohol use, particularly among older age groups, are also major risk factors for serious hypothermia and death by freezing, although increased risk of hypothermia can also occur among the young (e.g., [88]). Likely under-reported at a global level, alcohol's role in injuries arising from excessive cold nonetheless present ongoing challenges for cold climate countries during winter months, with many reporting alcohol's involvement in more than 40% of fatal cases (e.g., [89–91]).

3.7. Workplace Injuries

Despite growing use of alcohol and drug testing in the workplace, international research evidence for alcohol as a major contributor to workplace accidents and injuries (except impaired driving) is surprisingly under-developed. Although single studies have continued to support a causal relationship [92–94], almost three decades have passed since Stallones and Kraus' [95] review of epidemiological evidence regarding alcohol's role in workplace injuries.

3.8. Alcohol Poisoning and Other Injuries from Heavy Intoxication

Other key forms of injury arising from heavy intoxication include aspiration (i.e., choking) [49] and alcohol poisoning [96]. These are especially common among marginalised populations in intoxication-oriented and spirits drinking cultures, with—for example— rates substantially higher in Eastern Europe than the rest of Europe [97]. Additional injury risks arise when informal (e.g., home-made) or illegally produced counterfeit or adulterated products are consumed. These products often contain unknown quantities of pure alcohol (i.e., ethanol) and other toxic substances (e.g., methanol, ethylene glycol) not intended for human consumption (e.g., 'antifreeze', perfume, methylated spirits) that can cause blindness, brain injury, coma and death when ingested [98,99]. People with alcohol use disorders, low incomes and tourists appear to be at particular risk of injury from illicit alcohol. Large poisoning outbreaks have been documented in many countries, with fatality rates as high as 30% in some places (e.g., Uganda, Tunisia, Turkey, Pakistan, Norway, Nicaragua, Libya, Kenya, Indonesia, India, Estonia, Ecuador, Czech Republic, Cambodia) [100,101].

4. Effective Interventions and Policies

Broadly speaking, research evidence for effective interventions and policies aimed at reducing alcohol-related injuries can be grouped into two camps: (i) alcohol consumption-centred approaches oriented towards reducing use at a whole-of-population level, that may also have specific or more substantive effects on sub-populations (e.g., young people, heavy drinkers), and (ii) injury-centred approaches targeted at reducing the risk of specific types of injury (e.g., falls, assaults) or injuries that occur in specific situations (e.g., while driving, in the workplace). Though not intended to be exhaustive, the following sections summarise current research evidence for a wide range of policy and intervention options available to decision makers concerned with reducing alcohol-related injury.

4.1. Alcohol Consumption-Centred Approaches

Decades of research evidence clearly support policy approaches that reduce population-level alcohol consumption as having a central role to play in the reduction of alcohol-related injury overall. Whole-of-population consumption-centred approaches are highly cost-effective at reducing harmful alcohol use in general, alongside restrictions on marketing and brief interventions [102]. Furthermore, although most evidence in support of consumption-centred approaches has been derived from high-income countries, they are also highly effective in middle- and low-income countries when implemented appropriately [103].

Among consumption-centred approaches, research evidence in support of effective reduction of population-level alcohol use is arguably the strongest and most consistent for price-based interventions that influence alcohol's economic availability, i.e., retail price relative to disposable income [104–107]. A meta-analysis of 50 studies suggested that a doubling of alcohol taxes would reduce road injury deaths by 11%, violence by 2% and suicide by around 4% [108]. Price-based interventions have historically been delivered via governments raising alcohol taxes. However, there is growing evidence that raising the minimum price at which alcohol can be sold at retail (i.e., minimum unit pricing, MUP) is an effective strategy for reducing overall consumption, IV and drink-driving [109–112].

There is also very strong evidence that raising the legal minimum drinking age (e.g., from 18 to 21 years) leads to substantial reductions in road injuries among young people in the USA and elsewhere [113,114], and is most effective when supported by concerted enforcement efforts [115,116]. Benefits of higher legal minimum drinking age laws have also been shown to accrue to suicide [117], violence and morbidity from other accidental injuries [118] among young people.

Strategies which reduce alcohol's physical availability can substantially reduce injury rates, including in low- and middle-income countries [119]. Research evidence is particularly robust for reductions in permitted hours of sale of alcohol late at night and reduced violence and road injuries [15,120]. Links between physical density of outlets (i.e., both on- and off-trade access) and injury outcomes (including violence, road injury and self-harm [121]) have been demonstrated by many studies. However, uncertainties remain about the overall robustness of this literature [122] and confirmation is needed from new studies that incorporate information on alcohol sales with appropriate methods for studying geospatial data.

Of recent interest are potential impacts from alcohol's designation as an 'essential' product/service and liberalisation of off-trade alcohol sales by a large number of jurisdictions during the COVID-19 pandemic [123–126]. Expansion of off-trade sales often occurred simultaneously to the closure of workplaces, schools, childcare, leisure and physical activity centres [126]. Although evidence is still emerging, reports in the media and grey literature suggest increased drinking in the home in some countries [127–129] (largely off-setting reductions in on-premises drinking). These reports have appeared alongside several studies showing increased abusive head trauma among children [130] and family violence [131,132]. There is some concern that, pressured by commercial vested interests [133,134], governments will allow continuation of deregulatory changes originally intended as temporary, leading to increased risks of IV and trauma in the home [123,124,126].

Multi-component interventions that simultaneously implement a suite of strategies can markedly reduce injury, especially when supported by their target populations [135]. Although price increases and physical availability restrictions (e.g., reduced trading hours, limits on cheap high-risk beverage purchases) are considered central to the success of these programmes, they are often accompanied by supporting harm (e.g., mandatory server training, sobriety testing) and demand reduction strategies (e.g., advertising restrictions) [135,136]. Most recently, a series of price and availability restrictions in Lithuania reduced total mortality there by 3% [137], with the bulk of benefits arising from reductions in injury-related deaths [138]. Due to their relatively direct control of alcohol sales, advertising and promotion, jurisdictions with whole or partial alcohol monopolies (more common to Scandinavia and North America) are well-placed to implement multi-component interventions. All else being equal, alcohol monopolies have lower rates of alcohol-related injuries than those with free market systems [139]. Studies which have modelled potential impacts of disbanding retail monopolies in Sweden, for example, have estimated increases in alcohol-related injury deaths of between 18% and 28% annually [140].

4.2. Injury-Centred Approaches

From a global perspective, evidence-based strategies for reducing alcohol-related road injuries have undoubtedly received more government commitment to implementation than any other source of injury—and with striking results [58]. Of critical importance to minimizing the road toll in countries where alcohol is widely consumed are government laws prohibiting BACs exceeding 0.02% for probationary drivers and 0.05% for non-probationary drivers (WHO 2018). Current best-practice drink-driving laws should also be coupled with widely publicised, highly visible police enforcement and random breath testing [141]. At last count, 45 countries, covering less than a third of the world's population, had enacted such laws, with just 2% from low-income countries [58]. There is great scope, therefore, for governments of countries at all income levels to substantially reduce premature death and disability caused by alcohol-impaired road users among their citizenry.

Motor vehicle drivers who repeatedly drink and drive are often targeted for further preventative measures. Commonly referred to as ignition- or alcohol-interlocks, devices that detect breath alcohol can be retrofitted to motor vehicles of drink-driving offenders. When positive breath alcohol is detected, alcohol-interlocks incapacitate a vehicle by blocking engine ignition and are highly effective at reducing repeat drink-driving offences [7,142].

Interventions aimed at reducing alcohol-related IV have largely been focused around drinking venues and night-time entertainment precincts. While broad physical availability restrictions (e.g., reduced trading hours for licensed venues) have the clearest evidence, improving server training [143], venue security, environmental design and management have been shown in some settings to reduce violence and aggression [144]. Research into interventions aimed at alcohol-related domestic or family violence specifically is relatively scant, with systematic reviews finding few robust evaluations of alcohol policy interventions [145]. Individual-level interventions with offenders have shown generally poor results in terms of reducing reoffending [146], suggesting that upstream policy interventions should be the focus of work to reduce alcohol-related violence.

Research on interventions to reduce alcohol-related drowning or burns is scarce. A study of minimum-legal drinking age laws in the USA found no impact on young-adult drowning rates [147], while programs aiming to reduce alcohol consumption while boating and fishing remain largely unevaluated [148]. Similarly, evaluation studies of workplace alcohol and drug testing are poor, with only one relatively high-quality study finding an effect of testing in the transport industry [149].

5. Conclusions

Invitations to consider alcohol's role in the death, disability and distress that arises from human injury, and what actions could be taken to reduce the burden, can evoke a wide range of responses. To varying degrees, responses may be part of a broader agenda, compromised by politicisation, motivated by vested interests or simply reflect personal beliefs and experience. In contrast, scientific evidence in support of alcohol's causal and central role in injury has strengthened over time and is strikingly robust. Given the broad range of scientific disciplines and research approaches which have contributed to the evidence base, the many decades over which that evidence has accumulated and the variety of forms that alcohol-related injury can take, the high level of overall consistency among findings is remarkable.

Estimates of the human and economic costs of alcohol-related injury leave no doubt that the global burden is very large. In truth, it is probably larger still, and it may be many more decades (if at all) before the full extent of short- and long-term consequences, including chronic disease, mental health problems and reduced wellbeing, are fully understood and quantified across all counties. At the very least, the burden borne by low- and middle-income countries is likely far higher than current statistics imply.

Upward global trends in per capita alcohol consumption during the past thirty years or so are predicted to continue, increasing by more than a litre per person by 2030 (i.e., from 6.5 L in 2017 to 7.6 L in 2030) [10]. Increasing consumption combined with ageing populations and more drinking in the home (facilitated by pandemic-related liberalisation of off-trade sales) can be expected to bring about changes in the distribution and magnitude of alcohol-attributable injury in the next several decades. These changes may well add to challenges faced by governments already struggling to manage over-loaded healthcare systems, policing and social services [150].

Nonetheless, there is reason to be optimistic. The evidence is clear: population-level alcohol consumption-centred policies that reduce alcohol's economic and physical availability, especially when implemented in conjunction with each other, substantially reduce alcohol-related injury in its various forms. Strategies specifically targeted at reducing alcohol-impaired driving are also highly effective and indeed essential for addressing the world's leading cause of death and disability among people in their most productive years. There are major human capital and economic windfalls awaiting governments that adopt nation-wide, best-practice alcohol interventions. Manifestly underutilised, independently and collectively, the full potential of these strategies has been scarcely realised, though they offer evidence-based solutions for high- and low-income countries alike [151].

Funding: This work was supported by funding from the Australian Government under the Drug and Alcohol Program (NDRI core funder) and from Curtin University, Western Australia.

Conflicts of Interest: The authors declare no conflict of interest.

References

1. Institute for Health Metrics and Evaluation. *GBD Results Tool: Global Health Data Exchange*; University of Washington: Seattle, WA, USA, 2021.
2. Ye, Y.; Shield, K.; Cherpitel, C.J.; Manthey, J.; Korcha, R.; Rehm, J. Estimating alcohol-attributable fractions for injuries based on data from emergency department and observational studies: A comparison of two methods. *Addiction* **2018**, *114*, 462–470. [CrossRef]
3. White, A.M.; Slater, M.E.; Ng, G.; Hingson, R.; Breslow, R. Trends in Alcohol-Related Emergency Department Visits in the United States: Results from the Nationwide Emergency Department Sample, 2006 to 2014. *Alcohol. Clin. Exp. Res.* **2018**, *42*, 352–359. [CrossRef] [PubMed]
4. Canadian Substance Use Costs and Harms Scientific WORKING Group. Canadian Substance Use Costs and Harms Visualization Tool, Version 2.0.0 [Online Tool]. 2020. Available online: https://csuch.ca/explore-the-data/ (accessed on 23 June 2021).
5. Ranaweera, S.; Amarasinghe, H.; Chandraratne, N.; Thavorncharoensap, M.; Ranasinghe, T.; Karunaratna, S.; Kumara, D.; Santatiwongchai, B.; Chaikledkaew, U.; Abeykoon, P.; et al. Economic costs of alcohol use in Sri Lanka. *PLoS ONE* **2018**, *13*, e0198640. [CrossRef]
6. World Health Organisation. *The Global Health Observatory—Regional Prevalence, Aafs (15+), Road Traffic Crash Deaths*; World Health Organisation: Geneva, Switzerland, 2021.
7. Lefio, A.; Bachelet, V.C.; Jiménez-Paneque, R.; Gomolan, P.; Rivas, K. A systematic review of the effectiveness of interventions to reduce motor vehicle crashes and their injuries among the general and working populations. *Rev. Panam. Salud Públ.* **2018**, *42*, 1–8. [CrossRef] [PubMed]
8. Cummings, P.; Rivara, F.P.; Olson, C.M.; Smith, K.M. Changes in traffic crash mortality rates attributed to use of alcohol, or lack of a seat belt, air bag, motorcycle helmet, or bicycle helmet, United States, 1982–2001. *Inj. Prev.* **2006**, *12*, 148–154. [CrossRef] [PubMed]

9.   Sweedler, B.M.; Stewart, K. *Worldwide Trends in Alcohol and Drug Impaired Driving. Drugs, Driving and Traffic Safety*; Birkhäuser Basel Verlag AG: Basel, Switzerland, 2009; pp. 23–41.

10. Manthey, J.; Shield, K.D.; Rylett, M.; Hasan, O.M.; Probst, C.; Rehm, J. Global alcohol exposure between 1990 and 2017 and forecasts until 2030, A modelling study. *Lancet* **2019**, *393*, 2493–2502. [CrossRef]

11. Giancola, P.R.; Duke, A.A.; Ritz, K.Z. Alcohol, violence, and the alcohol myopia model: Preliminary findings and im-plications for prevention. *Addict. Behav.* **2011**, *36*, 1019–1022. [CrossRef]

12. Borges, G.; Cherpitel, C.; Orozco, R.; Bond, J.; Ye, Y.; Macdonald, S.; Rehm, J.; Poznyak, V. Multicentre study of acute alcohol use and non-fatal injuries: Data from the WHO collaborative study on alcohol and injuries. *Bull. World Health Organ.* **2006**, *84*, 453–460. [CrossRef]

13. Irwin, C.; Iudakhina, E.; Desbrow, B.; McCartney, D. Effects of acute alcohol consumption on measures of simulated driving: A systematic review and meta-analysis. *Accid. Anal. Prev.* **2017**, *102*, 248–266. [CrossRef]

14. Norström, T.; Ramstedt, M. Mortality and population drinking: A review of the literature. *Drug Alcohol Rev.* **2005**, *24*, 537–547. [CrossRef]

15. Nepal, S.; Kypri, K.; Tekelab, T.; Hodder, R.K.; Attia, J.; Bagade, T.; Chikritzhs, T.; Miller, P. Effects of Extensions and Restrictions in Alcohol Trading Hours on the Incidence of Assault and Unintentional Injury: Systematic Review. *J. Stud. Alcohol Drugs* **2020**, *81*, 5–23. [CrossRef]

16. Taylor, B.; Irving, H.; Kanteres, F.; Room, R.; Borges, G.; Cherpitel, C.; Greenfield, T.; Rehm, J. The more you drink, the harder you fall: A systematic review and meta-analysis of how acute alcohol consumption and injury or collision risk increase together. *Drug Alcohol Depend.* **2010**, *110*, 108–116. [CrossRef]

17. Zeisser, C.; Stockwell, T.; Chikritzhs, T.; Cherpitel, C.J.; Ye, Y.; Gardner, C. A Systematic Review and Meta-Analysis of Alcohol Consumption and Injury Risk as a Function of Study Design and Recall Period. *Alcohol. Clin. Exp. Res.* **2013**, *37*, E1–E8. [CrossRef] [PubMed]

18. Cherpitel, C.J.; Ye, Y.; Bond, J.; Borges, G.; Monteiro, M. Relative risk of injury from acute alcohol consumption: Modeling the dose-response relationship in emergency department data from 18 countries. *Addiction* **2015**, *110*, 279–288. [CrossRef]

19. Jiang, H.; Livingston, M.; Room, R. Alcohol Consumption and Fatal Injuries in Australia Before and After Major Traffic Safety Initiatives: A Time Series Analysis. *Alcohol. Clin. Exp. Res.* **2015**, *39*, 175–183. [CrossRef] [PubMed]

20. Norstrøm, T.-A.; Rossow, I. Alcohol Consumption as a Risk Factor for Suicidal Behavior: A Systematic Review of Associations at the Individual and at the Population Level. *Arch. Suicide Res.* **2016**, *20*, 489–506. [CrossRef] [PubMed]

21. Ramstedt, M. Population drinking and homicide in Australia: A time series analysis of the period 1950–2003. *Drug Alcohol Rev.* **2011**, *30*, 466–472. [CrossRef]

22. Bye, E.K. Alcohol and homicide in Eastern Europe: A time series analysis of six countries. *Homicide Stud.* **2008**, *12*, 7–27. [CrossRef]

23. Homel, R. Random breath testing in Australia: A complex deterrent. *Aust. Drug Alcohol Rev.* **1988**, *7*, 231–241. [CrossRef]

24. National Health and Medical Research Council. *Australian Guidelines to Reduce Health Risks from Drinking Alcohol*; NHMRC: Canberra, NSW, Australia, 2020.

25. Dry, M.J.; Burns, N.R.; Nettelbeck, T.; Farquharson, A.L.; White, J.M. Dose-related effects of alcohol on cognitive functioning. *PLoS ONE* **2012**, *7*, e50977. [CrossRef]

26. Davis, A.; Lipson, A. Central-Nervous-System Depression and High Blood Ethanol Levels. *Lancet* **1986**, *327*, 566. [CrossRef]

27. Exum, M.L. Alcohol and aggression: An integration of findings from experimental studies. *J. Crim. Justice* **2006**, *34*, 131–145. [CrossRef]

28. Bjork, J.M.; Gilman, J.M. The effects of acute alcohol administration on the human brain: Insights from neuroimaging. *Neuropharmacology* **2014**, *84*, 101–110. [CrossRef] [PubMed]

29. Zoethout, R.W.M.; Delgado, W.L.; Ippel, A.E.; Dahan, A.; Van Gerven, J.M.A. Functional biomarkers for the acute effects of alcohol on the central nervous system in healthy volunteers. *Br. J. Clin. Pharmacol.* **2011**, *71*, 331–350. [CrossRef]

30. Jacob, A.; Wang, P. Alcohol Intoxication and Cognition: Implications on Mechanisms and Therapeutic Strategies. *Front. Neurosci.* **2020**, *14*, 102. [CrossRef]

31. Van Skike, C.E.; Goodlett, C.; Matthews, D.B. Acute alcohol and cognition: Remembering what it causes us to forget. *Alcohol* **2019**, *79*, 105–125. [CrossRef] [PubMed]

32. Wells, S.; Flynn, A.; Tremblay, P.F.; Dumas, T.; Miller, P.; Graham, K. Linking masculinity to negative drinking conse-quences: The mediating roles of heavy episodic drinking and alcohol expectancies. *J. Stud. Alcohol Drugs* **2014**, *75*, 510–519. [CrossRef] [PubMed]

33. Attwood, A.S.; Munafò, M.R. Effects of acute alcohol consumption and processing of emotion in faces: Implications for understanding alcohol-related aggression. *J. Psychopharmacol.* **2014**, *28*, 719–732. [CrossRef] [PubMed]

34. Babor, T.; Caetano, R.; Casswell, S.; Edwards, G.; Giesbrecht, N.; Graham, K. *Alcohol: No Ordinary Commodity—Research and Public Policy*, 2nd ed.; Oxford University Press: Oxford, UK, 2010.

35. Cherpitel, C.J.; Witbrodt, J.; Korcha, R.A.; Ye, Y.; Monteiro, M.G.; Chou, P. Dose–Response Relationship of Alcohol and Injury Cause: Effects of Country-Level Drinking Pattern and Alcohol Policy. *Alcohol. Clin. Exp. Res.* **2019**, *43*, 850–856. [CrossRef]

36. Yin, X.; Li, D.; Zhu, K.; Liang, X.; Peng, S.; Tan, A.; Du, Y. Comparison of intentional and unintentional injuries among Chinese children and adolescents. *J. Epidemiol.* **2019**, *30*, 529–536. [CrossRef]

37. Cohen, L.; Miller, T.; Sheppard, M.A.; Gordon, E.; Gantz, T.; Atnafou, R. Bridging the gap: Bringing together intentional and unintentional injury prevention efforts to improve health and well being. *J. Saf. Res.* **2003**, *34*, 473–483. [CrossRef] [PubMed]
38. Violence Prevention Alliance. *Definition and Typology of Violence*; World Health Organisation: Geneva, Switzerland, 2021.
39. Shrivastava, S.R.; Shrivastava, P.S.; Ramasamy, J. Spotlight initiative: A step of united nations and European union to end violence against women. *Med. J. Dr. D.Y. Patil Vidyapeeth* **2018**, *11*, 380. [CrossRef]
40. Steele, C.M.; Southwick, L. Alcohol and social behavior: I. The psychology of drunken excess. *J. Per. Soc. Psychol.* **1985**, *48*, 18. [CrossRef]
41. Kuhns, J.B.; Exum, M.L.; Clodfelter, T.A.; Bottia, M.C. The prevalence of alcohol-involved homicide offending: A meta-analytic review. *Homicide Stud.* **2014**, *18*, 251–270. [CrossRef]
42. Duke, A.A.; Smith, K.M.Z.; Oberleitner, L.M.S.; Westphal, A.; McKee, S.A. Alcohol, drugs, and violence: A meta-meta-analysis. *Psychol. Violence* **2018**, *8*, 238–249. [CrossRef]
43. Kuhns, J.B.; Wilson, D.B.; Clodfelter, T.A.; Maguire, E.R.; Ainsworth, S.A. A meta-analysis of alcohol toxicology study findings among homicide victims. *Addiction* **2010**, *106*, 62–72. [CrossRef]
44. Cafferky, B.M.; Mendez, M.; Anderson, J.R.; Stith, S.M. Substance use and intimate partner violence: A meta-analytic review. *Psychol. Violence* **2018**, *8*, 110. [CrossRef]
45. Devries, K.M.; Child, J.; Bacchus, L.J.; Mak, J.; Falder, G.; Graham, K.; Watts, C.; Heise, L. Intimate partner violence victimization and alcohol consumption in women: A systematic review and meta-analysis. *Addiction* **2014**, *109*, 379–391. [CrossRef] [PubMed]
46. World Health Organization. *Preventing Child. Maltreatment: A Guide to Taking Action and Generating Evidence*; World Health Organization: Geneva, Switzerland, 2006.
47. Johannesen, M.; Lo Giudice, D. Elder abuse: A systematic review of risk factors in community-dwelling elders. *Age Ageing* **2013**, *42*, 292–298. [CrossRef]
48. World Health Organization. *Global Status Report on Violence Prevention 2014*; World Health Organization: Geneva, Switzerland, 2014.
49. Roth, G.A.; Abate, D.; Abate, K.H.; Abay, S.M.; Abbafati, C.; Abbasi, N.; Abbastabar, H.; Abd-Allah, F.; Abdela, J.; Abdelalim, A.; et al. Global, regional, and national age-sex-specific mortality for 282 causes of death in 195 countries and territories, 1980–2017, a systematic analysis for the Global Burden of Disease Study 2017. *Lancet* **2018**, *392*, 1736–1788. [CrossRef]
50. World Health Organisation. *Preventing Suicide: A Global Imperative*; World Health Organization: Geneva, Switzerland, 2014.
51. Darvishi, N.; Farhadi, M.; Haghtalab, T.; Poorolajal, J. Alcohol-Related Risk of Suicidal Ideation, Suicide Attempt, and Completed Suicide: A Meta-Analysis. *PLoS ONE* **2015**, *10*, e0126870. [CrossRef] [PubMed]
52. Borges, G.; Bagge, C.; Cherpitel, C.J.; Conner, K.; Orozco, R.; Rossow, I. A meta-analysis of acute use of alcohol and the risk of suicide attempt. *Psychol. Med.* **2017**, *47*, 949–957. [CrossRef]
53. Conner, K.R.; Huguet, N.; Caetano, R.; Giesbrecht, N.; McFarland, B.H.; Nolte, K.B.; Kaplan, M.S. Acute use of alcohol and methods of suicide in a US national sample. *Am. J. Public Health* **2014**, *104*, 171–178. [CrossRef]
54. Kaplan, M.S.; McFarland, B.; Huguet, N.; Conner, K.; Caetano, R.; Giesbrecht, N.; Nolte, K.B. Acute alcohol intoxication and suicide: A gender-stratified analysis of the National Violent Death Reporting System. *Inj. Prev.* **2012**, *19*, 38–43. [CrossRef]
55. Vos, T.; Lim, S.S.; Abbafati, C.; Abbas, K.M.; Abbasi, M.; Abbasifard, M.; Abbasi-Kangevari, M.; Abbastabar, H.; Abd-Allah, F.; Abdelalim, A.; et al. Global burden of 369 diseases and injuries in 204 countries and territories, 1990–2019: A systematic analysis for the Global Burden of Disease Study 2019. *Lancet* **2020**, *396*, 1204–1222. [CrossRef]
56. Voas, R.B.; Torres, P.; Romano, E.; Lacey, J.H. Alcohol-Related Risk of Driver Fatalities: An Update Using 2007 Data. *J. Stud. Alcohol Drugs* **2012**, *73*, 341–350. [CrossRef]
57. Adeloye, D.; Thompson, J.Y.; Akanbi, M.; Azuh, D.; Samuel, V.; Omoregbe, N.; Ayo, C.K. The burden of road traffic crashes, injuries and deaths in Africa: A systematic review and meta-analysis. *Bull. World Health Organ.* **2016**, *94*, 510–521A. [CrossRef] [PubMed]
58. World Health Organization. *Global Status Report on Road Safety 2018, Summary*; World Health Organization: Geneva, Switzerland, 2018.
59. Taylor, B.; Rehm, J. The relationship between alcohol consumption and fatal motor vehicle injury: High risk at low alco-hol levels. *Alcohol. Clin. Exp. Res.* **2012**, *36*, 1827–1834. [CrossRef]
60. Lira, M.C.; Sarda, V.; Heeren, T.C.; Miller, M.; Naimi, T.S. Alcohol Policies and Motor Vehicle Crash Deaths Involving Blood Alcohol Concentrations Below 0.08%. *Am. J. Prev. Med.* **2020**, *58*, 622–629. [CrossRef]
61. Zador, P.L.; Krawchuk, S.A.; Voas, R.B. Alcohol-related relative risk of driver fatalities and driver involvement in fatal crashes in relation to driver age and gender: An update using 1996 data. *J. Stud. Alcohol* **2000**, *61*, 387–395. [CrossRef]
62. Kerr, W.C.; Greenfield, T.K.; Midanik, L.T. How many drinks does it take you to feel drunk? Trends and predictors for subjective drunkenness. *Addiction* **2006**, *101*, 1428–1437. [CrossRef]
63. Norstrom, T. Assessment of the Impact of the 0.02 Percent BAC-Limit in Sweden. *Stud. Crime Crime Prev.* **1997**, *6*, 245–258.
64. Quinlan, K.; Shults, R.A.; Rudd, R.A. Child Passenger Deaths Involving Alcohol-Impaired Drivers. *Pediatrics* **2014**, *133*, 966–972. [CrossRef] [PubMed]
65. Lasota, D.; Goniewicz, M.; Kosson, D.; Ochal, A.; Krajewski, P.; Tarka, S.; Goniewicz, K.; Mirowska-Guzel, D. The effect of ethyl alcohol on the severity of injuries in fatal pedestrian victims of traffic crashes. *PLoS ONE* **2019**, *14*, e0221749. [CrossRef]
66. Hartung, B.; Mindiashvili, N.; Maatz, R.; Schwender, H.; Roth, E.H.; Ritz-Timme, S.; Moody, J.; Malczyk, A.; Daldrup, T. Regarding the fitness to ride a bicycle under the acute influence of alcohol. *Int. J. Leg. Med.* **2014**, *129*, 471–480. [CrossRef]

67. Harada, M.Y.; Gangi, A.; Ko, A.; Liou, D.Z.; Barmparas, G.; Li, T.; Hotz, H.; Stewart, D.; Ley, E.J. Bicycle trauma and alcohol intoxication. *Int. J. Surg.* **2015**, *24*, 14–19. [CrossRef]
68. Guitart, A.M.; Espelt, A.; Castellano, Y.; Suelves, J.M.; Villalbí, J.R.; Brugal, M.T. Injury-Related Mortality Over 12 Years in a Cohort of Patients with Alcohol Use Disorders: Higher Mortality Among Young People and Women. *Alcohol. Clin. Exp. Res.* **2015**, *39*, 1158–1165. [CrossRef]
69. Reid, M.C.; Boutros, N.N.; O'Connor, P.G.; Cadariu, A.; Concato, J. The health-related effects of alcohol use in older per-sons: A systematic review. *Subst. Abuse* **2002**, *23*, 149–164. [CrossRef]
70. Kool, B.; Ameratunga, S.; Jackson, R. The role of alcohol in unintentional falls among young and middle-aged adults: A systematic review of epidemiological studies. *Inj. Prev.* **2009**, *15*, 341–347. [CrossRef] [PubMed]
71. Cherpitel, C.J.; Witbrodt, J.; Ye, Y.; Korcha, R. A multi-level analysis of emergency department data on drinking patterns, alcohol policy and cause of injury in 28 countries. *Drug Alcohol Depend.* **2018**, *192*, 172–178. [CrossRef]
72. Cherpitel, C.J. Drinking Patterns and Problems and Drinking in the Event: An Analysis of Injury by Cause Among Casualty Patients. *Alcohol. Clin. Exp. Res.* **1996**, *20*, 1130–1137. [CrossRef]
73. Holton, A.; Boland, F.; Gallagher, P.; Fahey, T.; Moriarty, F.; Kenny, R.A.; Cousins, G. Potentially serious alcohol–medication in-teractions and falls in community-dwelling older adults: A prospective cohort study. *Age Ageing* **2019**, *48*, 824–831. [CrossRef]
74. Chandran, A.; Hyder, A.A.; Peek-Asa, C. The Global Burden of Unintentional Injuries and an Agenda for Progress. *Epidemiol. Rev.* **2010**, *32*, 110–120. [CrossRef]
75. Zaloshnja, E.; Miller, T.; Lawrence, B.A.; Romano, E. The costs of unintentional home injuries. *Am. J. Prev. Med.* **2005**, *28*, 88–94. [CrossRef]
76. Lutz, W.; Sanderson, W.; Scherbov, S. The coming acceleration of global population ageing. *Nat. Cell Biol.* **2008**, *451*, 716–719. [CrossRef] [PubMed]
77. Calvo, E.; Medina, J.T.; Ornstein, K.A.; Staudinger, U.M.; Fried, L.P.; Keyes, K.M. Cross-country and historical variation in alcohol consumption among older men and women: Leveraging recently harmonized survey data in 21 countries. *Drug Alcohol Depend.* **2020**, *215*, 108219. [CrossRef]
78. Han, B.H.; Moore, A.A.; Sherman, S.; Keyes, K.M.; Palamar, J. Demographic trends of binge alcohol use and alcohol use disorders among older adults in the United States, 2005–2014. *Drug Alcohol Depend.* **2017**, *170*, 198–207. [CrossRef]
79. Breslow, R.A.; Castle, I.J.P.; Chen, C.M.; Graubard, B.I. Trends in alcohol consumption among older Americans: Nation-al Health Interview Surveys, 1997 to 2014. *Alcohol. Clin. Exp. Res.* **2017**, *41*, 976–986. [CrossRef]
80. Livingston, M.; Callinan, S.; Dietze, P.; Stanesby, O.; Kuntsche, E. Is there gender convergence in risky drinking when tak-ing birth cohorts into account? Evidence from an Australian national survey 2001–2013. *Addiction* **2018**, *113*, 2019–2028. [CrossRef]
81. Hamilton, K.; Keech, J.J.; Peden, A.E.; Hagger, M.S. Alcohol use, aquatic injury, and unintentional drowning: A system-atic literature review. *Drug Alcohol Rev.* **2018**, *37*, 752–773. [CrossRef]
82. Driscoll, T.R.; Harrison, J.; Steenkamp, M. Review of the role of alcohol in drowning associated with recreational aquatic activity. *Inj. Prev.* **2004**, *10*, 107–113. [CrossRef]
83. Warda, L.; Tenenbein, M.; Moffatt, M.E.K. House fire injury prevention update. Part I. A review of risk factors for fatal and non-fatal house fire injury. *Inj. Prev.* **1999**, *5*, 145–150. [CrossRef]
84. Turner, S.L.; Johnson, R.D.; Weightman, A.; Rodgers, S.; Arthur, G.; Bailey, R.; Lyons, R.A. Risk factors associated with unintentional house fire incidents, injuries and deaths in high-income countries: A systematic review. *Inj. Prev.* **2017**, *23*, 131–137. [CrossRef]
85. Bruck, D.; Ball, M.; Thomas, I.R. Fire fatality and alcohol intake: Analysis of key risk factors. *J. Stud. Alcohol Drugs* **2011**, *72*, 731–736. [CrossRef]
86. Runyan, C.W.; Bangdiwala, S.I.; Linzer, M.A.; Sacks, J.J.; Butts, J. Risk Factors for Fatal Residential Fires. *New Engl. J. Med.* **1992**, *327*, 859–863. [CrossRef]
87. Ball, M.; Bruck, D. The effect of alcohol upon response to fire alarm signals in sleeping young adults. In Proceedings of the 3rd International Symposium on Human Behaviour in Fire, Belfast, Northern Ireland, 1–3 September 2004; pp. 291–302.
88. Schreurs, C.J.; Van Hoof, J.J.; Van Der Lely, N. Hypothermia and acute alcohol intoxication in Dutch adolescents: The relationship between core and outdoor temperatures. *J. Subst. Use* **2016**, *22*, 449–453. [CrossRef]
89. Brändström, H.; Eriksson, A.; Giesbrecht, G.; Ängquist, K.-A.; Haney, M. Fatal hypothermia: An analysis from a sub-arctic region. *Int. J. Circumpolar Health* **2012**, *71*, 1–7. [CrossRef]
90. Kosiński, S.; Darocha, T.; Gałązkowski, R.; Drwiła, R. Accidental hypothermia in Poland–estimation of prevalence, diagnostic methods and treatment. *Scand. J. Trauma Resusc. Emerg. Med.* **2015**, *23*, 1–6. [CrossRef] [PubMed]
91. Makela, P. Alcohol-related mortality by age and sex and its impact on life expectancy: Estimates based on the Finnish death register. *Eur. J. Public Health* **1998**, *8*, 43–51. [CrossRef]
92. McNeilly, B.; Ibrahim, J.; Bugeja, L.; Ozanne-Smith, J. The prevalence of work-related deaths associated with alcohol and drugs in Victoria, Australia, 2001–2006. *Inj. Prev.* **2010**, *16*, 423–428. [CrossRef]
93. Ramirez, M.; Bedford, R.; Sullivan, R.; Anthony, T.R.; Kraemer, J.; Faine, B.; Peek-Asa, C. Toxicology Testing in Fatally Injured Workers: A Review of Five Years of Iowa FACE Cases. *Int. J. Environ. Res. Public Health* **2013**, *10*, 6154–6168. [CrossRef]
94. Webb, G.R.; Redman, S.; Hennrikus, D.J.; Kelman, G.R.; Gibberd, R.W.; Sanson-Fisher, R.W. The relationships between high-risk and problem drinking and the occurrence of work injuries and related absences. *J. Stud. Alcohol* **1994**, *55*, 434–446. [CrossRef]

95. Stallones, L.; Kraus, J.F. The occurrence and epidemiologic features of alcohol-related occupational injuries. *Addiction* **1993**, *88*, 945–951. [CrossRef] [PubMed]

96. Yoon, Y.-H.; Stinson, F.S.; Yi, H.-Y.; Dufour, M.C. Accidental* Alcohol Poisoning Mortality in the United States, 1996–1998. *Alcohol Res. Health J. Natl. Inst. Alcohol Abus. Alcohol.* **2003**, *27*, 110–118.

97. Moskalewicz, J.; Razvodovsky, Y.; Wieczorek, Ł. East–west disparities in alcohol-related harm. *Alcohol. Drug Addict.* **2016**, *29*, 209–222. [CrossRef]

98. Brent, J. Current Management of Ethylene Glycol Poisoning. *Drugs* **2001**, *61*, 979–988. [CrossRef]

99. Kruse, J.A. Methanol and Ethylene Glycol Intoxication. *Crit. Care Clin.* **2012**, *28*, 661–711. [CrossRef] [PubMed]

100. World Health Organisation. *Methanol Poisoning Outbreaks*; World Health Organisation: Geneva, Switzerland, 2014.

101. Nekoukar, Z.; Zakariaei, Z.; Taghizadeh, F.; Musavi, F.; Banimostafavi, E.S.; Sharifpour, A.; Ghuchi, N.E.; Fakhar, M.; Tabaripour, R.; Safanavaei, S. Methanol poisoning as a new world challenge: A review. *Ann. Med. Surg.* **2021**, *66*, 102445. [CrossRef]

102. Chisholm, D.; Moro, D.; Bertram, M.; Pretorius, C.; Gmel, G.; Shield, K.; Rehm, J. Are the "Best Buys" for Alcohol Control Still Valid? An Update on the Comparative Cost-Effectiveness of Alcohol Control Strategies at the Global Level. *J. Stud. Alcohol Drugs* **2018**, *79*, 514–522. [CrossRef]

103. Rehm, J.; Babor, T.F.; Casswell, S.; Room, R. Heterogeneity in trends of alcohol use around the world: Do policies make a difference? *Drug Alcohol Rev.* **2021**, *40*, 345–349. [CrossRef] [PubMed]

104. Gallet, C.A. The demand for alcohol: A meta-analysis of elasticities. *Aust. J. Agric. Resour. Econ.* **2007**, *51*, 121–135. [CrossRef]

105. Wagenaar, A.C.; Salois, M.J.; Komro, K.A. Effects of beverage alcohol price and tax levels on drinking: A meta-analysis of 1003 estimates from 112 studies. *Addiction* **2009**, *104*, 179–190. [CrossRef] [PubMed]

106. Sornpaisarn, B.; Shield, K.D.; Cohen, J.E.; Schwartz, R.; Rehm, J. Elasticity of alcohol consumption, alcohol-related harms, and drinking initiation in low- and middle-income countries: A systematic review and meta-analysis. *Int. J. Alcohol Drug Res.* **2013**, *2*, 45–58. [CrossRef]

107. Nelson, J.P. Meta-analysis of alcohol price and income elasticities–with corrections for publication bias. *Health Econ. Rev.* **2013**, *3*, 1–10. [CrossRef]

108. Wagenaar, A.C.; Tobler, A.L.; Komro, K.A. Effects of alcohol tax and price policies on morbidity and mortality: A systematic review. *Am. J. Public Health* **2010**, *100*, 2270–2278. [CrossRef]

109. Taylor, N.; Miller, P.; Coomber, K.; Livingston, M.; Scott, D.; Buykx, P.; Chikritzhs, T. The impact of a minimum unit price on wholesale alcohol supply trends in the Northern Territory, Australia. *Aust. N. Z. J. Public Health* **2021**, *45*, 26–33. [CrossRef]

110. Stockwell, T.; Zhao, J.; Marzell, M.; Gruenewald, P.J.; Macdonald, S.; Ponicki, W.R.; Martin, G. Relationships Between Minimum Alcohol Pricing and Crime During the Partial Privatization of a Canadian Government Alcohol Monopoly. *J. Stud. Alcohol Drugs* **2015**, *76*, 628–634. [CrossRef]

111. O'Donnell, A.; Anderson, P.; Jané-Llopis, E.; Manthey, J.; Kaner, E.; Rehm, J. Immediate impact of minimum unit pricing on alcohol purchases in Scotland: Controlled interrupted time series analysis for 2015–18. *BMJ* **2019**, *366*, l5274. [CrossRef]

112. Zhao, J.; Stockwell, T.; Martin, G.; Macdonald, S.; Vallance, K.; Treno, A.; Ponicki, W.R.; Tu, A.; Buxton, J. The relationship between minimum alcohol prices, outlet densities and alcohol-attributable deaths in British Columbia, 2002–2009. *Addiction* **2013**, *108*, 1059–1069. [CrossRef]

113. Jiang, H.; Livingston, M.; Manton, E. The effects of random breath testing and lowering the minimum legal drinking age on traffic fatalities in Australian states. *Inj. Prev.* **2014**, *21*, 77–83. [CrossRef]

114. Wagenaar, A.C.; Toomey, T.L. Effects of minimum drinking age laws: Review and analyses of the literature from 1960 to 2000. *J. Stud. Alcohol Suppl.* **2002**, *S206*, 206–225. [CrossRef] [PubMed]

115. Scribner, R.; Cohen, D. The Effect of Enforcement on Merchant Compliance with the Minimum Legal Drinking Age Law. *J. Drug Issues* **2001**, *31*, 857–866. [CrossRef]

116. Wagenaar, A.C.; Wolfson, M. Enforcement of the Legal Minimum Drinking Age in the United States. *J. Public Health Policy* **1994**, *15*, 37–53. [CrossRef]

117. Xuan, Z.; Naimi, T.S.; Kaplan, M.S.; Bagge, C.L.; Few, L.R.; Maisto, S.; Saitz, R.; Freeman, R. Alcohol Policies and Suicide: A Review of the Literature. *Alcohol. Clin. Exp. Res.* **2016**, *40*, 2043–2055. [CrossRef] [PubMed]

118. Carpenter, C.; Dobkin, C. The minimum legal drinking age and morbidity in the United States. *Rev. Econ. Stat.* **2017**, *99*, 95–104. [CrossRef]

119. Cook, W.K.; Bond, J.; Greenfield, T.K. Are alcohol policies associated with alcohol consumption in low- and middle-income countries? *Addiction* **2014**, *109*, 1081–1090. [CrossRef]

120. Wilkinson, C.; Livingston, M.; Room, R. Impacts of changes to trading hours of liquor licences on alcohol-related harm: A systematic review 2005–2015. *Public Health Res. Pr.* **2016**, *26*, e2641644. [CrossRef]

121. Fitterer, J.L.; Nelson, T.A.; Stockwell, T. A Review of Existing Studies Reporting the Negative Effects of Alcohol Access and Positive Effects of Alcohol Control Policies on Interpersonal Violence. *Front. Public Health* **2015**, *3*, 253. [CrossRef] [PubMed]

122. Gmel, G.; Holmes, J.; Studer, J. Are alcohol outlet densities strongly associated with alcohol-related outcomes? A critical review of recent evidence. *Drug Alcohol Rev.* **2015**, *35*, 40–54. [CrossRef]

123. Reynolds, J.; Wilkinson, C. Accessibility of 'essential' alcohol in the time of COVID-19, casting light on the blind spots of licensing? *Drug Alcohol Rev.* **2020**, *39*, 305–308. [CrossRef]

124. Hobin, E.; Smith, B. Is another public health crisis brewing beneath the COVID-19 pandemic? *Can. J. Public Health* **2020**, *111*, 392–396. [CrossRef]
125. Braillon, A. Alcohol control and the COVID-19 crisis on the other side of the Atlantic. *Can. J. Public Health* **2020**, *111*, 1–2. [CrossRef]
126. Andreasson, S.; Chikritzhs, T.; Dangardt, F.; Holder, H.; Naimi, T.; Sherk, A.; Stockwell, T. *Alcohol and Society 2021, Alcohol and the Coronavirus Pandemic: Individual, Societal and Policy Perspectives*; Swedish Society of Nursing; SFAM; SAFF; CERA; The Swedish Society of Addiction Medicine; SIGHT, Movendi International & IOGT-NTO: Stockholm, Sweden, 2021.
127. Callinan, S.; Mojica-Perez, Y.; Wright, C.J.C.; Livingston, M.; Kuntsche, S.; Laslett, A.-M.; Room, R.; Kuntsche, E. Purchasing, consumption, demographic and socioeconomic variables associated with shifts in alcohol consumption during the COVID-19 pandemic. *Drug Alcohol Rev.* **2021**, *40*, 183–191. [CrossRef] [PubMed]
128. Ryerson, N.C.; Wilson, O.W.; Pena, A.; Duffy, M.; Bopp, M. What happens when the party moves home? The effect of the COVID-19 pandemic on US college student alcohol consumption as a function of legal drinking status using longitudinal data. *Transl. Behav. Med.* **2021**, *11*, 772–774. [CrossRef]
129. Calina, D.; Hartung, T.; Mardare, I.; Mitroi, M.; Poulas, K.; Tsatsakis, A.; Rogoveanu, I.; Docea, A.O. COVID-19 pandemic and alcohol consumption: Impacts and interconnections. *Toxicol. Rep.* **2021**, *8*, 529–535. [CrossRef]
130. Sidpra, J.; Abomeli, D.; Hameed, B.; Baker, J.; Mankad, K. Rise in the incidence of abusive head trauma during the COVID-19 pandemic. *Arch. Dis. Child.* **2021**, *106*, e14. [CrossRef] [PubMed]
131. Piquero, A.R.; Jennings, W.G.; Jemison, E.; Kaukinen, C.; Knaul, F.M. Domestic violence during the COVID-19 pandemic-Evidence from a systematic review and meta-analysis. *J. Crim. Justice* **2021**, *74*, 101806–101815. [CrossRef]
132. Mclay, M.M. When "Shelter-in-Place" Isn't Shelter that's safe: A rapid analysis of domestic violence case differences during the COVID-19 pandemic and stay-at-home orders. *J. Fam. Violence* **2021**, *7*, 1–10.
133. Keric, D.; Stafford, J. Alcohol industry arguments for putting profit before health in the midst of a pandemic: The Western Australian experience. *Drug Alcohol Rev.* **2021**, *40*, 201–204. [CrossRef] [PubMed]
134. Opp, S.M.; Mosier, S.L. Liquor, marijuana, and guns: Essential services or political tools during the Covid-19 pandemic? *Policy Des. Pr.* **2020**, *3*, 297–311. [CrossRef]
135. Chikritzhs, T.; Gray, D.; Lyons, Z.; Saggers, S. *Restrictions on the Sale and Supply of Alcohol: Evidence and Outcomes*; National Drug Research Institute: Perth, WA, Australia, 2007.
136. Porthé, V.; García-Subirats, I.; Ariza, C.; Villalbí, J.R.; Bartroli, M.; Júarez, O.; Díez, E. Community-Based Interventions to Reduce Alcohol Consumption and Alcohol-Related Harm in Adults. *J. Commun. Health* **2021**, *46*, 565–576. [CrossRef]
137. Štelemėkas, M.; Manthey, J.; Badaras, R.; Casswell, S.; Ferreira-Borges, C.; Kalėdienė, R.; Lange, S.; Neufeld, M.; Petkevičienė, J.; Radišauskas, R.; et al. Alcohol control policy measures and all-cause mortality in Lithuania: An interrupted time–series analysis. *Addiction* **2021**. [CrossRef]
138. Stumbrys, D.; Telksnys, T.; Jasilionis, D.; Liutkutė Gumarov, V.; Galkus, L.; Goštautaitė Midttun, N.; Štelemėkas, M. Alcohol-related male mortality in the context of changing alcohol control policy in Lithuania 2000–2017. *Drug Alcohol Rev.* **2020**, *39*, 818–826. [CrossRef]
139. Miller, T.; Snowden, C.; Birckmayer, J.; Hendrie, D. Retail alcohol monopolies, underage drinking, and youth impaired driving deaths. *Accid. Anal. Prev.* **2006**, *38*, 1162–1167. [CrossRef] [PubMed]
140. Stockwell, T.; Sherk, A.; Norström, T.; Angus, C.; Ramstedt, M.; Andréasson, S.; Chikritzhs, T.; Gripenberg, J.; Holder, H.; Holmes, J.; et al. Estimating the public health impact of disbanding a government alcohol monopoly: Application of new methods to the case of Sweden. *BMC Public Health* **2018**, *18*, 1400. [CrossRef] [PubMed]
141. Shults, R.A.; Elder, R.W.; Sleet, D.A.; Nichols, J.L.; Alao, M.O.; Carande-Kulis, V.G.; Zaza, S.; Sosin, D.M.; Thompson, R.S. Reviews of evidence regarding interventions to reduce alcohol-impaired driving. *Am. J. Prev. Med.* **2001**, *21*, 66–88. [CrossRef]
142. Kaufman, E.J.; Wiebe, D.J. Impact of state ignition interlock laws on alcohol-involved crash deaths in the United States. *Am. J. Public Health* **2016**, *106*, 865–871. [CrossRef]
143. Wallin, E.; Norström, T.; Andréasson, S. Alcohol prevention targeting licensed premises: A study of effects on violence. *J. Stud. Alcohol* **2003**, *64*, 270–277. [CrossRef]
144. Graham, K.; Homel, R. *Raising the Bar*; Taylor & Francis: Abingdon, UK, 2008.
145. Wilson, I.M.; Graham, K.; Taft, A. Alcohol interventions, alcohol policy and intimate partner violence: A systematic review. *BMC Public Health* **2014**, *14*, 881. [CrossRef]
146. McMurran, M. *Alcohol-Related Violence: Prevention and Treatment*; John Wiley & Sons: Hoboken, NJ, USA, 2012.
147. Howland, J.; Birckmayer, J.; Hemenway, D.; Cote, J. Did changes in minimum age drinking laws affect adolescent drowning (1970–1990)? *Inj. Prev.* **1998**, *4*, 288–291. [CrossRef] [PubMed]
148. Leavy, J.E.; Crawford, G.; Portsmouth, L.; Jancey, J.; Leaversuch, F.; Nimmo, L.; Hunt, K. Recreational Drowning Prevention Interventions for Adults, 1990–2012: A Review. *J. Community Health* **2015**, *40*, 725–735. [CrossRef] [PubMed]
149. Pidd, K.; Roche, A.M. How effective is drug testing as a workplace safety strategy? A systematic review of the evidence. *Accid. Anal. Prev.* **2014**, *71*, 154–165. [CrossRef] [PubMed]
150. Stockwell, T.; Andreasson, S.; Cherpitel, C.; Chikritzhs, T.; Dangardt, F.; Holder, H.; Naimi, T.; Sherk, A. The burden of alcohol on health care during COVID-19. *Drug Alcohol Rev.* **2021**, *40*, 3–7. [CrossRef] [PubMed]
151. Medina-Mora, M.E.; Monteiro, M.; Rafful, C.; Samano, I. Comprehensive analysis of alcohol policies in the Latin America and the Caribbean. *Drug Alcohol Rev.* **2021**, *40*, 385–401. [CrossRef] [PubMed]

nutrients

Review

'Joining the Dots': Individual, Sociocultural and Environmental Links between Alcohol Consumption, Dietary Intake and Body Weight—A Narrative Review

Mackenzie Fong *, Stephanie Scott, Viviana Albani, Ashley Adamson and Eileen Kaner

Population Health Sciences Institute, Newcastle University, Newcastle-upon-Tyne NE1 4LP1, UK;
Stephanie.scott@newcastle.ac.uk (S.S.); viviana.albani@newcastle.ac.uk (V.A.);
ashley.adamson@newcastle.ac.uk (A.A.); Eileen.kaner@newcastle.ac.uk (E.K.)
* Correspondence: mackenzie.fong@newcastle.ac.uk

Abstract: Alcohol is energy-dense, elicits weak satiety responses relative to solid food, inhibits dietary fat oxidation, and may stimulate food intake. It has, therefore, been proposed as a contributor to weight gain and obesity. The aim of this narrative review was to consolidate and critically appraise the evidence on the relationship of alcohol consumption with dietary intake and body weight, within mainstream (non-treatment) populations. Publications were identified from a PubMed keyword search using the terms 'alcohol', 'food', 'eating', 'weight', 'body mass index', 'obesity', 'food reward', 'inhibition', 'attentional bias', 'appetite', 'culture', 'social'. A snowball method and citation searches were used to identify additional relevant publications. Reference lists of relevant publications were also consulted. While limited by statistical heterogeneity, pooled results of experimental studies showed a relatively robust association between acute alcohol intake and greater food and total energy intake. This appears to occur via metabolic and psychological mechanisms that have not yet been fully elucidated. Evidence on the relationship between alcohol intake and weight is equivocal. Most evidence was derived from cross-sectional survey data which does not allow for a cause-effect relationship to be established. Observational research evidence was limited by heterogeneity and methodological issues, reducing the certainty of the evidence. We found very little qualitative work regarding the social, cultural, and environmental links between concurrent alcohol intake and eating behaviours. That the evidence of alcohol intake and body weight remains uncertain despite no shortage of research over the years, indicates that more innovative research methodologies and nuanced analyses are needed to capture what is clearly a complex and dynamic relationship. Also, given synergies between 'Big Food' and 'Big Alcohol' industries, effective policy solutions are likely to overlap and a unified approach to policy change may be more effective than isolated efforts. However, joint action may not occur until stronger evidence on the relationship between alcohol intake, food intake and weight is established.

Keywords: alcohol; body weight; obesity; eating dietary intake; drinking pattern

Citation: Fong, M.; Scott, S.; Albani, V.; Adamson, A.; Kaner, E. 'Joining the Dots': Individual, Sociocultural and Environmental Links between Alcohol Consumption, Dietary Intake and Body Weight—A Narrative Review. *Nutrients* **2021**, *13*, 2927. https://doi.org/10.3390/nu13092927

Academic Editor: Peter Anderson

Received: 4 August 2021
Accepted: 23 August 2021
Published: 24 August 2021

Publisher's Note: MDPI stays neutral with regard to jurisdictional claims in published maps and institutional affiliations.

1. Introduction

Excess body weight and heavy alcohol consumption remain two of the most intractable public health challenges globally. Worldwide, around 39% of adults have overweight and 13% have obesity [1], while one in four adults in England and Scotland regularly consume over 14 units of alcohol per week; the maximum number of units considered 'low risk' based on the Chief Medical Officer's guidelines in the UK [2]. One in five drinkers in Great Britain binge drink on their heaviest drinking day (>8 units for men and >6 units for women) [2]. Obesity and alcohol misuse are leading causes of disability and disease, and disease burden attributable to elevated body mass index (BMI) and alcohol-related mortality and morbidity are greater in socioeconomically disadvantaged populations compared with more advantaged counterparts [3–5]. The phenomenon whereby people of lower socioeconomic position (SEP) consume the same (or less) alcohol as those from higher SEP yet experience greater alcohol-related harm is described as the alcohol-harm paradox [6]. This relationship is consistent internationally and across varying measures of SEP such as education, car ownership, income, employment, and housing tenure [7].

Like other nutritive beverages (mainly sugar-sweetened beverages (SSBs)), alcohol has been proposed as a causal factor of weight gain and obesity. Beverages elicit weak satiety signals relative to solid food [8–10] resulting in insufficient dietary compensation and greater daily energy intake [11,12]. However, compared to other nutritive beverages, alcohol has a very high energy density (29 kJ/g), providing more energy for volume than most (e.g., wine = 280 kJ/100 mL vs. cola drink = 177 kJ/100 mL). It is estimated that alcohol contributes around 10% of an adult drinker's weekly energy intake [13] which, in the UK, roughly equates to 690–770 kJ based on data from the National Diet and Nutrition Survey showing that daily energy intake of adults 19–64 years and 65 years and over was 7690 and 6900 kJ/day, respectively [14]. This is significant given that a reduction of 400 kJ/day may attenuate population-level weight gain [15]. Further, unlike other beverages, alcohol metabolism inhibits dietary fat oxidation [16], thereby promoting body fat storage. The pharmacological properties of alcohol are posited to stimulate food intake during the drinking occasion ostensibly through integrated cognitive and hedonic mechanisms [17]. Taken together, the nutritional properties, and metabolic and psychological effects of alcohol make it a logical driver of weight gain and obesity.

As with most research attempting to disentangle the relationship of weight with a single food/beverage item, evidence on the relationship between alcohol and weight is drawn mostly from heterogeneous cross-sectional studies, and findings are inconsistent [18]. To unpack this relationship further, researchers have investigated the role of several moderators, with sex, drinking pattern and type of alcoholic drink garnering most interest. Briefly, the hypotheses are that the relationship of alcohol and body weight is stronger: (a) in females; (b) in heavy drinkers; and (c) for beer/spirit intake. There are well-documented social and cultural differences in how men and women engage with alcohol [19,20], as well as biological sex differences in body composition, genetic factors, absorption, and metabolism [21,22]. Hypotheses regarding heavy drinking may be related to socioeconomic factors; people of lower SEP tend to binge drink more frequently than their counterparts from more advantaged backgrounds [23,24]. Similarly, drink preference may be socially patterned; wine is preferred by the middle class and is considered to be a source of cultural capital and distinction [25,26]. Wine drinking has been associated with more healthful dietary intake relative to beer drinking [27,28]. Further, standard serving sizes generally differ between drink types. In the UK, the standard serving size for beer is a 570 mL 'pint', providing ~880 kJ (full strength beer). For wine, the standard serving size is a 175 mL ('medium') glass, providing ~630 kJ. Spirits are often served with nutritive beverage mixers which contribute to greater energy content.

Given the clear public health importance of these two interrelated issues, and the breadth of available literature, it is important to synthesise and appraise this literature to understand the knowledge base and current thinking, identify areas of further research, and inform future interventions. Thus, the aim of this narrative review was to consolidate and critically appraise the evidence on the relationship between alcohol consumption, dietary intake, and body weight within mainstream (non-treatment) populations. We acknowledge that the relationship of alcohol and weight in people with alcohol dependency (AD) is distinct. This group is more likely to experience undernutrition and have a low BMI for reasons including substitution of food for alcohol [29], changes in appetite [29,30], and interferences with nutrient digestion, absorption, and metabolism [30]. While this review focuses on general (non-treatment) populations, we discuss diverse population groups (including people with AD) in the Section 4 of this paper. As intake of both alcohol and food occurs within complex socioecological contexts, we also aimed to explore how individual, sociocultural, and environmental factors shape consumption behaviours.

2. Materials and Methods

We undertook a narrative review to consolidate and critically appraise the literature exploring the impact of alcohol on food intake, the relationship of alcohol with body weight, and the individual-level mechanisms and sociocultural and environmental influences of alcohol and food intake. Publications were identified from a PubMed search using combinations of the keywords: 'alcohol', 'food', 'eating', 'weight', 'body mass index', 'obesity', 'food reward', 'inhibition', 'attentional bias', 'appetite', 'culture', 'social'. We used the snowball method and citation searches and consulted articles' reference lists to identify additional publications. Results are grouped thematically and examined below as follows: (1) The relationship between alcohol and dietary intake, and alcohol and body weight; (2) Individual-level mechanisms; (3) Social, cultural, and environmental influences on alcohol and food intake.

3. Results

3.1. Alcohol and Dietary Intake

A recent systematic review and meta-analysis identified 22 experimental trials investigating the effect of acute alcohol intake on ad libitum food energy intake (participants are invited to eat as much or as little food as they wish) and total energy intake (energy from beverages and food) [31]. While narrative synthesis conveyed that the effect of alcohol on food energy intake was inconsistent and dependent on the comparator condition (no/negligible energy non-alcoholic drink vs. energy containing non-alcoholic drink) and the alcohol dose, meta-analysis found that food energy intake was significantly greater in the alcohol condition compared to pooled effect of energy and non-energy containing non-alcoholic beverage comparators ($n = 12$; WMD 343 kJ; 95% CI 161, 525 kJ). With regards to total energy intake, 7/8 studies found that compared to no beverage and no/negligible-energy beverage comparators, alcoholic beverages significantly increased total energy intake. Comparisons with energy-containing beverages were mixed. Meta-analysis found that total energy intake was greater in the alcohol condition compared to the energy and non-energy containing non-alcoholic/no beverage conditions ($n = 8$; WMD 1072 kJ; 95% CI 820, 1323 kJ). In keeping with these findings, analyses of nationally representative survey data in Australian adults found that energy intake for those that reported alcohol consumption on the day of the dietary recall was significantly greater than those who did not. Between-participant, age-adjusted mean daily energy intake on alcohol vs. no alcohol days was 11,409 kJ vs. 9944 kJ ($p < 0.0001$) for men, and 8303 kJ vs. 7070 kJ ($p < 0.0001$) for women, respectively. This was a mean increase of 1514 kJ (462) for men and 1227 kJ (424) for women [32].

There are suggestions that alcohol may impact macronutrient intake, and this may have implications on weight as macronutrients may differentially affect energy metabolism [33]. Cummings and colleagues [34] systematically reviewed the evidence on the relationship of alcohol and macronutrient intake (refined and unrefined carbohydrate/fat/protein) intake. The 18 experimental studies identified in the search yielded mixed results. While there was a trend for a single occasion of light and moderate drinking to promote fat (8/18 studies) and protein intake (5/12 studies), most studies did not detect differences in macronutrient intake in response to these drinking behaviours. Similarly, findings from 12 observational studies, most of which used nationally representative cross-sectional datasets, were also inconsistent. The most consistent findings were of the inverse association between frequent heavy drinking and refined carbohydrate intake from foods such as candies, cereal (unspecified), and chocolate (reported in 90.9% of studies), but also intake of unrefined carbohydrate from foods such as fruits, grains, and vegetables (reported in 90% of studies). Meanwhile, Parekh et al. [35] recently conducted a nuanced analysis of the Framingham Heart Offspring Cohort study (1971–2008) and explored the relationship of drinking patterns with dietary intake, and how these shifted over time as participants aged. Across all data collection points, binge drinkers consumed less fruits and vegetables and wholegrains, and had greater total fat intake than non-binge drinkers. As participants aged, total fat intake increased in binge drinkers only.

3.2. Alcohol and Body Weight

Given there is relatively robust evidence of the effect of alcohol on food and total energy intake compared to non-alcoholic comparator beverages, it is plausible that, without sufficient energy compensation i.e., reduction in dietary intake and/or greater physical activity, alcohol intake can cause positive energy balance. Over time, depending on the frequency and intensity of alcohol consumption, this may lead to weight gain and contribute to obesity development (Figure 1). Very few experimental studies have investigated the longer-term effect of alcohol on weight. In a cross-over trial [36], 23 healthy men with abdominal obesity drank 450 mL of alcoholic red wine and de-alcoholised wine daily for 4 weeks, randomised by sequence. There was no significant between-condition difference in body weight nor deposition of subcutaneous, abdominal, or liver fat, although a positive trend was observed in the red-wine condition for the latter ($p = 0.09$). An earlier crossover trial found that daily consumption of two 135 mL servings of red wine over six weeks did not result in differences in body weight, energy intake or macronutrient composition in free-living, healthy males compared to a 6-week abstinence phase [37]. Similarly, a crossover study in women with overweight found no significant effects of alcohol consumption (two 135 mL servings of red wine on five days/week) vs. abstinence on body weight or dietary intake over 10 weeks [38].

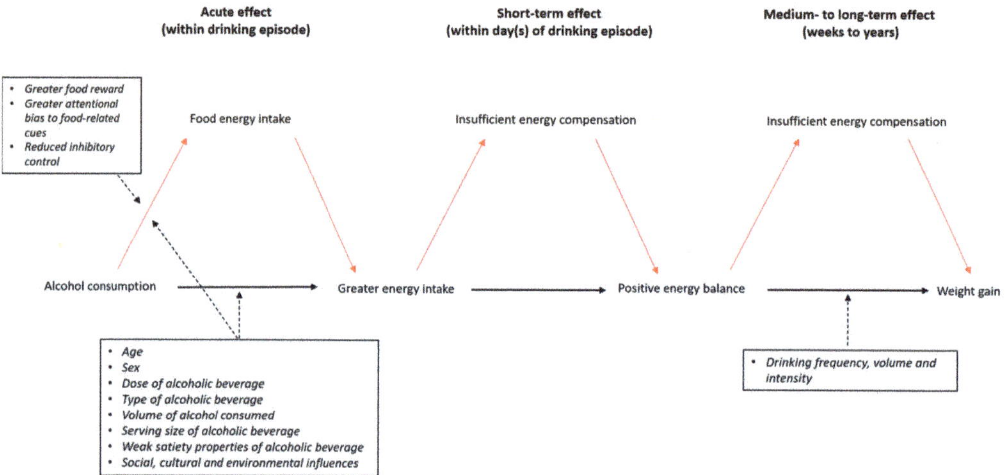

Figure 1. The hypothesised causal pathway from alcohol intake to weight gain. The red lines represent a proposed mediational relationship, and the dotted black lines represent proposed moderation. Body weight and energy compensation are influenced by a significant number of factors; these are beyond the scope of this review and have intentionally been omitted from this figure.

Sayon-Orea et al. [39] conducted the first systematic review of observational evidence on the association between alcohol consumption (all drink types) with body weight and other measures of adiposity. Of the 14 cross-sectional studies in adults, 9/14 studies (seven in men; two in women) reported a significant positive association of alcohol intake with BMI or weight gain, with a stronger positive relationship observed for heavy or binge drinking. However, a significant negative association was also observed in 9/14 studies (seven in women; two in men). For prospective cohort studies, 5/9 (three in men; two in women) showed a positive association of alcohol intake and weight gain or BMI. However, the absolute magnitude of differences was clinically important in just two studies. Associations of mixed direction were observed for cohort studies that reported measures of abdominal adiposity, although subgroup analyses in larger cohorts revealed only negative associations with wine drinking despite positive associations with total alcohol intake. One cohort study exhibited a U-shaped association between wine consumption and waist circumference, and three exhibited a J-shaped association, such that heavy (male and female) alcohol drinkers ($\geq$28 drinks/week) showed greatest weight gain, BMI or waist circumference. Travesy and Chaput [18] conducted an updated narrative review which included additional cross-sectional and longitudinal evidence. Attention was drawn to more recent evidence that examined the relationship of alcohol intake pattern, especially heavy drinking, with adiposity outcomes. For instance, Shelton and Knott [40] analysed data from the Health Survey for England and calculated % recommended dietary intake (RDA) from alcohol on participants' heaviest drinking day over the previous seven days. The risk of obesity was approximately 70% greater in the heaviest ($\geq$75%RDA) vs. lightest (0–25% RDA) groups. Concordant with Sayon-Orea et al., the authors concluded that while the literature is heterogeneous and findings are inconsistent, there is a trend for the relationship between alcohol and body weight to be non-linear (J-shaped), and that drinking intensity (heavy drinking) may be a more salient behaviour than drinking frequency.

Bendsen et al. [41] focused on beer intake specifically; their systematic review included 12 experimental studies testing the effect of beer intake vs. intake of low- or non-alcoholic beer/water/no substitute beverage on anthropometric outcomes. Experimental interventions lasted between 21–126 days and, for most, participants were prescribed a certain volume of beer to drink per day or per week. Random effects pooling found that beer consumption did not increase body weight compared to the no-substitute beverage comparator (mean difference 0.54 kg; 95% CI-1.00, 2.08; $p = 0.49$; $I^2 = 0$%). In contrast, consumption of alcoholic beer was found to increase body weight compared with no and low-alcohol beer (mean difference 0.73 kg; 95% CI 0.53, 0.92; $p < 0.0001$; $I^2 = 0$%). Among the ten prospective cohort studies and 25 cross-sectional studies included in the review, most showed a positive relationship or no relationship between beer intake and markers of obesity; in women a negative relationship was observed in several studies. Twenty-one studies reported obesity outcomes by level of beer intake, although heterogeneity in outcome measures and beer intake e.g., frequency vs. amount, prohibited quantitative data synthesis. The cumulative raw data did not indicate a dose-response at lower or moderate beer intake (~500 mL/day), but higher beer intake (>4 L/week) may be associated positively with abdominal obesity, particularly among men. The authors noted that most studies controlled for at least some potential confounders i.e., age, education, physical activity, and smoking, but that the degree of statistical adjustment varied widely.

Most recently, Golzarand et al. [42] conducted meta-analyses in 127 observational studies investigating the association between alcohol intake and markers of adiposity. Meta-analyses of cohort studies revealed no significant association between alcohol drinking and risk of overweight (HR 0.93; 95% CI 0.46, 1.89; $I^2 = 97.7$; $p = 0.84$), obesity (HR 0.84, 95% CI 0.52, 1.37; $I^2 = 90.7$; $p = 0.48$), overweight/obesity (HR 1.15; 95% CI 0.84, 1.58; $I^2 = 87.0$; $p = 0.37$), nor abdominal obesity (HR 1.13; 95% CI 0.90, 1.41; $I^2 = 61.0$; $p = 0.28$). In cross sectional studies, alcohol intake was associated with greater odds of having overweight (OR 1.11; 95% CI 1.05; 1.18; $I^2 = 87.7$; $p = 0.001$), but not obesity (OR 1.03; 95%CI 0.95, 1.12; $I^2 = 95.1$; $p = 0.48$). Subgroup analysis by alcohol dose revealed that heavy drinking (>28 g/day), but not light (<14 g/day), nor moderate drinking (14–28 g/day), was associated positively with overweight (OR 1.12; 95% CI 1.01, 1.24; $p = 0.02$). Regarding obesity, moderate drinkers had 16% lower odds of having obesity, but neither light nor heavy drinking were associated with greater odds of obesity. Those in the highest category of alcohol intake had 19% increased odds of abdominal obesity compared to those in the lowest category (OR 1.19; 95% CI 1.09, 1.29; $p < 0.001$). Heterogeneity between studies was very high and the authors rated most studies as being of very low quality/certainty based on the GRADE working group grades of evidence (i.e., the true effect is probably markedly different from the estimated effect) [43].

3.3. Alcohol and Weight Loss

A smaller number of studies have investigated the effect of alcohol intake on weight loss outcomes. In one randomised controlled trial (RCT), adults with overweight or obesity were prescribed hypo-energetic diets of 1500 cal/day (6300 kJ/day), of which 10% was to be consumed as either grape juice or white wine (200 mL). At 3 months follow-up, there were no significant differences in weight loss nor other anthropometric measures between groups [44]. Secondary analyses of a weight loss RCT found that while energy on no-alcohol days was significantly less than days alcohol was consumed, alcohol consumption was not associated with weekly weight loss [45]. Further, analyses of a large multicentre RCT (the Look AHEAD (Action for Health in Diabetes) study) identified that weight loss between participants who did and did not abstain from alcohol were similar during the first year regardless of study group. However, at Year 4, participants assigned to the Intensive Lifestyle Intervention (ILI) who abstained from alcohol lost 1.6% more weight relative to individuals who drank alcohol at any time during the intervention [46]. Kase and colleagues found that alcohol intake was not associated with weight before or after 26-week behavioural weight loss intervention, nor was change in alcohol intake related to change in weight [47]. However, reduction in alcohol was more relevant to weight loss for certain personality traits; specifically, those with higher levels of impulsivity experienced greater weight loss from reducing alcohol consumption.

3.4. Indiviudal-Level Mechanisms Linking Alcohol and Food Intake

3.4.1. Appetite, Hunger and Satiety

It is proposed that alcohol stimulates appetite; a phenomenon referred to as the aperitif effect [48]. However, several studies examining the effect of alcohol on levels of ghrelin (predominant hunger-stimulating hormone), have yielded mixed findings, with some even observing a reduction in ghrelin [18,49]. It is speculated that alcohol may enhance subjective appetite, however, most studies do not indicate a significant increase in self-reported appetite following an aperitif [48]. One of the more recent studies involved participants consuming either an alcohol (0.6 g/kg) or lemonade primer, rested for twenty minutes (to allow time for alcohol absorption) and then consumed snacks ad libitum for ten minutes [50]. Compared to the lemonade group, the alcohol groups experienced greater snack urge between baseline and the rest period, and experienced a smaller decline in appetite following snack consumption.

As with other nutritive beverages, alcohol is proposed to elicit weak satiety signals relative to solid food [8–10,51]. Compared to food, beverages have faster gastric transit times [52,53], demand less oral processing [54,55], reduce ghrelin suppression [56,57] and elicit lower cognitive perception of anticipated satiety [56,58]. The weaker satiety responses elicited by alcohol may not stimulate sufficient feelings of fullness needed to inhibit further energy intake.

3.4.2. Inhibitory Control

Inhibitory control is an umbrella term that describes the suppression of goal-irrelevant stimuli and behavioural responses [59]. Several studies have shown that alcohol can cause deficits in inhibitory control in relation to suppression of autonomic appetitive responses [60]. A study in female undergraduate students showed that consumption of cookies following an alcohol primer (compared to an alcohol-free placebo) was mediated by inhibitory control (assessed through performance on a Stroop task), such that poorer inhibitory control was associated with greater cookie consumption [61]. Impairment of inhibitory control may be more pronounced in people who have higher trait disinhibition [50] and dietary restraint [62], although evidence is conflicted [63]. Notably, even alcohol-related cues (e.g., alcohol odour, memory elicitation of drinking) in the absence of actual consumption has been shown to reduce inhibitory control [64].

3.4.3. Food-Related Reward and Attentional-Bias

Alcohol may stimulate greater food intake by enhancing its reward value. Studies using explicit self-report measures to assess the effect of alcohol on indices of food reward showed that intake of an alcohol primer increased: appetite, snack urge, ad libiutm intake and explicit liking of high-fat savoury foods [50,65,66].

It is posited that alcohol may increase attentional bias (AB) (i.e., selective attention) to food cues through classical conditioning [67] e.g., associations between an 'aperitif' before meals, 'drinks and nibbles', wine and cheese pairings etc. As AB is thought to indicate underlying appetitive motivational processes, several studies have used AB to food cues as an implicit measure of food reward. Between [68] and within-subjects [66] studies found that low dose alcohol primers (0.3–0.4 mg/kg) did not affect AB towards energy-dense foods relative to placebo. Gough et al. [66] observed that AB to food cues increased following a higher dose alcohol primer (0.6 mg/kg) compared to placebo, suggesting that these effects may only occur at higher doses. A smaller study in 23 young adults found that while low- (0.3 g/kg) and high-dose (0.65 g/kg) alcohol primers reduced alcohol-cue related AB relative to placebo, AB towards food-cues were sustained across all doses [69]. Recent evidence has indicated that alcohol (beer) odour increased attentional bias for food-cues, even in the absence of actual alcohol consumption [70].

3.5. Social, Cultural and Environmental Influences on Food and Alcohol Intake

3.5.1. The Interconnected Role of Food and Alcohol in Social and Cultural Life

Food and alcohol 'products' are a source of pleasure and a valued component of social spaces (especially within the night-time economy), as well as emotional and cultural life. A qualitative synthesis of 62 papers, identified that alcohol and food were both used to overcome personal problems, facilitate fun experiences, exercise control and restraint, and demonstrate a sense of identity in 10–17-year-olds [71]. However, we have a poorer understanding of how drinking and eating behaviours interact; just a handful of studies have explored this relationship in any depth, with most data focusing on young adult populations. The first (and, to our knowledge, only) qualitative study to examine the interconnectedness of food and alcohol consumption for UK young adults, found that sociocultural, physical and emotional links between food and alcohol were an unquestioned norm for 18–25-year-olds [72]. For interviewees, eating patterns whilst drinking alcohol were tied not only to hunger, but also to sociability, traditions, and identity. Further, young adults in this study conceptualised and calculated acute risks to weight, appearance, and social status, rather than risks to their long-term health. In a study focused upon multiple health behaviours (physical activity, nutrition, weight gain), Nelson et al. [73] found that US college students suggested their weight gain resulted, in part, from drinking alcohol and alcohol-related eating, including both eating late at night after alcohol was consumed as well as eating before going out to allow themselves to consume more alcohol.

3.5.2. Role of the Environment

Physical, economic, and political environments drive the availability and affordability of food and alcohol products which, in turn, influence their consumption. In developed countries, people have virtually unfettered access to alcohol and food products high in fat, sugar and salt (HFSS). Research has demonstrated an association between alcohol outlet density (a proxy marker of physical availability) and higher alcohol consumption [74,75], and between density of hot food takeaway outlets and obesity [3,76]. Importantly, both density of fast food [77,78] and alcohol outlets [79] are associated with greater socioeconomic deprivation. The recent proliferation of digital 'on demand' food and alcohol delivery services has enhanced their availability and may facilitate greater consumption [80–82], however, this requires more investigation. Regarding affordability, both HFSS food and alcohol products are relatively inexpensive. In the UK, affordability of alcohol has increased over time and is 13% more affordable than in 2008 [83]. High strength white ciders and spirits are especially cheap and may cost as little as 19 p per unit, meaning that 14 units (the maximum number of units considered 'low risk' based on the UK Chief Medical Officer's guidelines) are available for just £2.68 [84].

3.5.3. The Role of Food and Alcohol Industries

As corporate industries, 'Big Food' and 'Big Alcohol' use similar marketing and lobbying tactics with high degrees of co-operation [85–88]. For example, both industries have engaged in the use of 'framing'; whereby actors use discursive strategies to 'frame' a debate to benefit their corporate agenda [89]. Recent work by Rinaldi et al. [90] identified that different stakeholders in alcohol control (including industry) use framing discredit public health policy solutions; whilst similar studies have explored the use of framing in relation to the sugar tax [91] and food industry [92]. Petticrew et al. [93] identified the existence of a 'cross-industry' playbook represented by two incongruous 'frames': (i) aetiology of public health issues is complex, therefore, individual products cannot be blamed; and (ii) population health measures are 'too simple' to address complex public health problems.

Both industries also use strategic advertising and marketing to promote acceptability, and even glamorisation, of convenience food and alcohol consumption. Recent research demonstrated that marketing messages by 'Big Food' and 'Big Alcohol' were both adapted in the early stages of the pandemic. Martino et al. [94] analysed the extent and nature of online marketing by leading alcohol and food/beverage brands (and their parent companies) in Australia over a 4-month period. They found that nearly 80% of posts from brands studies related to COVID-19, with quick service restaurants, food and alcohol delivery companies, alcohol brands and bottle shops most active.

4. Discussion

4.1. Appraising the Evidence

We found good evidence from pooled laboratory studies showing that acute intake of alcoholic beverages consistently and significantly increased food and total energy intake within the drinking episode. This energy appears to be additive to energy from other sources, resulting in greater daily energy intake relative to alcohol-free days. However, the cumulative findings of the four previous reviews on alcohol intake and body weight are equivocal and do not support a fully consistent relationship. That the acute effect of alcohol on food intake did not translate to a robust relationship between alcohol and weight may speak to the poor external validity of laboratory studies; responses to alcohol manifest differently in a laboratory vs. real life setting [95]. Also, Kwok et al. noted that the findings of their meta-analyses of the effect of alcohol on food intake were only generalisable to younger adults aged 18–37 years, and that significant statistical heterogeneity and small effect sizes were observed for some outcomes.

Neither experimental nor observational evidence supported a clear association between alcohol and weight. Experimental studies were conducted in small samples and studies were likely to be underpowered to detect intervention effects. Similarly, the longest trial was conducted over ten weeks, possibly not long enough for group differences to emerge. An unclear relationship between alcohol and weight in observational evidence can be partially explained by significant heterogeneity among studies i.e., differences in assessment of drinking behaviour, cut-offs for levels of drinking, cut offs for abdominal obesity. Most epidemiological evidence was obtained from cross-sectional studies which carry the inherent limitation of not being able to determine a cause-effect relationship. Assessment of alcohol intake relies largely on self-report surveys e.g., Alcohol Use Disorders Identification Test (AUDIT-C) [96]. While these measures have adequate psychometric properties [97], they are prone to recall [98,99] and social desirability biases [100], and people tend to overestimate the amount of alcohol that constitutes a standard unit [101]. Taken together, there is a real risk for alcohol intake to be underreported which reduces confidence in studies' findings. Recall methods e.g., the 'Yesterday Method'—collection of detailed information about alcohol consumed the previous day—are also more likely to overestimate abstention [102]. Relating average alcohol intake at a single time point e.g., units per week, may lead to spurious findings, particularly if drinking patterns (i.e., binge) are not considered. Further, surveys may only capture current drinking behaviour, and 'non-drinkers' may include former drinkers who have previously experienced alcohol-related weight change; potentially reducing the validity of studies that compared drinkers to 'non-drinkers'.

There is also the significant potential for various factors to confound the relationship between alcohol and weight. Aside from the review by Bendsen et al. [41], others did not report the extent to which included studies statistically adjusted for confounders. Particularly pertinent is the adjustment for physical activity which could mitigate alcohol-induced energy intake. Systematic reviews have found that alcohol intake is positively (linearly or curvilinearly) associated with physical activity in young people, college-students, and the general adult population [103,104]. A cohort study in US college students found that weight motives mediated the positive relationship between heavy episodic drinking and vigorous (but not moderate) physical activity, suggesting that drinkers may engage in physical activity to compensate for additional energy intake or neutralise alcohol-related harms in line with the compensatory health behaviours model [105]. In the weight loss RCT by Carels et al. [45], duration of exercise was greater on days that alcohol was consumed, and participants who consumed alcohol more frequently had higher energy expenditure than those who drank less frequently. This may also explain why experimental studies in free-living participants found no relationship between alcohol and weight change. Another important confounding factor is 'delayed' dietary compensation. Studies in free-living participants have shown that corrective dietary compensation in response to deviations in daily energy intake was observed over a 24-hour period [12] and even up to three to four days later [106]. Therefore, while alcohol intake can increase daily energy intake, this additional energy may be offset in the following days. Distinctly, dietary compensation can be problematic when it manifests as 'drunkorexia' or 'alcorexia', characterised by dietary restriction, purging and/or excessive exercise to enable consumption of large quantities of alcohol and avoid weight gain and/or enhance intoxication [107–110].

Regarding the effect of moderating variables, the collective evidence gleaned from the four previous reviews of observational evidence does not support the hypotheses that drink type or sex moderates the relationship between alcohol and weight. That the effect of these moderators was inconsistent is expected given the heterogeneity and limitations mentioned previously. Notwithstanding these limitations, evidence of a J-shaped relationship between alcohol intake and weight-related outcomes (such that the positive association was strongest for heavy/binge drinking) were relatively consistent and is feasible given the substantial amount of energy consumed during a binge drinking episode. In analyses of a large national sample from England and Scotland in 18–25 year olds, Albani and colleagues [111] observed a significant positive association between alcohol consumption and BMI observed at Very High levels of intake (>75% RDA energy) in men and High to Very High intakes (>50% RDA energy) in women; equating to >1875 calories (7838 kJ) and >1000 calories (4180 kJ), respectively. This is a substantial amount of energy and would require significant dietary restriction and/or increased physical activity to mitigate the risk of weight gain. Further, drinking intensity is socially patterned and heavy/binge drinking is more prevalent in deprived groups [112–114]; SEP may moderate the relationship between heavy drinking and weight. This may be related to observations that those with lower SEP are more likely to engage in multiple health-risk behaviours (e.g., alcohol use, smoking, poor diet, lower physical activity) [115]. It is also suggested that heavy and/or binge drinking is linked to other behaviours implicated in weight gain. For instance, binge drinking is associated with impulsivity [116,117], which is also associated positively with energy intake and snacking, and negatively associated with diet quality [117].

4.2. Strengths and Limitations

This review consolidates an expansive body of literature on the relationship between alcohol intake, dietary intake, and body weight. We explored the topic from physiological, psychological, and socioecological perspectives, and included experimental, epidemiological, and qualitative evidence. Taking this comprehensive and holistic approach enabled us to 'join the dots' and present a nuanced and contextualised exploration of a complex issue. Given their inextricable link, we integrated evidence on food and alcohol intake to build a narrative that was more reflective of real life. In terms of limitations, neither the literature search nor article selection were conducted systematically and, therefore, selection of publications included in the review may have been biased. It is possible that some seminal papers may have been omitted unintentionally. Also, as several reviews on this topic having been conducted previously, we focused on their cumulative findings rather than those of individual studies. Therefore, there is a risk that biases, misreporting and/or misinterpretation from previous reviews may have been carried forward here.

4.3. Policy and Practice Implications

Our findings have several important policy and practice implications. First, alcohol and obesity policies are typically developed and implemented independently. Given significant commonalities between these two industries (as above), a unified approach to lobbying for policy change may be more effective than siloed efforts. Also, given synergies across these two industries, effective policy solutions are likely to overlap. Knowledge exchange and collaboration between policy makers may help to optimise the effectiveness of policies and interventions. Table 1 summarises population-level food and alcohol policies/interventions across shared targets, along with selected references for further reading. We acknowledge that reducing obesity and alcohol use require a whole-systems approach and no single intervention presents a panacea. We also acknowledge that a unified approach to policy change may not occur until stronger evidence on the relationship between alcohol use, food intake and weight is available (see Section 4.4 for recommendations for future research).

Table 1. Population-level food and alcohol policies across shared targets/points of intervention.

Target of Policy/Strategy	Foods and Non-Alcoholic Beverages	Alcoholic Beverages
Fiscal policy	• Taxation of sugar content in SSBs [118,119], and dietary fat content of foods [120]	• Minimum unit pricing [121–123] • Taxation of alcohol content [124,125]
Mass media and marketing	• Restrictions on advertising [126] e.g., UK government's 9 pm watershed on television advertising of food and drink products high in fat, sugar and salt (HFSS) [127] • Regulation of marketing e.g., HFSS food companies' sponsorship of, and advertising in sport environments [128,129]	• Restrictions on advertising [130,131] • Regulation of marketing e.g., alcohol sponsorship of, and advertising in sport environments [132,133]
Sales availability	• Regulation of takeaway outlet density e.g., exclusion zones around schools [134], planning regulations [135,136] • Regulation of digital on demand food delivery services [137]	• Regulation of alcohol outlet density [138,139] • Regulation of licensing hours [138,140] • Regulation of digital on demand alcohol delivery services [141]
Product server setting	• Information-based cue at point-of-purchase e.g., grocery store [142,143] • Regulation of volume-based price promotions e.g., '2 for 1 deals' [144]	• Information-based cue at point-of-purchase e.g., bar [145] • Regulation of price and volume promotions e.g., '2 for 1 deals' [146,147]
Product reformulation	• Production of reduced fat and sugar product varieties [148,149]	• Production of low- and no-alcohol beverage alternatives [123,150]
Product labelling	• Use of on-pack nutrition labelling [142] • Use of on-pack health warning labels [151,152]	• Use of on-bottle energy content labelling [153] • Use of on-bottle health warning labels [152,154]
Standard serving sizes	• Reduction of standard serving sizes [155,156]	• Reduction of standard serving sizes [157,158]

Second, there are only a handful of studies examining strategies that specifically aim to reduce energy intake from alcohol. Evidence of the effect of product reformulation strategies i.e., low- and no-alcohol drinks, is limited. A recent study found that low-alcohol products are perceived to target non-traditional consumers (pregnant women) and occasions (weekday lunchtimes), suggesting potential challenges to their uptake [159]. In a recent systematic review and meta-analysis, Robinson et al. [153] found moderate evidence that consumers are unaware of the energy content of alcoholic drinks, and that consumers support energy labelling. However, most studies found no effect of energy-labelling on actual or intended alcohol consumption; studies were generally of poor methodological quality and none were conducted in a real-life setting. Displaying the energy content of alcohol may result in complex public health messaging and disordered 'drunkorexia' or 'alcorexia' behaviours e.g., binge drinking, restriction of dietary intake/purging. These behaviours increase the likelihood of intoxication, result in blood alcohol levels rising sharply affecting the brain and subsequent behaviour, which in turn steeply increases the risk of acute harm such as from accidents. Some [160–162] but not all [163] studies suggest that such weight control behaviours are particularly prevalent amongst females and can lead to wider repercussions for health and wellbeing. For example, amongst UK women, consumption of alcohol without food is associated with higher risk of liver cirrhosis [164]. Meanwhile, in a prospective cohort study of UK Biobank (UKB) participants, Jani et al. [165] found a 10% higher risk of mortality with alcohol drinking without food compared to alcohol drinking with food amongst 38–73 year-olds.

Third, policy and practice strategies are more challenging to implement in marginalised or at-risk populations, such as those in alcohol recovery or experiencing homelessness or food-insecurity. Here, the evidence-base becomes sparser, with a small pool of quantitative research and a dearth of qualitative data. Again, most studies do not focus explicitly on this relationship in marginalised/at risk groups; usually this is discussed in the context of wider aims/objectives or with one product given a peripheral mention in the context of the other. Thus, Puddephatt et al. [166] found that a small minority of UK food bank clients reported using alcohol and other illicit substances to suppress hunger and to cope with the stress of their access to food. Meanwhile, Reitzel et al. [167] identified that, among US homeless adults, 28.4% of the sample had probable alcohol dependence, 25% were heavy drinkers, and 78.4% were food insecure. Further, heavy drinking and probable alcohol dependence/abuse were each associated with increased odds of food insecurity. Similar associations were identified by Bergmans et al. [168]. Tan and Johns [169] did specifically focus on the links between alcohol use and eating disorders amongst those alcohol dependent or in early recovery, and found the 'alcohol wheel' to be triggering for those more susceptible to an eating disorder in alcohol services. Further, Thomson and Paudel (2018) highlight the possibility of addiction swap (substitution of sugar for alcohol addiction) for those in alcohol recovery.

4.4. Future Research

Despite no shortage of observational research, the relationship between alcohol and weight remains uncertain. To genuinely improve our understanding of this relationship, rather than gratuitously generating further cross-sectional analyses, resources should be prioritised for research that uses innovative methodologies to make a novel and meaningful contribution to the literature, and that considers the complexities of the relationship and important variables linking food and alcohol consumption (regular drinking pattern and previous drinking status, physical activity levels, personality traits, health motives, SEP). Rather than examining the relationship of alcohol intake and weight which may be an oversimplification even with statistical adjustment, a latent class analysis may be more illuminating, given that weight-related behaviours and food/beverage consumption patterns tend to cluster. Methodologies that sample behaviour in the field such as ecological momentary assessment (EMA) [170] or diary studies [106] that are conducted over several weeks would provide valuable, richly detailed data on alcohol use and energy balance behaviours in real life. Similarly, more qualitative work is needed to improve our understanding around consumers' knowledge, perceptions and attitudes regarding alcohol, food, weight, and weight-related behaviours across the life course. This would help to identify and develop targeted interventions that are informed by real life experiences. Also, youth alcohol consumption has declined steadily and significantly over the last 15 years in many high-income countries including the UK, with those young people who do drink starting to do so at a later age, less often and in smaller quantities [171,172]. What, then, are the implications of this on eating behaviours and body weight? How consumers use no- and low-alcohol products requires further research i.e., who uses them, how are they used and for what reasons, are they substitutive or additive to the whole diet, what are facilitators and barriers to their uptake, do they lead to self-licensing (e.g., I had a low-alcohol beer, therefore, I can eat more chips). More generally, all future research must adequately represent people from marginalised and disadvantaged communities who are disproportionally affected by alcohol misuse and obesity.

5. Conclusions

Alcohol appears to increase food and energy intake within a drinking episode, and this energy increases daily energy intake. However, as with other research that attempts to disentangle the relationship of weight with a single nutrient/food/beverage/meal, evidence is equivocal and does not support a fully consistent relationship. Most evidence is derived from cross-sectional survey data and is limited by methodological weaknesses and heterogeneity. That the evidence remains uncertain despite no shortage of observational research indicates that more nuanced analyses and creative research methodologies are needed to capture what is clearly a complex and dynamic relationship. To meaningfully advance our understanding of the relationship between alcohol, food and weight, future research must account for relevant behaviours and variables and, importantly, how they cluster with patterns of alcohol intake e.g., binge drinking, and how they shift across the life course. Qualitative work and sampling weight-related behaviour and alcohol use in the field (e.g., using EMA protocols or diary studies) would add much needed richness and nuance to the literature. Given synergies between 'Big Food' and 'Big Alcohol' industries, effective policy solutions are likely to overlap and a unified approach to policy change may be more effective than isolated efforts. However, joint action may not occur until stronger evidence on the relationship between alcohol intake, food intake and weight is established.

Author Contributions: M.F. and S.S.: determined the direction of the review, designed the search strategy, conducted the literature search, synthesised search results, appraised the evidence, and drafted and refined the manuscript. A.A., E.K., V.A.: critically reviewed the manuscript. All authors have read and agreed to the published version of the manuscript.

Funding: This research received no external funding.

Institutional Review Board Statement: Not applicable.

Informed Consent Statement: Not applicable.

Data Availability Statement: Not applicable.

Acknowledgments: This research is supported by the National Institute of Health Research (NIHR) Applied Research Collaboration (ARC) for the North East and North Cumbria (NENC). The National Institute for Health Research (NIHR) is the nation's largest funder of health and care research and provides the people, facilities and technology that enables research to thrive. Find out more about the NIHR.

Conflicts of Interest: The authors declare no conflict of interest.

References

1. World Health Organisation. Overweight and Obesity 2021. Available online: https://www.who.int/news-room/fact-sheets/detail/obesity-and-overweight (accessed on 2 April 2021).
2. Office for National Statistics. Adult Drinking Habits in Great Britain 2018. Available online: https://www.ons.gov.uk/peoplepopulationandcommunity/healthandsocialcare/drugusealcoholandsmoking/datasets/adultdrinkinghabits (accessed on 7 June 2021).
3. Burgoine, T.; Forouhi, N.G.; Griffin, S.J.; Brage, S.; Wareham, N.J.; Monsivais, P. Does neighborhood fast-food outlet exposure amplify inequalities in diet and obesity? A cross-sectional study. *Am. J. Clin. Nutr.* **2016**, *103*, 1540–1547. [CrossRef] [PubMed]
4. Katikireddi, S.V.; Whitley, E.; Lewsey, J.; Gray, L.; Leyland, A.H. Socioeconomic status as an effect modifier of alcohol consumption and harm: Analysis of linked cohort data. *Lancet Public Health* **2017**, *2*, e267–e276. [CrossRef]
5. Dai, H.; Alsalhe, T.A.; Chalghaf, N.; Riccò, M.; Bragazzi, N.L.; Wu, J. The global burden of disease attributable to high body mass index in 195 countries and territories, 1990–2017: An analysis of the Global Burden of Disease Study. *PLoS Med.* **2020**, *17*, e1003198. [CrossRef]
6. Boyd, J.; Bambra, C.; Purshouse, R.C.; Holmes, J. Beyond Behaviour: How Health Inequality Theory Can Enhance Our Understanding of the 'Alcohol-Harm Paradox'. *Int. J. Environ. Res. Public Health* **2021**, *18*, 6025. [CrossRef]
7. Beard, E.; Brown, J.; West, R.; Angus, C.; Brennan, A.; Holmes, J.; Kaner, E.; Meier, P.; Michie, S. Deconstructing the Alcohol Harm Paradox: A Population Based Survey of Adults in England. *PLoS ONE* **2016**, *11*, e0160666. [CrossRef]
8. Mourao, D.M.; Bressan, J.; Campbell, W.W.; Mattes, R.D. Effects of food form on appetite and energy intake in lean and obese young adults. *Int. J. Obes.* **2007**, *31*, 1688–1695. [CrossRef]
9. Houchins, J.A.; Tan, S.Y.; Campbell, W.W.; Mattes, R.D. Effects of fruit and vegetable, consumed in solid vs beverage forms, on acute and chronic appetitive responses in lean and obese adults. *Int. J. Obes.* **2012**, *37*, 1109–1115. [CrossRef] [PubMed]
10. Wiessing, K.R.; Xin, L.; Budgett, S.C.; Poppitt, S.D. No evidence of enhanced satiety following whey protein- or sucrose-enriched water beverages: A dose response trial in overweight women. *Eur. J. Clin. Nutr.* **2015**, *69*, 1238–1243. [CrossRef] [PubMed]
11. Fong, M.; Li, A.; Hill, A.J.; Cunich, M.; Skilton, M.R.; Madigan, C.D.; Caterson, I.D. Modelling the Association between Core and Discretionary Energy Intake in Adults with and without Obesity. *Nutrients* **2019**, *11*, 683. [CrossRef]
12. McKiernan, F.; Hollis, J.; Mattes, R. Short-term dietary compensation in free-living adults. *Physiol. Behav.* **2008**, *93*, 975–983. [CrossRef] [PubMed]
13. Arif, A.A.; Rohrer, J.E. Patterns of alcohol drinking and its association with obesity: Data from the third national health and nutrition examination survey, 1988–1994. *BMC Public Health* **2005**, *5*, 126. [CrossRef]
14. Public Health England. National Diet and Nutrition Survey (NDNS) from Years 1 to 9: Data Tables 201. Available online: https://www.gov.uk/government/statistics/ndns-time-trend-and-income-analyses-for-years-1-to-9 (accessed on 7 June 2021).
15. Hill, J.O.; Wyatt, H.R.; Reed, G.W.; Peters, J.C. Obesity and the environment: Where do we go from here? *Science* **2003**, *299*, 853–855. [CrossRef] [PubMed]
16. Sozio, M.; Crabb, D.W. Alcohol and lipid metabolism. *Am. J. Physiol. Endocrinol. Metab.* **2008**, *295*, E10–E16. [CrossRef]
17. Yeomans, M.R. Alcohol, appetite and energy balance: Is alcohol intake a risk factor for obesity? *Physiol. Behav.* **2010**, *100*, 82–89. [CrossRef]
18. Traversy, G.; Chaput, J.-P. Alcohol Consumption and Obesity: An Update. *Curr. Obes. Rep.* **2015**, *4*, 122–130. [CrossRef] [PubMed]
19. Erol, A.; Karpyak, V.M. Sex and gender-related differences in alcohol use and its consequences: Contemporary knowledge and future research considerations. *Drug Alcohol Depend.* **2015**, *156*, 1–13. [CrossRef] [PubMed]
20. Wilsnack, R.W.; Wilsnack, S.C. *Gender and Alcohol: Consumption and Consequences. Alcohol: Science, Policy, and Public Health*; Oxford University Press: New York, NY, USA, 2013; pp. 153–160.
21. Baraona, E.; Abittan, C.S.; Dohmen, K.; Moretti, M.; Pozzato, G.; Chayes, Z.W.; Schaefer, C.; Lieber, C.S. Gender differences in pharmacokinetics of alcohol. *Alcohol Clin. Exp. Res.* **2001**, *25*, 502–507. [CrossRef] [PubMed]
22. Thomasson, H.R. Gender Differences in Alcohol Metabolism. In *Recent Developments in Alcoholism*; Springer: Boston, MA, USA, 2002; Volume 12.
23. Fone, D.L.; Farewell, D.; White, J.; Lyons, R.A.; Dunstan, F.D. Socioeconomic patterning of excess alcohol consumption and binge drinking: A cross-sectional study of multilevel associations with neighbourhood deprivation. *BMJ Open* **2013**, *3*, e002337. [CrossRef] [PubMed]
24. Beard, E.; Brown, J.; West, R.; Kaner, E.; Meier, P.; Michie, S. Associations between socio-economic factors and alcohol consumption: A population survey of adults in England. *PLoS ONE* **2019**, *14*, e0209442.
25. Scott, S.; Shucksmith, J.; Baker, R.; Kaner, E. 'Hidden Habitus': A Qualitative Study of Socio-Ecological Influences on Drinking Practices and Social Identity in Mid-Adolescence. *Int. J. Environ. Res. Public Health* **2017**, *14*, 611. [CrossRef]
26. Brierley-Jones, L.; Ling, J.; McCabe, K.E.; Wilson, G.B.; Crosland, A.; Kaner, E.F.; Haighton, C.A. Habitus of home and traditional drinking: A qualitative analysis of reported middle-class alcohol use. *Soc. Health Illn.* **2014**, *36*, 1054–1076. [CrossRef] [PubMed]
27. Sánchez-Villegas, A.; Toledo, E.; Bes-Rastrollo, M.; Martín-Moreno, J.; Tortosa, A.; Martinez-Gonzalez, M.A. Association between dietary and beverage consumption patterns in the SUN (Seguimiento Universidad de Navarra) cohort study. *Public Health Nutr.* **2008**, *12*, 351–358. [CrossRef] [PubMed]
28. Ruidavets, J.-B.; Bataille, V.; Dallongeville, J.; Simon, C.; Bingham, A.; Amouyel, P.; Arveiler, M.; Ducimetière, P.; Ferrières, J. Alcohol intake and diet in France, the prominent role of lifestyle. *Eur. Heart J.* **2004**, *25*, 1153–1162. [CrossRef]

29. De Timary, P.; Cani, P.D.; Duchemin, J.; Neyrinck, A.M.; Gihousse, D.; Laterre, P.-F.; Badaoui, A.; Leclercq, S.; Delzenne, N.M.; Stärkel, P. The loss of metabolic control on alcohol drinking in heavy drinking alcohol-dependent subjects. *PLoS ONE* **2012**, *7*, e38682. [CrossRef]

30. Ross, L.J.; Wilson, M.; Banks, M.; Rezannah, F.; Daglish, M. Prevalence of malnutrition and nutritional risk factors in patients undergoing alcohol and drug treatment. *Nutrition* **2012**, *28*, 738–743. [CrossRef] [PubMed]

31. Kwok, A.; Dordevic, A.L.; Paton, G.; Page, M.J.; Truby, H. Effect of alcohol consumption on food energy intake: A systematic review and meta-analysis. *Br. J. Nutr.* **2019**, *121*, 481–495. [CrossRef]

32. Grech, A.; Rangan, A.; Allman-Farinelli, M. Increases in Alcohol Intakes Are Concurrent with Higher Energy Intakes: Trends in Alcohol Consumption in Australian National Surveys from 1983, 1995 and 2012. *Nutrients* **2017**, *9*, 944. [CrossRef] [PubMed]

33. Ebbeling, C.B.; Swain, J.F.; Feldman, H.A.; Wong, W.W.; Hachey, D.L.; Garcia-Lago, E.; Ludwig, D.S. Effects of Dietary Composition on Energy Expenditure During Weight-Loss Maintenance. *JAMA* **2012**, *307*, 2627–2634. [CrossRef] [PubMed]

34. Cummings, J.R.; Gearhardt, A.N.; Ray, L.A.; Choi, A.K.; Tomiyama, A.J. Experimental and observational studies on alcohol use and dietary intake: A systematic review. *Obes Rev.* **2020**, *21*, e12950. [CrossRef] [PubMed]

35. Parekh, N.; Lin, Y.; Chan, M.; Juul, F.; Makarem, N. Longitudinal dimensions of alcohol consumption and dietary intake in the Framingham Heart Study Offspring Cohort (1971–2008). *Br. J. Nutr.* **2020**, *125*, 685–694. [CrossRef] [PubMed]

36. Beulens, J.W.; van Beers, R.M.; Stolk, R.P.; Schaafsma, G.; Hendriks, H.F. The effect of moderate alcohol consumption on fat dis-tribution and adipocytokines. *Obesity* **2006**, *14*, 60–66. [CrossRef]

37. Cordain, L.; Bryan, E.D.; Melby, C.L.; Smith, M.J. Influence of moderate daily wine consumption on body weight regulation and metabolism in healthy free-living males. *J. Am. Coll. Nutr.* **1997**, *16*, 134–139. [CrossRef] [PubMed]

38. Cordain, L.; Melby, C.L.; Hamamoto, A.E.; O'Neill, D.; Cornier, M.-A.; Barakat, H.A.; Israel, R.G.; Hill, J.O. Influence of moderate chronic wine consumption on insulin sensitivity and other correlates of syndrome X in moderately obese women. *Metabolism* **2000**, *49*, 1473–1478. [CrossRef] [PubMed]

39. Sayon-Orea, C.; Martinez-Gonzalez, M.A.; Bes-Rastrollo, M. Alcohol consumption and body weight: A systematic review. *Nutr. Rev.* **2011**, *69*, 419–431. [CrossRef] [PubMed]

40. Shelton, N.J.; Knott, C.S. Association Between Alcohol Calorie Intake and Overweight and Obesity in English Adults. *Am. J. Public Health* **2014**, *104*, 629–631. [CrossRef]

41. Bendsen, N.T.; Christensen, R.; Bartels, E.M.; Kok, F.J.; Sierksma, A.; Raben, A.; Astrup, A. Is beer consumption related to measures of abdominal and general obesity? A systematic review and meta-analysis. *Nutr. Rev.* **2012**, *71*, 67–87. [CrossRef]

42. Golzarand, M.; Salari-Moghaddam, A.; Mirmiran, P. Association between alcohol intake and overweight and obesity: A sys-tematic review and dose-response meta-analysis of 127 observational studies. *Crit. Rev. Food Sci. Nutr.* **2021**, 1–21. [CrossRef]

43. Guyatt, G.H.; Oxman, A.D.; Vist, G.E.; Kunz, R.; Falck-Ytter, Y.; Alonso-Coello, P.; Schünemann, H.J. GRADE: An emerging consensus on rating quality of evidence and strength of recommendations. *BMJ* **2008**, *336*, 924–926. [CrossRef]

44. Flechtner-Mors, M.; Biesalski, H.K.; Jenkinson, C.P.; Adler, G.; Ditschuneit, H.H. Effects of moderate consumption of white wine on weight loss in overweight and obese subjects. *Int. J. Obes.* **2004**, *28*, 1420–1426. [CrossRef]

45. Carels, R.A.; Young, K.M.; Coit, C.; Clayton, A.M.; Spencer, A.; Wagner, M. Skipping meals and alcohol consumption: The regulation of energy intake and expenditure among weight loss participants. *Appetite* **2008**, *51*, 538–545. [CrossRef]

46. Chao, A.M.; Wadden, T.A.; Tronieri, J.S.; Berkowitz, R.I. Alcohol Intake and Weight Loss During Intensive Lifestyle Intervention for Adults with Overweight or Obesity and Diabetes. *Obesity* **2018**, *27*, 30–40. [CrossRef]

47. Kase, C.A.; Piers, A.D.; Schaumberg, K.; Forman, E.M.; Butryn, M.L. The relationship of alcohol use to weight loss in the context of behavioral weight loss treatment. *Appetite* **2016**, *99*, 105–111. [CrossRef]

48. Eiler, W.J., 2nd; Džemidžić, M.; Case, K.R.; Soeurt, C.M.; Armstrong, C.L.; Mattes, R.D.; O'Connor, S.J.; Harezlak, J.; Acton, A.J.; Considine, R.V.; et al. The apéritif effect: Alcohol's effects on the brain's response to food aromas in women. *Obesity* **2015**, *23*, 1386–1393. [CrossRef]

49. Koopmann, A.; Schuster, R.; Kiefer, F. The impact of the appetite-regulating, orexigenic peptide ghrelin on alcohol use disorders: A systematic review of preclinical and clinical data. *Biol. Psychol.* **2016**, *131*, 14–30. [CrossRef]

50. Rose, A.; Hardman, C.; Christiansen, P. The effects of a priming dose of alcohol and drinking environment on snack food intake. *Appetite* **2015**, *95*, 341–348. [CrossRef] [PubMed]

51. Mattes, R.D.; Campbell, W.W. Effects of Food Form and Timing of Ingestion on Appetite and Energy Intake in Lean Young Adults and in Young Adults with Obesity. *J. Am. Diet Assoc.* **2009**, *109*, 430–437. [CrossRef]

52. Glasbrenner, B.; Pieramico, O.; Brecht-Krau, D.; Baur, M.; Malfertheiner, P. Gastric emptying of solids and liquids in obesity. *J. Mol. Med.* **1993**, *71*. [CrossRef]

53. Hoad, C.; Rayment, P.; Spiller, R.C.; Marciani, L.; Alonso, B.D.C.; Traynor, C.; Mela, D.; Peters, H.P.F.; Gowland, P.A. In Vivo Imaging of Intragastric Gelation and Its Effect on Satiety in Humans. *J. Nutr.* **2004**, *134*, 2293–2300. [CrossRef] [PubMed]

54. Lavin, J.; French, S.; Ruxton, C.; Read, N. An investigation of the role of oro-sensory stimulation in sugar satiety? *Int. J. Obes.* **2002**, *26*, 384–388. [CrossRef]

55. French, S.J.; Cecil, J.E. Oral, gastric and intestinal influences on human feeding. *Physiol. Behav.* **2001**, *74*, 729–734. [CrossRef]

56. Cassady, B.A.; Considine, R.V.; Mattes, R.D. Beverage consumption, appetite, and energy intake: What did you expect? *Am. J. Clin. Nutr.* **2012**, *95*, 587–593. [CrossRef] [PubMed]

57. Leidy, H.J.; Apolzan, J.; Mattes, R.D.; Campbell, W.W. Food Form and Portion Size Affect Postprandial Appetite Sensations and Hormonal Responses in Healthy, Nonobese, Older Adults. *Obesity* **2010**, *18*, 293–299. [CrossRef] [PubMed]
58. DiMeglio, D.; Mattes, R. Liquid versus solid carbohydrate: Effects on food intake and body weight. *Int. J. Obes.* **2000**, *24*, 794–800. [CrossRef] [PubMed]
59. Diamond, A. Executive Functions. *Ann. Rev. Psychol.* **2013**, *64*, 135–168. [CrossRef]
60. Field, M.; Wiers, R.; Christiansen, P.; Fillmore, M.T.; Verster, J.C. Acute Alcohol Effects on Inhibitory Control and Implicit Cognition: Implications for Loss of Control Over Drinking. Alcohol. *Clin. Exp. Res.* **2010**, *34*, 1346–1352. [CrossRef]
61. Christiansen, P.; Rose, A.; Randall-Smith, L.; Hardman, C.A. Alcohol's acute effect on food intake is mediated by inhibitory control impairments. *Health Psychol.* **2016**, *35*, 518–522. [CrossRef]
62. Caton, S.J.; Nolan, L.J.; Hetherington, M.M. Alcohol, Appetite and Loss of Restraint. *Curr. Obes. Rep.* **2015**, *4*, 99–105. [CrossRef]
63. Luce, K.H.; Crowther, J.H.; Leahey, T.; Buchholz, L.J. Do restrained eaters restrict their caloric intake prior to drinking alcohol? *Eat Behav.* **2013**, *14*, 361–365. [CrossRef]
64. Jones, A.; Robinson, E.; Duckworth, J.; Kersbergen, I.; Clarke, N.; Field, M. The effects of exposure to appetitive cues on inhibitory control: A meta-analytic investigation. *Appetite* **2018**, *128*, 271–282. [CrossRef]
65. Schrieks, I.C.; Stafleu, A.; Griffioen-Roose, S.; de Graaf, C.; Witkamp, R.; Boerrigter-Rijneveld, R.; Hendriks, H.F. Moderate alcohol consumption stimulates food intake and food reward of savoury foods. *Appetite* **2015**, *89*, 77–83. [CrossRef]
66. Gough, T.; Christiansen, P.; Rose, A.K.; Hardman, C.A. The effect of alcohol on food-related attentional bias, food reward and intake: Two experimental studies. *Appetite* **2021**, *162*, 105173. [CrossRef]
67. Castellanos, E.H.; Charboneau, E.; Dietrich, M.S.; Park, S.; Bradley, B.; Mogg, K.; Cowan, R.L. Obese adults have visual attention bias for food cue images: Evidence for altered reward system function. *Int. J. Obes.* **2009**, *33*, 1063–1073. [CrossRef] [PubMed]
68. Adams, S.; Wijk, E. Effects of Acute Alcohol Consumption on Food Intake and Pictorial Stroop Response to High-Calorie Food Cues. *Alcohol Alcohol.* **2020**, *56*, 275–283. [CrossRef] [PubMed]
69. Monem, R.; Fillmore, M.T. Alcohol administration reduces attentional bias to alcohol-related but not food-related cues: Evi-dence for a satiety hypothesis. *Psychol. Addict. Behav.* **2019**, *33*, 677–684. [CrossRef]
70. Karyadi, K.A.; Cyders, M.A. Preliminary support for the role of alcohol cues in food cravings and attentional biases. *J. Health Psychol.* **2017**, *24*, 812–822. [CrossRef]
71. Scott, S.; Elamin, W.; Giles, E.L.; Hillier-Brown, F.; Byrnes, K.; Connor, N.; Newbury-Birch, D.; Ells, L. Socio-Ecological Influences on Adolescent (Aged 10–17) Alcohol Use and Unhealthy Eating Behaviours: A Systematic Review and Synthesis of Qualitative Studies. *Nutrients* **2019**, *11*, 1914. [CrossRef] [PubMed]
72. Scott, S.; Muir, C.; Stead, M.; Fitzgerald, N.; Kaner, E.; Bradley, J.; Wrieden, W.; Power, C.; Adamson, A. Exploring the links between unhealthy eating behaviour and heavy alcohol use in the social, emotional and cultural lives of young adults (aged 18–25): A qualitative research study. *Appetite* **2019**, *144*, 104449. [CrossRef] [PubMed]
73. Nelson, M.C.; Kocos, R.; Lytle, L.A.; Perry, C.L. Understanding the Perceived Determinants of Weight-related Behaviors in Late Adolescence: A Qualitative Analysis among College Youth. *J. Nutr. Educ. Behav.* **2009**, *41*, 287–292. [CrossRef]
74. Ayuka, F.; Barnett, R.; Pearce, J. Neighbourhood availability of alcohol outlets and hazardous alcohol consumption in New Zealand. *Health Place* **2014**, *29*, 186–199. [CrossRef]
75. Weitzman, E.R.; Folkman, A.; Folkman, M.K.L.; Wechsler, H. The relationship of alcohol outlet density to heavy and frequent drinking and drinking-related problems among college students at eight universities. *Health Place* **2002**, *9*, 1–6. [CrossRef]
76. Burgoine, T.; Sarkar, C.; Webster, C.J.; Monsivais, P. Examining the interaction of fast-food outlet exposure and income on diet and obesity: Evidence from 51,361 UK Biobank participants. *Int. J. Behav. Nutr. Phys. Activity* **2018**, *15*, 71. [CrossRef] [PubMed]
77. Maguire, E.R.; Burgoine, T.; Monsivais, P. Area deprivation and the food environment over time: A repeated cross-sectional study on takeaway outlet density and supermarket presence in Norfolk, UK, 1990–2008. *Health Place* **2015**, *33*, 142–147. [CrossRef] [PubMed]
78. Public Health England. Obesity and the Environment. Density of Fast Food Outlets at 31 December 2017. Available on-line: https://assets.publishing.service.gov.uk/government/uploads/system/uploads/attachment_data/file/741555/Fast_Food_map.pdf (accessed on 23 July 2021).
79. Berke, E.M.; Tanski, S.E.; Demidenko, E.; Alford-Teaster, J.; Shi, X.; Sargent, J.D. Alcohol Retail Density and Demographic Predictors of Health Disparities: A Geographic Analysis. *Am. J. Public Health* **2010**, *100*, 1967–1971. [CrossRef]
80. Huckle, T.; Parker, K.; Romeo, J.S.; Casswell, S. Online alcohol delivery is associated with heavier drinking during the first New Zealand COVID -19 pandemic restrictions. *Drug Alcohol Rev.* **2020**, *40*, 826–834. [CrossRef] [PubMed]
81. Mojica-Perez, Y.; Callinan, S.; Livingston, M. Alcohol Home Delivery Services: An Investigation of Use and Risk 2019. Available online: https://fare.org.au/wp-content/uploads/Alcohol-home-delivery-services.pdf (accessed on 23 July 2021).
82. Keeble, M.; Adams, J.; Sacks, G.; Vanderlee, L.; White, C.M.; Hammond, D.; Burgoine, T. Use of Online Food Delivery Services to Order Food Prepared Away-From-Home and Associated Sociodemographic Characteristics: A Cross-Sectional, Multi-Country Analysis. *Int. J. Environ. Res. Public Health* **2020**, *17*, 5190. [CrossRef]
83. NHS Digital. Expenditure and Affordability 2020. Available online: https://digital.nhs.uk/data-and-information/publications/statistical/statistics-on-alcohol/2020/part-7 (accessed on 23 July 2021).
84. Alcohol Health Alliance. Small Change: Alcohol at Pocket Money Prices. *AHA Pricing Survey*. 2020. Available online: https://ahauk.org/wp-content/uploads/2021/04/Small-change-Final-04_2021-compressed.pdf (accessed on 23 July 2021).

85. Petticrew, M.; Maani, N.; Pettigrew, L.; Rutter, H.; VAN Schalkwyk, M.C. Dark Nudges and Sludge in Big Alcohol: Behavioral Economics, Cognitive Biases, and Alcohol Industry Corporate Social Responsibility. *Milbank Q.* **2020**, *98*, 1290–1328. [CrossRef]
86. Smith, R.; Kelly, B.; Yeatman, H.; Boyland, E. Food Marketing Influences Children's Attitudes, Preferences and Consumption: A Systematic Critical Review. *Nutrients* **2019**, *11*, 875. [CrossRef]
87. Viacava, K.R.; Weydmann, G.J.; De Vasconcelos, M.F.; Jaboinski, J.; Batista, G.D.; De Almeida, R.M.M.; Bizarro, L. It is pleasant and heavy: Convergence of visual contents in tobacco, alcohol and food marketing in Brazil. *Health Promot. Int.* **2015**, *31*, 674–683. [CrossRef]
88. Miller, D.; Harkins, C. Corporate strategy, corporate capture: Food and alcohol industry lobbying and public health. *Crit. Soc. Policy* **2010**, *30*, 564–589. [CrossRef]
89. Livonen, K. Defensive responses to strategic sustainability paradoxes: Have your coke and drink it too! *J. Bus. Ethics* **2018**, *148*, 309–327. [CrossRef]
90. Rinaldi, C.; van Schalkwyk, M.C.; Egan, M.; Petticrew, M. A Framing Analysis of Consultation Submissions on the WHO Global Strategy to Reduce the Harmful Use of Alcohol: Values and Interests. *Int. J. Health Policy Manag.* **2021**, in press. [CrossRef]
91. Campbell, N.; Mialon, M.; Reilly, K.; Browne, S.; Finucane, F.M. How are frames generated? Insights from the industry lobby against the sugar tax in Ireland. *Soc. Sci. Med.* **2020**, *264*, 113215. [CrossRef]
92. Lauber, K.; Ralston, R.; Mialon, M.; Carriedo, A.; Gilmore, A. Non-communicable disease governance in the era of the sustainable development goals: A qualitative analysis of food industry framing in WHO consultations. *Glob. Health* **2020**, *16*, 76. [CrossRef]
93. Petticrew, M.; Katikireddi, S.V.; Knai, C.; Cassidy, R.; Maani Hessari, N.; Thomas, J.; Weishaar, H. Nothing can be done until everything is done: The use of complexity arguments by food, beverage, alcohol and gambling industries. *J. Epidemiol. Community Health* **2017**, *71*, 1078–1083.
94. Martino, F.; Brooks, R.; Browne, J.; Carah, N.; Zorbas, C.; Corben, K.; Saleeba, E.; Martin, J.; Peeters, A.; Backholer, K. The Nature and Extent of Online Marketing by Big Food and Big Alcohol During the COVID-19 Pandemic in Australia: Content Analysis Study. *JMIR Public Health Surveill.* **2021**, *7*, e25202. [CrossRef]
95. Schoenmakers, T.; Wiers, R. Craving and Attentional Bias Respond Differently to Alcohol Priming: A Field Study in the Pub. *Eur. Addict. Res.* **2010**, *16*, 9–16. [CrossRef]
96. Bush, K.; Kivlahan, D.R.; McDonell, M.B.; Fihn, S.D.; Bradley, K.A. The AUDIT Alcohol Consumption Questions (AUDIT-C) An Effective Brief Screening Test for Problem Drinking. *Arch. Intern. Med.* **1998**, *158*, 1789–1795. [CrossRef]
97. McKenna, H.; Treanor, C.; O'Reilly, D.; Donnelly, M. Evaluation of the psychometric properties of self-reported measures of alcohol consumption: A COSMIN systematic review. *Subst. Abus. Treat. Prev. Policy* **2018**, *13*, 6. [CrossRef]
98. Gmel, G.; Daeppen, J.B. Recall bias for seven-day recall measurement of alcohol consumption among emergency department patients: Implications for case-crossover designs. *J. Stud. Alcohol Drugs.* **2007**, *68*, 303–310. [CrossRef]
99. Cherpitel, C.J.; Ye, Y.; Stockwell, T.; Vallance, K.; Chow, C. Recall bias across 7 days in self-reported alcohol consumption prior to injury among emergency department patients. *Drug Alcohol Rev.* **2017**, *37*, 382–388. [CrossRef]
100. Davis, C.G.; Thake, J.; Vilhena, N. Social desirability biases in self-reported alcohol consumption and harms. *Addict. Behav.* **2010**, *35*, 302–311. [CrossRef]
101. Devos-Comby, L.; Lange, J.E. My Drink is Larger than Yours? A Literature Review of Self-Defined Drink Sizes and Standard Drinks. *Curr. Drug Abus. Rev.* **2008**, *1*, 162–176. [CrossRef]
102. Gilligan, C.; Anderson, K.G.; Ladd, B.O.; Yong, Y.M.; David, M. Inaccuracies in survey reporting of alcohol consumption. *BMC Public Health* **2019**, *19*, 1639. [CrossRef]
103. Dodge, T.; Clarke, P.; Dwan, R. The Relationship Between Physical Activity and Alcohol Use Among Adults in the United States. *Am. J. Health Promot.* **2016**, *31*, 97–108. [CrossRef] [PubMed]
104. Piazza-Gardner, A.K.; Barry, A.E. Examining physical activity levels and alcohol consumption: Are people who drink more active? *Am. J. Health Promot.* **2012**, *26*, e95–e104. [CrossRef]
105. Matley, F.A.I.; Davies, E.L. Resisting temptation: Alcohol specific self-efficacy mediates the impacts of compensatory health beliefs and behaviours on alcohol consumption. *Psychol. Health Med.* **2017**, *23*, 259–269. [CrossRef]
106. Bray, G.A.; Flatt, J.-P.; Volaufova, J.; Delany, J.P.; Champagne, C.M. Corrective responses in human food intake identified from an analysis of 7-d food-intake records. *Am. J. Clin. Nutr.* **2008**, *88*, 1504–1510. [CrossRef] [PubMed]
107. Eisenberg, M.; Fitz, C.C. "Drunkorexia": Exploring the Who and Why of a Disturbing Trend in College Students' Eating and Drinking Behaviors. *J. Am. Coll. Health* **2014**, *62*, 570–577. [CrossRef]
108. Knight, A.; Castelnuovo, G.; Pietrabissa, G.; Manzoni, G.M.; Simpson, S. Drunkorexia: An Empirical Investigation among Australian Female University Students. *Aust. Psychol.* **2017**, *52*, 414–423. [CrossRef]
109. Lupi, M.; Martinotti, G.; Di Giannantonio, M. Drunkorexia: An emerging trend in young adults. *Eat. Weight. Disord. Stud. Anorex. Bulim. Obes.* **2017**, *22*, 619–622. [CrossRef]
110. Wilkerson, A.H.; Hackman, C.L.; Rush, S.E.; Usdan, S.L.; Smith, C.S. "Drunkorexia": Understanding eating and physical activity be-haviors of weight conscious drinkers in a sample of college students. *J. Am. Coll. Health* **2017**, *65*, 492–501. [CrossRef]
111. Albani, V.; Bradley, J.; Wrieden, W.L.; Scott, S.; Muir, C.; Power, C.; Fitzgerald, N.; Stead, M.; Kaner, E.; Adamson, A.J. Examining Associations between Body Mass Index in 18–25 Year-Olds and Energy Intake from Alcohol: Findings from the Health Survey for England and the Scottish Health Survey. *Nutrients* **2018**, *10*, 1477. [CrossRef]

112. Thern, E.; Landberg, J. Understanding the differential effect of alcohol consumption on the relation between socio-economic position and alcohol-related health problems: Results from the Stockholm Public Health Cohort. *Addiction* **2020**, *116*, 799–808. [CrossRef]
113. Boniface, S.; Lewer, D.; Hatch, S.; Goodwin, L. Associations between interrelated dimensions of socio-economic status, higher risk drinking and mental health in South East London: A cross-sectional study. *PLoS ONE* **2020**, *15*, e0229093. [CrossRef] [PubMed]
114. Sadler, S.; Angus, C.; Gavens, L.; Gillespie, D.; Holmes, J.; Hamilton, J.; Brennan, A.; Meier, P. Understanding the alcohol harm paradox: An analysis of sex- and condition-specific hospital admissions by socio-economic group for alcohol-associated conditions in England. *Addiction* **2017**, *112*, 808–817. [CrossRef]
115. Bellis, M.A.; Hughes, K.; Nicholls, J.; Sheron, N.; Gilmore, I.; Jones, L. The alcohol harm paradox: Using a national survey to explore how alcohol may disproportionately impact health in deprived individuals. *BMC Public Health* **2016**, *16*, 111. [CrossRef] [PubMed]
116. Adan, A.; Forero, D.; Navarro, J.F. Personality Traits Related to Binge Drinking: A Systematic Review. *Front. Psychiatry* **2017**, *8*, 134. [CrossRef] [PubMed]
117. Bénard, M.; Bellisle, F.; Kesse-Guyot, E.; Julia, C.; Andreeva, V.A.; Etilé, F.; Reach, G.; Dechelotte, P.; Tavolacci, M.-P.; Hercberg, S.; et al. Impulsivity is associated with food intake, snacking, and eating disorders in a general population. *Am. J. Clin. Nutr.* **2018**, *109*, 117–126. [CrossRef]
118. World Obesity. Sugar-Sweetened Beverage Tax: Systematic Reviews 2019. Available online: https://www.worldobesity.org/resources/policy-dossiers/pd-1/systematic-reviews (accessed on 23 July 2021).
119. World Obesity. Sugar-Sweetened Beverage Tax: Case Studies 2019. Available online: https://www.worldobesity.org/resources/policy-dossiers/pd-1/case-studies (accessed on 23 July 2021).
120. Lhachimi, S.K.; Pega, F.; Heise, T.L.; Fenton, C.; Gartlehner, G.; Griebler, U.; Sommer, I.; Bombana, M.; Katikireddi, S. Taxation of the fat content of foods for reducing their consumption and preventing obesity or other adverse health outcomes. *Cochrane Database Syst Rev.* **2020**, *9*, CD012415. [CrossRef]
121. Boniface, S.; Scannell, J.W.; Marlow, S. Evidence for the effectiveness of minimum pricing of alcohol: A systematic review and assessment using the Bradford Hill criteria for causality. *BMJ Open* **2017**, *7*, e013497. [CrossRef]
122. Anderson, P.; O'Donnell, A.; Kaner, E.; Llopis, E.J.; Manthey, J.; Rehm, J. Impact of minimum unit pricing on alcohol purchases in Scotland and Wales: Controlled interrupted time series analyses. *Lancet Public Health* **2021**, *6*, e557–e565. [CrossRef]
123. Llopis, E.J.; O'Donnell, A.; Anderson, P. Impact of price promotion, price, and minimum unit price on household purchases of low and no alcohol beers and ciders: Descriptive analyses and interrupted time series analysis of purchase data from 70, 303 British households, 2015–2018 and first half of 2020. *Soc. Sci. Med.* **2021**, *270*, 113690. [CrossRef] [PubMed]
124. Nelson, J.P.; McNall, A.D. Alcohol prices, taxes, and alcohol-related harms: A critical review of natural experiments in alcohol policy for nine countries. *Health Policy* **2016**, *120*, 264–272. [CrossRef] [PubMed]
125. Elder, R.W.; Lawrence, B.; Ferguson, A.; Naimi, T.; Brewer, R.D.; Chattopadhyay, S.K.; Toomey, T.L.; Fielding, J.E. The Effectiveness of Tax Policy Interventions for Reducing Excessive Alcohol Consumption and Related Harms. *Am. J. Prev. Med.* **2010**, *38*, 217–229. [CrossRef] [PubMed]
126. Boyland, E.J.; Nolan, S.N.; Kelly, B.; Tudur-Smith, C.; Jones, A.; Halford, J.C.; Robinson, E. Advertising as a cue to consume: A systematic review and meta-analysis of the effects of acute exposure to unhealthy food and nonalcoholic beverage advertising on intake in children and adults. *Am. J. Clin. Nutr.* **2016**, *103*, 519–533. [CrossRef]
127. Obesity Health Alliance. The Need for a Comprehensive 9pm Watershed on All Junk Food Adverts Shown on All Types of Media 2021. Available online: http://obesityhealthalliance.org.uk/wp-content/uploads/2019/03/The-need-for-a-comprehensive-9pm-watershed-Briefing.pdf (accessed on 23 July 2021).
128. Ireland, R.; Chambers, S.; Bunn, C. Exploring the relationship between Big Food corporations and professional sports clubs: A scoping review. *Public Health Nutr.* **2019**, *22*, 1888–1897. [CrossRef]
129. Smith, M.; Signal, L.; Edwards, R.; Hoek, J. Children's and parents' opinions on the sport-related food environment: A systematic review. *Obes. Rev.* **2017**, *18*, 1018–1039. [CrossRef]
130. Stautz, K.; Brown, K.G.; King, S.E.; Shemilt, I.; Marteau, T.M. Immediate effects of alcohol marketing communications and media portrayals on consumption and cognition: A systematic review and meta-analysis of experimental studies. *BMC Public Health* **2016**, *16*, 465. [CrossRef]
131. Courtney, A.L.; Rapuano, K.M.; Sargent, J.D.; Heatherton, T.F.; Kelley, W.M. Reward System Activation in Response to Alcohol Advertisements Predicts College Drinking. *J. Stud Alcohol Drugs* **2018**, *79*, 29–38. [CrossRef]
132. Westberg, K.; Stavros, C.; Smith, A.C.; Munro, G.; Argus, K. An examination of how alcohol brands use sport to engage consumers on social media. *Drug Alcohol Rev.* **2016**, *37*, 28–35. [CrossRef]
133. Jones, S. When does alcohol sponsorship of sport become sports sponsorship of alcohol? A case study of developments in sport in Australia. *Int. J. Sports Mark. Spons.* **2010**, *11*, 67–78. [CrossRef]
134. Williams, J.; Scarborough, P.; Matthews, A.; Cowburn, G.; Foster, C.; Roberts, N.; Rayner, M. A systematic review of the influence of the retail food environment around schools on obesity-related outcomes. *Obes. Rev.* **2014**, *15*, 359–374. [CrossRef] [PubMed]
135. Keeble, M.; Burgoine, T.; White, M.; Summerbell, C.; Cummins, S.; Adams, J. How does local government use the planning system to regulate hot food takeaway outlets? A census of current practice in England using document review. *Health Place* **2019**, *57*, 171–178. [CrossRef] [PubMed]

136. Keeble, M.; Adams, J.; White, M.; Summerbell, C.; Cummins, S.; Burgoine, T. Correlates of English local government use of the planning system to regulate hot food takeaway outlets: A cross-sectional analysis. *Int. J. Behav. Nutr. Phys. Act.* **2019**, *16*, 127. [CrossRef] [PubMed]

137. Goffe, L.; Uwamahoro, N.S.; Dixon, C.J.; Blain, A.P.; Danielsen, J.; Kirk, D.; Adamson, A.J. Supporting a Healthier Takeaway Meal Choice: Creating a Universal Health Rating for Online Takeaway Fast-Food Outlets. *Int. J. Environ. Res. Public Health* **2020**, *17*, 9260. [CrossRef]

138. Popova, S.; Giesbrecht, N.; Bekmuradov, D.; Patra, J. Hours and Days of Sale and Density of Alcohol Outlets: Impacts on Alcohol Consumption and Damage: A Systematic Review. *Alcohol Alcohol.* **2009**, *44*, 500–516. [CrossRef]

139. Campbell, C.A.; Hahn, R.A.; Elder, R.; Brewer, R.; Chattopadhyay, S.; Fielding, J.; Naimi, T.; Toomey, T.; Lawrence, B.; Middleton, J.C. The Effectiveness of Limiting Alcohol Outlet Density as a Means of Reducing Excessive Alcohol Consumption and Alcohol-Related Harms. *Am. J. Prev. Med.* **2009**, *37*, 556–569. [CrossRef]

140. Hahn, R.A.; Kuzara, J.L.; Elder, R.; Brewer, R.; Chattopadhyay, S.; Fielding, J.; Naimi, T.S.; Toomey, T.; Middleton, J.C.; Lawrence, B. Effectiveness of policies restricting hours of alcohol sales in preventing excessive alcohol consumption and related harms. *Am. J. Prev Med.* **2010**, *39*, 590–604. [CrossRef]

141. Colbert, S.; Wilkinson, C.; Thornton, L.; Feng, X.; Richmond, R. Online alcohol sales and home delivery: An international policy review and systematic literature review. *Health Policy* **2021**, in press. [CrossRef]

142. Gupta, A.; Billich, N.; George, N.A.; Blake, M.R.; Huse, O.; Backholer, K.; Boelsen-Robinson, T.; Peeters, A. The effect of front-of-package labels or point-of-sale signage on consumer knowledge, attitudes and behavior regarding sugar-sweetened beverages: A systematic review. *Nutr. Rev.* **2020**, in press. [CrossRef]

143. Chan, J.; McMahon, E.; Brimblecombe, J. Point-of-sale nutrition information interventions in food retail stores to promote healthier food purchase and intake: A systematic review. *Obes. Rev.* **2021**, in press. [CrossRef]

144. Watt, T.L.S.; Beckert, W.; Smith, R.D.; Cornelsen, L. Reducing consumption of unhealthy foods and beverages through banning price promotions: What is the evidence and will it work? *Public Health Nutr.* **2020**, *23*, 2228–2233. [CrossRef]

145. Carter, P.; Bignardi, G.; Hollands, J.; Marteau, T.M. Information-based cues at point of choice to change selection and consumption of food, alcohol and tobacco products: A systematic review. *BMC Public Health* **2018**, *18*, 418. [CrossRef]

146. Meier, P.; Booth, A.; Stockwell, T.; Sutton, A.; Wilkerson, A.; Wong, R. *Independent Review of the Effects of Alcohol Pricing and Promotion*; University of Sheffield: Sheffield, UK, 2008.

147. Jackson, R.; Johnson, M.; Campbell, F.; Messina, J.; Guillaume, L.; Meier, P.; Goyder, E.; Chilcott, J.; Payne, N. *Interventions on Control of Alcohol Price, Promotion and Availability for Prevention of Alcohol Use Disorders in Adults and Young People*; University of Sheffield: Sheffield, UK, 2010.

148. Gressier, M.; Swinburn, B.; Frost, G.; Segal, A.B.; Sassi, F. What is the impact of food reformulation on individuals' behaviour, nutrient intakes and health status? A systematic review of empirical evidence. *Obes. Rev.* **2020**, *22*, e13139. [CrossRef]

149. Hashem, K.M.; He, F.J.; A MacGregor, G. Effects of product reformulation on sugar intake and health—A systematic review and meta-analysis. *Nutr. Rev.* **2019**, *77*, 181–196. [CrossRef]

150. Rehm, J.; Lachenmeier, D.W.; Llopis, E.J.; Imtiaz, S.; Anderson, P. Evidence of reducing ethanol content in beverages to reduce harmful use of alcohol. *Lancet Gastroenterol. Hepatol.* **2016**, *1*, 78–83. [CrossRef]

151. Rosenblatt, D.H.; Bode, S.; Dixon, H.; Murawski, C.; Summerell, P.; Ng, A.; Wakefield, M. Health warnings promote healthier dietary decision making: Effects of positive versus negative message framing and graphic versus text-based warnings. *Appetite* **2018**, *127*, 280–288. [CrossRef] [PubMed]

152. Clarke, N.; Pechey, E.; Kosīte, D.; König, L.M.; Mantzari, E.; Blackwell, A.K.; Marteau, T.M.; Hollands, G.J. Impact of health warning labels on selection and consumption of food and alcohol products: Systematic review with meta-analysis. *Health Psychol. Rev.* **2020**, 1–24. [CrossRef] [PubMed]

153. Robinson, E.; Humphreys, G.; Jones, A. Alcohol, calories, and obesity: A rapid systematic review and meta-analysis of consumer knowledge, support, and behavioral effects of energy labeling on alcoholic drinks. *Obes Rev.* **2021**, *22*, e13198. [CrossRef] [PubMed]

154. Stautz, K.; Marteau, T.M. Viewing alcohol warning advertising reduces urges to drink in young adults: An online experiment. *BMC Public Health* **2016**, *16*, 530. [CrossRef] [PubMed]

155. Steenhuis, I.H.M.; Vermeer, W.M. Portion size: Review and framework for interventions. *Int. J. Behav. Nutr. Phys. Activity* **2009**, *6*, 58. [CrossRef]

156. Marteau, T.M.; Hollands, G.; Shemilt, I.; Jebb, S.A. Downsizing: Policy options to reduce portion sizes to help tackle obesity. *BMJ* **2015**, *351*, h5863. [CrossRef] [PubMed]

157. Kersbergen, I.; Oldham, M.; Jones, A.; Field, M.; Angus, C.; Robinson, E. Reducing the standard serving size of alcoholic beverages prompts reductions in alcohol consumption. *Addiction* **2018**, *113*, 1598–1608. [CrossRef] [PubMed]

158. Pilling, M.; Clarke, N.; Pechey, R.; Hollands, G.; Marteau, T. The effect of wine glass size on volume of wine sold: A mega-analysis of studies in bars and restaurants. *Addiction* **2020**, *115*, 1660–1667. [CrossRef] [PubMed]

159. Vasiljevic, M.; Couturier, D.-L.; Marteau, T.M. What are the perceived target groups and occasions for wines and beers labelled with verbal and numerical descriptors of lower alcohol strength? An experimental study. *BMJ Open* **2019**, *9*, e024412. [CrossRef] [PubMed]

160. Cummings, J.R.; Ray, L.A.; Tomiyama, A.J. Food-alcohol competition: As young females eat more food, do they drink less alcohol? *J. Health Psychol.* **2017**, *22*, 674–683. [CrossRef] [PubMed]
161. Farhat, T.; Iannotti, R.J.; Simons-Morton, B.G. Overweight, obesity, youth, and health-risk behaviors. *Am. J. Prev. Med.* **2010**, *38*, 258–267. [CrossRef] [PubMed]
162. Pasch, K.E.; Nelson, M.C.; Lytle, L.A.; Moe, S.G.; Perry, C.L. Adoption of Risk-Related Factors Through Early Adolescence: Associations with Weight Status and Implications for Causal Mechanisms. *J. Adolesc. Health* **2008**, *43*, 387–393. [CrossRef] [PubMed]
163. Peralta, R.L.; Barr, P.B. Gender orientation and alcohol-related weight control behavior among male and female college students. *J. Am. Coll. Health* **2016**, *65*, 229–242. [CrossRef]
164. Simpson, R.F.; Hermon, C.; Liu, B.; Green, J.; Reeves, G.K.; Beral, V.; Floud, S. Alcohol drinking patterns and liver cirrhosis risk: Analysis of the prospective UK Million Women Study. *Lancet Public Health* **2018**, *4*, e41–e48. [CrossRef]
165. Jani, B.D.; McQueenie, R.; Nicholl, B.I.; Field, R.; Hanlon, P.; Gallacher, K.I.; Mair, F.S.; Lewsey, J. Association between patterns of alcohol consumption (beverage type, frequency and consumption with food) and risk of adverse health outcomes: A prospective cohort study. *BMC Med.* **2021**, *19*, 8. [CrossRef] [PubMed]
166. Puddephatt, J.-A.; Keenan, G.; Fielden, A.; Reaves, D.; Halford, J.C.; Hardman, C.A. 'Eating to survive': A qualitative analysis of factors influencing food choice and eating behaviour in a food-insecure population. *Appetite* **2019**, *147*, 104547. [CrossRef] [PubMed]
167. Reitzel, L.R.; Chinamuthevi, S.; Daundasekara, S.S.; Hernandez, D.C.; Chen, T.-A.; Harkara, Y.; Obasi, E.M.; Kendzor, D.E.; Businelle, M.S. Association of Problematic Alcohol Use and Food Insecurity among Homeless Men and Women. *Int. J. Environ. Res. Public Health* **2020**, *17*, 3631. [CrossRef] [PubMed]
168. Bergmans, R.S.; Coughlin, L.; Wilson, T.; Malecki, K. Cross-sectional associations of food insecurity with smoking cigarettes and heavy alcohol use in a population-based sample of adults. *Drug Alcohol Depend.* **2019**, *205*, 107646. [CrossRef] [PubMed]
169. Tan, J.; Johns, G. *Drinking and Eating*; Alcohol Change UK: London, UK, 2019.
170. Wray, T.B.; Merrill, J.E.; Monti, P.M. Using Ecological Momentary Assessment (EMA) to Assess Situation-Level Predictors of Alcohol Use and Alcohol-Related Consequences. *Alcohol Res. Curr. Rev.* **2014**, *36*, 19–27.
171. Oldham, M.; Callinan, S.; Whitaker, V.; Fairbrother, H.; Curtis, P.; Meier, P.; Livingston, M.; Holmes, J. The decline in youth drinking in England—is everyone drinking less? A quantile regression analysis. *Addiction* **2019**, *115*, 230–238. [CrossRef]
172. Vashishtha, R.; Pennay, A.; Dietze, P.; Marzan, M.B.; Room, R.; Livingston, M. Trends in adolescent drinking across 39 high-income countries: Exploring the timing and magnitude of decline. *Eur. J. Public Health* **2020**, *31*, 424–431. [CrossRef]

Review

Production, Consumption, and Potential Public Health Impact of Low- and No-Alcohol Products: Results of a Scoping Review

Peter Anderson [1,2,*]**, Daša Kokole** [1] **and Eva Jané Llopis** [1,3,4]

[1]   Department of Health Promotion, CAPHRI Care and Public Health Research Institute, Maastricht University, 6200 MD Maastricht, The Netherlands; d.kokole@maastrichtuniversity.nl (D.K.); eva.jane@esade.edu (E.J.L.)
[2]   Population Health Sciences Institute, Newcastle University, Newcastle upon Tyne NE2 4AX, UK
[3]   ESADE Business School, University Ramon Llull, 08034 Barcelona, Spain
[4]   Institute for Mental Health Policy Research, Centre for Addiction and Mental Health (CAMH), Toronto, ON M5S 2S1, Canada
*   Correspondence: peteranderson.mail@gmail.com

Abstract: Switching from higher strength to low- and no-alcohol products could result in consumers buying and drinking fewer grams of ethanol. We undertook a scoping review with systematic searches of English language publications between 1 January 2010 and 17 January 2021 using PubMed and Web of Science, covering production, consumption, and policy drivers related to low- and no-alcohol products. Seventy publications were included in our review. We found no publications comparing a life cycle assessment of health and environmental impacts between alcohol-free and regular-strength products. Three publications of low- and no-alcohol beers found only limited penetration of sales compared with higher strength beers. Two publications from only one jurisdiction (Great Britain) suggested that sales of no- and low-alcohol beers replaced rather than added to sales of higher strength beers. Eight publications indicated that taste, prior experiences, brand, health and wellbeing issues, price differentials, and overall decreases in the social stigma associated with drinking alcohol-free beverages were drivers of the purchase and consumption of low- and no-alcohol beers and wines. Three papers indicated confusion amongst consumers with respect to the labelling of low- and no-alcohol products. One paper indicated that the introduction of a minimum unit price in both Scotland and Wales favoured shifts in purchases from higher- to lower-strength beers. The evidence base for the potential beneficial health impact of low- and no-alcohol products is very limited and needs considerable expansion. At present, the evidence base could be considered inadequate to inform policy.

Keywords: no-alcohol products; low-alcohol products; production; consumption; health impact

Citation: Anderson, P.; Kokole, D.; Llopis, E.J. Production, Consumption, and Potential Public Health Impact of Low- and No-Alcohol Products: Results of a Scoping Review. *Nutrients* **2021**, *13*, 3153. https://doi.org/10.3390/nu13093153

Academic Editor: Joerg Koenigstorfer

Received: 28 July 2021
Accepted: 8 September 2021
Published: 10 September 2021

Publisher's Note: MDPI stays neutral with regard to jurisdictional claims in published maps and institutional affiliations.

1. Introduction

Ethanol in alcoholic beverages is toxic to human health. Whilst consumption of up to 30 g of ethanol a day may be associated with a reduced risk of ischemic heart disease compared with no consumption [1], ethanol is genotoxic and a carcinogen, with no level of risk-free consumption [2].

Alcohol is a risk factor for early death. At an individual level, forty-year-olds who drink more than 350 g of alcohol per week (about five drinks a day) lose four to five years of life compared with those who drink 100 g of alcohol or less per week (approximately one and a half drinks a day) [3]. At a global level, alcohol is the cause of approximately 3 million deaths each year [4].

Reducing alcohol consumption reduces the risk of dying prematurely and the likelihood of a wide range of conditions, including cancer, elevated blood pressure, stroke, liver disease, mental health disorders, and accidents and injuries [5].

There are many strategies that enable people to drink less alcohol. For example, the WHO SAFER initiative calls on governments at all levels to (i) strengthen restrictions on alcohol availability; (ii) advance and enforce drinking and driving countermeasures; (iii) facilitate access to screening, brief interventions, and treatment; (iv) enforce bans or comprehensive restrictions on alcohol advertising, sponsorship, and promotion; and (v) raise prices on alcohol through excise taxes and pricing policies [6].

As an additional strategy, there is a growing discourse around the potential public health benefit of low- and no-alcohol products (alcohol-free and low-alcohol versions of alcoholic drinks such as beer, wine, and spirits) [7,8]. The WHO's global alcohol strategy called on the alcohol industry to contribute to reducing the harmful use of alcohol by addressing its products [9], for example, by reducing the amount of alcohol they contain. In its consultation document, 'Advancing our health: prevention in the 2020s,' the UK Government made a commitment to work with the alcohol industry to deliver a significant increase in the availability of alcohol-free and low-alcohol products by 2025; [10].

Low- and no-alcohol products can only be of public health benefit if they replace rather than add to existing consumption of higher strength products. In addition, any such potential health benefits resulting from the production and replacement consumption of low- and no-alcohol products should be offset against any environmental external costs (for example, due to extra steps in production), which can be assessed through life cycle assessments.

To inform the discourse on low- and no-alcohol products, we have undertaken a scoping review on their production, consumption, and potential health impact. In the review, we identify five research questions, for which we aim to synthesise knowledge about low- and no-alcohol products related to:

1. Production, including life cycle assessment compared with production of regular strength products
2. Prevalence of purchase and consumption
3. Potential health impact
4. Consumer perceptions and preferences
5. Policy drivers of purchase and consumption.

2. Methods

2.1. Design and Registration

We used scoping review methodology because of the breadth of the research questions and the lack of clarity on the amount and nature of existing research on this topic. Scoping reviews are used to map the main concepts in research areas and present a broad overview of the existing evidence, including the identification of research gaps, regardless of the study quality [11]. The design was guided by the methodological framework of Arksey and O'Malley [12]. The review protocol was pre-registered at https://osf.io/kv3rj/ (accessed on 1 September 2021).

2.2. Eligibility Criteria

To be included in the review, papers had to include topics related to the production, consumption, and health impact of low- and no-alcohol products. There was no restriction on research design. To delineate the scope of our research, we only focused on peer-reviewed literature rather than grey literature. Original articles and reviews in English published in 2011 or later were included. For production, only reviews were included, as a detailed examination of production methods was not the focus of the paper. Papers not specific to low- and no-alcohol products were excluded.

2.3. Information Sources and Search Strategy

Two databases (PubMed and Web of Science) were searched with the abovementioned restrictions on language and dates (English language, published between 1 January 2011 and 17 January 2021). The search strategy contained blocks with terms related to low- and no- alcoholic products ('Low alcohol' or 'No alcohol' or 'Zero alcohol' or 'Alcohol-free' or 'Alcohol free' or Reformulation or Reduc* ethanol content or Reduc* ethanol strength or

Reduc* 'alcohol strength' or Reduc* 'alcohol content' or Low strength alcohol OR 'non-alcoholic') and (beer or cider or wine or spirits or ready to drink or fortified wine or fermented beverages or intermediate products) in combination with other blocks related to production, consumption, and impact, producing several searches (see Tables S1 and S2 for a description of blocks and full search strategy). Database searches were complemented with Google Scholar inspection and reference searches.

2.4. Study Selection and Summary

Study selection was performed by two researchers (DK and PA) on the basis of the abovementioned inclusion and exclusion criteria. After the removal of duplicates, the studies were first screened by title and abstract by one researcher, followed by full-text examination and final article selection. Any doubts were discussed and resolved by consensus. The selected articles were grouped according to their main themes. A data extraction form was prepared to collect information on the authors, year of publication, paper objectives, studied topic, and key findings or conclusions. No quality appraisal of studies was undertaken, as the purpose of the review was to map all the available literature. For each theme, the main findings are presented in a narrative manner.

3. Results

In total, 3024 papers were identified across the eight conducted searches, and five papers were identified from other sources (Figure 1). After the removal of duplicates across the two databases and between the searches, 1121 papers remained for title and abstract screening. Ninety papers were selected for full-text inspection, and 70 of them were selected for final qualitative synthesis (see Table S3).

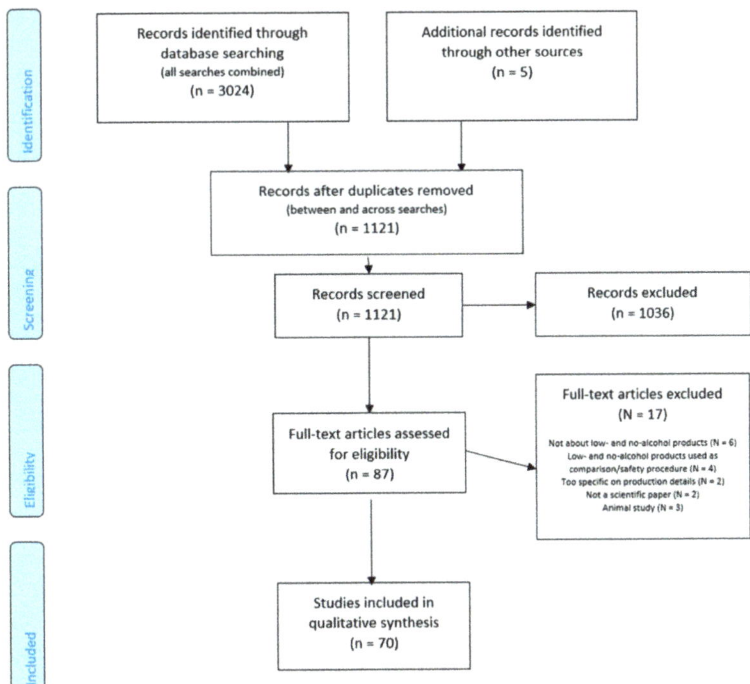

Figure 1. PRISMA flow diagram.

We matched the thematic analysis of each paper's topic to the five main research questions: production and life-cycle assessment, consumption and purchase, the potential impact on health, perception and preferences, and policy drivers of purchase and consumption. The results are presented within these categories. Additionally, several reviews were found which encompassed broader overviews of the selected topics and sometimes covered several themes at once [13–20]; thus, their findings are presented where appropriate. Along with the key findings of the papers, we also examined the definition of low- and no-alcohol products and the funding sources for the research (see Table S4).

3.1. Production and Life Cycle Assessment

In terms of the production of low- and non-alcoholic beverages, the majority of the research focused on beer and wine. For non-alcoholic or low-alcohol beer, the reason given for the importance of improving production used in many papers is flavour is one of the main problems with non-alcoholic beer acceptance. Thus, much of the production research has focused on how to make the flavour more acceptable and similar to alcoholic beer and whether biological (limiting ethanol formation during beer fermentation) or physical (removing ethanol from regular beer) approaches are more suitable to achieve this goal [21–27]. Most of the reviews refer to regulations when defining beer products; for example, EU regulations require 0.5% alcohol by volume (ABV) or less for non-alcoholic beer and 1.2% ABV or less for low-alcohol beer, but they acknowledge that these regulations differ by country. Other directions of research are related to low alcoholic or alcohol-free beers within the niche of craft beer [13] and dry hopping as a technique to produce non-alcoholic beer [14].

In the field of wine, the de-alcoholisation research has two directions: one objective is to decrease the alcohol strength in wine, and the second is to produce new low-alcohol beverages [28]. With regard to the first objective, the main issue that the wine industry faces is the increase in wine ABV due to climate change conditions (e.g., in Australian red wines, alcohol concentration has increased approximately 1% ABV per decade since the 1980s [29]). Thus, research has focused on how to reduce the elevated alcohol content in wine without leading to a taste that is not as well-accepted by consumers (e.g., [29–31]). For the second objective, the issue is also to overcome the loss of desirable sensory properties appreciated by consumers when alcohol content is reduced/removed from wine, with research focusing on identifying the most appropriate methods to achieve this goal (e.g., [32–35]). In terms of the definition of lower-alcohol wine products, the reviews have acknowledged country differences in regulations: one review mentioned the definitions adopted by the International Organisation of Vine and Wine: 'Beverages obtained by wine dealcoholization' for beverages with 0.5% ABV or lower, and 'Beverages obtained by partial wine dealcoholization' for those in the range of 0.5–8.5% ABV [28]; other reviews have defined wine products as lower-alcohol with an ABV of up to 11.0% [33,34].

While the comparison of various production methods is difficult, and thus it is difficult to define the best processes [22], some attempts at life cycle assessment have been carried out for alcohol-free bitter extracts as aperitifs [36,37] as well as for partial de-alcoholisation of wines [38]. We found no published life cycle assessments comparing the production of zero-alcohol beers or wines with the production of regular beers or wines.

As mentioned previously, no research was found on alcohol-free or low-alcohol spirits, but another review examined the field of traditional low-alcoholic and non-alcoholic fermented beverages [39], such as kefir, boz, or kvass. Finally, we found one study of ethanol production in non-alcoholic beer across the storage period, which found that storage temperature and packaging can have significant effects on ethanol production during the storage period, although the excess production does not exceed the allowed amount of 0.5% ABV [40].

3.2. Consumption and Purchase

Relatively little research in peer-reviewed literature has focused on low- and no-alcohol drink consumption and purchase trends, and all studies have focused on non-alcoholic or low-alcohol beer (none on wine or spirits). One study [41,42] examined the introduction of new low- and no-alcohol beers (defined as beers with 3.5% ABV or less) and reformulated beers in Great Britain and found that the volume of purchases of new low- and no-alcohol beer products (2.6% of the volume of all beers purchased in 2018) and new reformulated beer products (6.9% of the volume of all beers purchased in 2018) was very small. More widely, an analysis of official data from ethanol beer sales in Australia and New Zealand [43] showed that in Australia, the consumption of ethanol in mid-strength beer (3.01–3.5% ABV) increased, whereas consumption of low-strength beer (<3% ABV) decreased between 2000 and 2016. In New Zealand, the consumption of mid-strength beer (2.501–4.35% ABV) decreased substantially.

3.3. Impact on Health

Overall, only one public health-oriented review has been conducted to examine the evidence base of the reduction of ethanol content of alcoholic beverages as a means to reduce the harmful use of alcohol [20]. The review concluded that the literature is still too scarce to draw conclusions, although some mechanisms for how this might occur have been proposed: first, current drinkers may replace standard alcoholic beverages with similar beverages of lower alcoholic strength, and second, current drinkers may switch to no-alcohol alternatives some of the time. On the other hand, lowering alcoholic strength could reduce the threshold and initiate alcohol use in current abstainers, especially in adolescents.

The study [41] that examined the impact of the introduction of new low- and no-alcohol beers (defined as beers with 3.5% ABV or less) and reformulated beers in Great Britain on the average alcoholic strength of beer and the number of grams of alcohol purchased by households obtained the following results: a combined associated impact of both events with relative reductions of alcohol by volume of beer between 1.2% and 2.3%, purchases of grams of alcohol within beer between 7.1% and 10.2%, and purchases of grams of alcohol as a whole between 2.6% and 3.9%. Another study investigated the reformulation of products by one company and found that the mean ABV of its beer products dropped from 4.69 in 2015 to 4.55 in 2018, and these changes were associated with reduced purchases of grams of alcohol within its beer products [42].

Several individual studies have focused on the impact of low- and no-alcohol products on health-related topics. All but one focused on beer. The topics covered have included the impact of non-alcoholic beer on anxiety [44] and sleep quality [45,46]; the impact of alcohol-free beer enriched with isomaltulose on insulin resistance in diabetic patients with overweight or obesity [47]; the effect of non-alcoholic beer compared with improved diet and exercise on nutritional status, endothelial function, and quality of life in patients with cirrhosis [48]; the impact of non-alcoholic beer [49–51] or alcohol-free wine [52] on cardiovascular health; and the relationship between non-alcoholic beer and breastfeeding in terms of whether supplementing with non-alcoholic beer improves the oxidative stress and antioxidant content of breast milk [53] as well as how much ethanol in non-alcoholic beer may reach the breastfed child [54]. All these studies concluded that the impact of the tested drink was in a favourable direction, but they focused on relatively short-term effects (days to months) and the studies were conducted on small samples (ranging between 7 and 60 participants). The only study examining effects over a longer term (two years) found that in Australian older women, the frequency of drinking low-alcohol beer was positively associated with bone mass density in the lumbar spine but not in the hip [55]. Additionally, among the studies mentioned in this section, only two of twelve [52,55] were not funded by the alcohol industry or industry-related organisations. The majority of the studies also did not specifically define the possible alcohol content in the drinks, although they used terms in line with those used in regulations (alcohol-free, non-alcoholic, and low-alcohol). Additionally, two reviews examined the health properties of low-alcohol and alcohol-free beer [15,16] but focused more on the theorised nutritional benefits of beer and how to retain them in low- and non-alcoholic alternatives.

Another prominent topic in terms of study focus is the evaluation of non-alcoholic or low-alcohol beer in the context of sports/exercise, studying these products' effect on decreasing post-race inflammation and upper respiratory tract infection incidence among marathon runners [56] or fluid retention after exercise, with mixed results [57–60]. Finally, one study investigate chemical and physical properties to determine whether beer can be considered an isotonic drink [61] and concluded that only yeast-clouded alcohol-free beer (but not regular beer or clear alcohol-free beer) could be declared and promoted as isotonic, as it matches the Codex Alimentarius threshold values. However, none of the tested beverages matched the EC recommendation for sodium content.

3.4. Perceptions and Preferences

A relatively large number of papers focused on examining individual perceptions and preferences related to low- and non-alcoholic beverages, although the range of examined topics was rather heterogeneous. The majority of studies focused on beers, although regarding health impact, more studies focused on low-alcohol wine, and one considered alcohol-free spirits.

Several studies examined responses to low- and non-alcoholic beverages (beer and wine). One study found that, while the alcohol content affected participants' sensory expectations, it had no significant effect on expected liking; it also found that describing the sensory qualities of beer using a sensory descriptor had a larger effect than labelled alcohol content and label colour [62]. Another study [63] investigated whether non-alcoholic beer induced a conditioned response even when participants know that the beer is non-alcoholic. The study found that non-alcoholic beer produced a conditioned response in older drinkers because these drinkers more strongly associated the alcohol conditioned stimulus with the unconditioned stimulus than did younger participants [63]. One study [64] investigated how people evaluate low-alcohol wine (8% ABV content) and if the reduction in alcohol and the information that a wine is low in alcohol influenced consumption. The study found no difference in liking and consumption between low-alcohol and standard-alcohol wines [64]; however, participants were willing to pay more for standard wine compared with lower-alcohol wine. Two studies used functional magnetic resonance approaches to compare peoples' reactions to lower-alcohol alternatives and higher-alcohol alternatives: one [65] found no differences between acute brain rewards in the consumption of beer with and without alcohol when presented in a context in which regular alcoholic beer is expected; the second study [66] found greater activation in brain regions that are sensitive to taste intensity in low-alcohol compared with high-alcohol wines, although the definition of low-alcohol wine in the study was 13–13.5% ABV. Two studies examined whether consumers were able to discriminate between alcoholic and non-alcoholic products; both found that consumers were able to distinguish between mock and real sparkling wine [67] and between alcohol-free spirits and alcoholic spirits [68].

A number of studies investigated preferences and found that taste is an important preference driver in Australian wine consumers [69] and that that innovative wine attributes, including alcohol-free wine, were ranked among the least important attributes in a sample of Italian wine consumers [70]. Alcoholic aroma and flavour contributed as positive preference drivers for the acceptance of non-alcoholic beer in a sample of beer consumers [71]. Consumers perceived light beer (not clearly defined by % ABV) as healthier but less tasty than regular beer, and the preference for light beer was driven mostly by taste, prior experience, and brand [72]. A study of regular beer or wine consumers from the UK found that participants perceived pregnant women, athletes, and those aged 6–13 years old as target groups, and they perceived weekday lunches as the target occasions for drinking wine and beer labelled as lower-strength [73]. Another survey of Australian wine consumers found that taste was an important driver of consumption and considered 'low-alcohol wine' to contain around 3–8% alcohol [69]. The perceived reasons for preferring a low-alcohol wine included driving after drinking, lessening the adverse effects of alcohol, and being able to consume more without the effects of higher-alcohol wine [69]. Finally, in a sample of Dutch and Portuguese respondents, non-alcoholic beer was conceptualized as useful when alcohol was not convenient and functional as a substitute for regular beer, with the consumption of non-alcoholic beer driven by health and wellbeing issues, price differentials, and overall decreases in the social stigma associated with drinking alcohol-free beverages [74,75].

It is important to note that perceptions and preferences of low- and no-alcohol products might differ by culture and country. Two reviews examined the cultural context of non-alcoholic or low-alcohol beer consumption in the US [19] and the Netherlands and Portugal [17]. Another review pointed out that much low alcohol wine research originates from Australia [18].

3.5. Policy Drivers

We identified a number of policy driver-related studies, such as marketing [76] and labelling of lower-strength beer and wine [77–79] as well as the price of low- and non-alcoholic beer [80–82], with funding either received by governmental bodies or not received at all. In terms of marketing, one study conducted a content analysis of how the low- or lower-strength equivalents of beer and wine were marketed in an online context in the UK [76] and found that they were more often marketed in association with occasions deemed to be suitable for their consumption, including lunchtime (for wine), outdoor events/barbeques (for beer), and on sport/fitness occasions (for beer). Compared with regular-strength wines and beers, low- and lower-strength equivalents were more frequently marketed with images or text associated with health or information about low alcohol content and appeared to be marketed not as substitutes for higher-strength products but as products that can be consumed on additional occasions with an added implication of healthiness.

Two studies considered the labelling of low- and non-alcoholic products among regular drinkers from the UK. One study found that products with verbal descriptors denoting lower strength (low and super-low) had a lower appeal than regular-strength products, with appeal decreasing as % ABV decreased [79]. The second study found that the total amount of drink consumed increased as the label on the drink denoted successively lower alcohol strengths [78]. A related study also found that 17 of 18 verbal descriptors for lower-strength products were perceived as denoting products far higher in strength than the currently legislated cap (in the UK) of 1.2% ABV for low-alcohol products [77].

Related to price, one study [80] found that the introduction of minimum unit price in Scotland and Wales shifted purchases from higher- to lower-strength products, more so for ciders than for beers. In Australia, one analysis [82] found that after the varying nominal rates of tax were introduced for beer products according to three alcohol content levels (low-, mid-, and high-strength) in 2000/01, the relatively higher nominal tax rates for two beer categories (mid- and high-strength off-premises) had a significant negative effect on their consumption. Another Australian study examined price elasticity for several alcoholic beverages, including low-alcohol beer, but did not report on this specific category [81].

4. Discussion

The main finding of this scoping review was that there is only a relatively small and incoherent scientific literature on the production, consumption, and potential health impact of low- and no-alcohol products. The evidence base could be considered insufficient to inform policy.

Producers have a responsibility to report on their health and environmental impacts. Despite European guidance on undertaking life cycle assessments for beer production [83], such assessments are voluntary, with no requirement for public reporting. Whilst alcohol-free products, for example, may have a potential impact in reducing the harm inflicted by alcohol, and such potential benefits should be weighed against any environmental external costs, which may be present because there are extra steps in the production processes. Such scientific assessments have not been published.

There are hardly any published scientific data on the extent of purchase and consumption of low- and no-alcohol products. Published British data, which is restricted to household purchases of beer, suggest only low penetration in terms of volumes purchased.

In terms of health impact, there are a number of disparate studies on potential low quality, from which it is difficult to draw firm conclusions. British household purchase data suggest that the introduction of new low- and no-alcohol beers as well as product reformulation of existing beers to contain less alcohol result in households purchasing fewer grams of ethanol. However, these analyses refer to only one product (beer) and only one jurisdiction (Great Britain).

There have been several studies on perceptions and preferences for the uptake of no- and low-alcohol products, primarily beer. These indicate the importance of taste, prior experiences, brand (same brand as regular-strength beers), health and wellbeing issues, price differentials, and overall decreases in the social stigma associated with drinking alcohol-free beverages as important drivers of the purchase and consumption of low- and no-alcohol beers.

With respect to policy drivers, some research suggests that low- and no-alcohol products are marketed as products that can be used on additional occasions or in additional circumstances rather than as substitutes for higher-strength products. Although concern has been expressed concern on this issue [7], the scoping review found no published studies to indicate whether or not the marketing of low- and no-alcohol brands is specifically used to market higher-strength products. Several studies on labelling have suggested that the labelling of low- and no-alcohol products is sometimes inconsistent and not always as clear as it should be. Finally, some evidence from several jurisdictions suggests that pricing policy, such as the introduction of a minimum unit price, can favour shifts in purchases and consumption from higher- to lower-strength alcohol products.

4.1. Limitations

Our review is subject to the general limitations of scoping reviews: we did not appraise studies for quality, and the conclusions are still somewhat broad and qualitative. The results of our review are constrained by the relatively small size of published studies, which were not necessarily coherent within the five main groups that we used. In addition, as we started with very broad research questions, we decided to limit certain methodological criteria (such as using only two search databases and focusing on peer-reviewed articles in English only) in order to rapidly find and appraise the most relevant literature, but this means that findings from grey literature in other languages (such as government reports) are not included in this review.

4.2. Implications

Despite its limitations, our review indicates large research gaps and identifies policy actions to address these gaps, as proposed in the Table 1.

Table 1. Research and policy implications.

Future Research

There needs to be a major investment and expansion in scientific research on low and no alcohol products that covers, at least, the following topic areas:

- ○ Production: Full life cycle assessments comparing the health and environmental impacts of the production of low- and no-alcohol products compared with the production of the same or similar branded regular strength products
- ○ Purchase and Consumption:
  - Detailed analyses of the purchase and consumption of low and no-alcohol products across different product categories (e.g., beers, wines, and spirits) in a range of jurisdictions and over time.
  - Detailed analyses of the socio-demographic characteristics of who buys and drinks low- and no-alcohol products in terms of gender, age, income, educational level, occupational group, index of residential deprivation, and geographical area
- ○ Perceptions and Preferences: at the individual level, what drives the purchase and consumption of low- and no-alcohol products, including drinking occasion and location; taste, health, and well-being concerns; and previous experiences and loyalty to brands (of regular strength products)
- ○ Health Impact:
  - Does the introduction of new low- and no-alcohol products and the new purchase and consumption of such products result in consumers drinking fewer grams of alcohol (sustainable amounts) over time?
  - If less ethanol is purchased and consumed, how does this differ by socioeconomic characteristics of consumers?
- ○ Policy Drivers:
  - From an evidence perspective, what are the most appropriate definitions of low- and no-alcohol products across different categories (e.g., beers, wines, and spirits)?
  - How should low- and no-alcohol products be labelled to adequately and accurately inform consumers?
  - How should low- and no-alcohol products be placed in stores to best promote their purchase at the expense of higher strength products?
  - How should the marketing of low and no-alcohol products be regulated to prevent any negative impact of marketing in leading to increased consumption of ethanol?
  - To what extent can pricing policy, including a minimum unit price, encourage the purchase of low- and no-alcohol products at the expense of higher strength products.

Policy Implications

- ○ Clear standards and definitions need to be put in place regarding the definitions of low- and no-alcohol products across the different categories of beers, wines, and spirits, recognising, for example, that low-alcohol equivalents of spirits cannot be currently classified as spirits.
- ○ On one hand, there is a dearth of scientific information to adequately inform policy
- ○ On the other hand, there is a need for natural policy experiments in which low- and no-alcohol products are produced, marketed, and supported by relevant policy in the domains of marketing requirements, labelling, and pricing
- ○ The proviso is that all initiatives and natural policy experiments are fully and adequately subject to independent evaluation
- ○ Published life cycle assessments should become a mandatory requirement of all alcohol product development, building on existing guidance, and specifically comparing the production of no-alcohol products with the same branded regular strength products

5. Conclusions

At present, the published scientific literature on low- and no-alcohol products is too scarce and incoherent to adequately inform policy. Although analyses from one jurisdiction (Great Britain) suggest that at least low- and no-alcohol beers might be associated with reduced purchases of ethanol overall, there needs to be a rapid extension of published research across different jurisdictions and different product categories. Such research requires a series of natural experiments on new product development and availability, along with new policy drivers that promote shifts from higher- to lower-strength products that are fully and independently evaluated for their impact on drinking less ethanol, thus reducing ill-health and premature death. Such research should be a required price tag for all relevant natural experiments led by producers and governments.

Supplementary Materials: The following are available online at https://www.mdpi.com/article/10.3390/nu13093153/s1.

Author Contributions: Conceptualization: P.A. and D.K.; methodology: D.K. and P.A.; investigation: D.K. and P.A.; data analysis: D.K. and P.A.; supervision: E.J.L.; writing—original draft: D.K. and P.A.; writing—review and editing: D.K., P.A. and E.J.L. All authors have read and agreed to the published version of the manuscript.

Funding: This work was supported by the European Health and Digital Executive Agency (HaDEA) (previously Consumers, Health, Agriculture and Food Executive Agency (CHAFEA)) acting under the mandate from the European Commission (EC) specifically by the project ALHAMBRA (EU Health Programme 2014–2020 under service contract 2019 71 05). The views expressed in this article are those of the authors only and do not necessarily reflect the views of the EC or HaDEA.

Institutional Review Board Statement: Not applicable.

Data Availability Statement: Extracted data is presented in the Supplementary Materials.

Conflicts of Interest: P.A., D.K., and E.J.L. received funds from the European Health and Digital Executive Agency under a service contract. Within the previous five years, PA has received funds from the AB InBev Foundation.

References

1. Rehm, J.; Rovira, P.; Shield, K.D. Dose-response relationships between levels of alcohol use and risks of mortality for all people, and for different ages, by sex, and for groups defined by risk factors. *Nutrients* **2021**, *13*, 2652. [CrossRef] [PubMed]
2. Rumgay, H.; Murphy, N.; Ferrari, P.; Soerjomataram, I. Alcohol and cancer: Epidemiology and biological mechanisms. *Nutrients* **2021**, in press.
3. Wood, A.; Kaptoge, S.; Butterworth, A.; Paul, D.; Burgess, S.; Sweeting, M.; Bell, S.; Astle, W.; Willeit, P.; Bolton, T.; et al. Risk thresholds for alcohol consumption: Combined analysis of individual-participant data for 599,912 current drinkers in 83 prospective studies. *Lancet* **2018**, *391*, 1513–1523. [CrossRef]
4. Shield, K.; Manthey, J.; Rylett, M.; Probst, C.; Wettlaufer, A.; Parry, C.D.H.; Rehm, J. National, regional, and global burdens of disease from 2000 to 2016 attributable to alcohol use: A comparative risk assessment study. *Lancet Public Health* **2020**, *5*, e51–e61. [CrossRef]
5. World Health Organization (WHO). Global Status Report on Alcohol and Health. 2018. Available online: https://apps.who.int/iris/bitstream/handle/10665/274603/9789241565639-eng.pdf?ua=1&ua=1 (accessed on 1 September 2021).
6. World Health Organization. SAFER, Alcohol Control Initiative. 2020. Available online: https://www.who.int/substance_abuse/safer/en/ (accessed on 1 September 2021).
7. European Health and Digital Executive Agency. Service Contract for the Provision of 2019—Support to Member States in Studies and Capacity Building Activities to Reduce Alcohol Related Harm. Available online: https://etendering.ted.europa.eu/cft/cft-document.html?docId=65290 (accessed on 1 September 2021).
8. Corfe, S.; Hyde, R.; Shepherd, J. Alcohol-Free and Low-Strength Drinks Understanding Their Role in Reducing Alcohol-Related Harms. Available online: https://www.smf.co.uk/publications/no-low-alcohol-harms/ (accessed on 1 September 2021).
9. World Health Organization. *Global Strategy to Reduce the Harmful Use of Alcohol*; World Health Organization: Geneva, Switzerland, 2010.
10. UK Government. Advancing our Health: Prevention in the 2020s—Consultation Document. 2019. Available online: https://www.gov.uk/government/consultations/advancing-our-health-prevention-in-the-2020s/advancing-our-health-prevention-in-the-2020s-consultation-document (accessed on 1 September 2021).
11. Arksey, H.; O'Malley, L. Scoping studies: Towards a methodological framework. *Int. J. Soc. Res. Methodol.* **2005**, *8*, 19–32. [CrossRef]
12. Tricco, A.C.; Lillie, E.; Zarin, W.; O'Brien, K.; Colquhoun, H.; Kastner, M.; Levac, D.; Ng, C.; Sharpe, J.P.; Wilson, K.; et al. A scoping review on the conduct and reporting of scoping reviews. *BMC Med. Res. Methodol.* **2016**, *16*, 1–10. [CrossRef] [PubMed]
13. Baiano, A. Craft beer: An overview. *Compr. Rev. Food Sci. Food Saf.* **2020**, *20*, 1829–1856. [CrossRef] [PubMed]
14. Hagemann, M.H.; Bogner, K.; Marchioni, E.; Braun, S. Chances for dry-hopped non-alcoholic beverages? Part 1: Concept and market prospects. *Brew. Sci.* **2016**, *69*, 50–55.
15. Hagemann, M.H.; Schmidt-Cotta, V.; Marchioni, E.; Braun, S. Chance for Dry-hopped Non-alcoholic Beverages? Part 2: Health Properties and Target Consumers. *Brew. Sci.* **2017**, *70*, 118–123. [CrossRef]

16. Mellor, D.D.; Hanna-Khalil, B.; Carson, R. A Review of the Potential Health Benefits of Low Alcohol and Alcohol-Free Beer: Effects of Ingredients and Craft Brewing Processes on Potentially Bioactive Metabolites. *Beverages* **2020**, *6*, 25. [CrossRef]

17. Silva, A.P.; Jager, G.; Van Zyl, H.; Voss, H.-P.; Pintado, M.M.; Hogg, T.; De Graaf, C. Cheers, proost, saúde: Cultural, contextual and psychological factors of wine and beer consumption in Portugal and in the Netherlands. *Crit. Rev. Food Sci. Nutr.* **2017**, *57*, 1340–1349. [CrossRef] [PubMed]

18. Bucher, T.; Deroover, K.; Stockley, C. Low-Alcohol Wine: A Narrative Review on Consumer Perception and Behaviour. *Beverages* **2018**, *4*, 82. [CrossRef]

19. Hoalst-Pullen, N.; Patterson, M.W. *The Geography of Beer: Culture and Economics*; Springer International Publishing: Berlin/Heidelberg, Germany, 2020; ISBN 9783030416546.

20. Rehm, J.; Lachenmeier, D.W.; Llopis, E.J.; Imtiaz, S.; Anderson, P. Evidence of reducing ethanol content in beverages to reduce harmful use of alcohol. *Lancet Gastroenterol. Hepatol.* **2016**, *1*, 78–83. [CrossRef]

21. Blanco, C.A.; Andrés-Iglesias, C.; Montero, O. Low-alcohol Beers: Flavor Compounds, Defects, and Improvement Strategies. *Crit. Rev. Food Sci. Nutr.* **2016**, *56*, 1379–1388. [CrossRef] [PubMed]

22. Brányik, T.; da Silva, D.P.; Baszczyński, M.; Lehnert, R.; Silva, J.B.D.A.E. A review of methods of low alcohol and alcohol-free beer production. *J. Food Eng.* **2012**, *108*, 493–506. [CrossRef]

23. Catarino, M.; Mendes, A. Non-alcoholic beer—A new industrial process. *Sep. Purif. Technol.* **2011**, *79*, 342–351. [CrossRef]

24. Jackowski, M.; Trusek, A. Non-alcoholic beer production—An overview. *Pol. J. Chem. Technol.* **2018**, *20*, 32–38. [CrossRef]

25. Muller, C.; Neves, L.E.; Gomes, L.; Guimarães, M.; Ghesti, G. Processes for alcohol-free beer production: A review. *Food Sci. Technol.* **2020**, *40*, 273–281. [CrossRef]

26. Mangindaan, D.; Khoiruddin, K.; Wenten, I.G. Beverage dealcoholization processes: Past, present, and future. *Trends Food Sci. Technol.* **2018**, *71*, 36–45. [CrossRef]

27. Salanta, L.C.; Coldea, T.E.; Ignat, M.V.; Pop, C.R.; Tofana, M.; Mudura, E.; Zhao, H.F. Non-Alcoholic and Craft Beer Pro-duction and Challenges. *Processes* **2020**, *8*, 1382. [CrossRef]

28. Liguori, L.; Albanese, D.; Crescitelli, A.; Di Matteo, M.; Russo, P. Impact of dealcoholization on quality properties in white wine at various alcohol content levels. *J. Food Sci. Technol.* **2019**, *56*, 3707–3720. [CrossRef] [PubMed]

29. Varela, C.; Dry, P.R.; Kutyna, D.R.; Francis, I.L.; Henschke, P.A.; Curtin, C.D.; Chambers, P.J. Strategies for reducing alcohol concentration in wine. *Aust. J. Grape Wine Res.* **2015**, *21*, 670–679. [CrossRef]

30. Dequin, S.; Escudier, J.-L.; Bely, M.; Noble, J.; Albertin, W.; Masneuf-Pomarède, I.; Marullo, P.; Salmon, J.-M.; Sablayrolles, J.M. How to adapt winemaking practices to modified grape composition under climate change conditions. *OENO One* **2017**, *51*, 205. [CrossRef]

31. Ozturk, B.; Anli, E. Different techniques for reducing alcohol levels in wine: A review. *BIO Web Conf.* **2014**, *3*, 02012. [CrossRef]

32. Longo, R.; Blackman, J.W.; Torley, P.J.; Rogiers, S.Y.; Schmidtke, L.M. Changes in volatile composition and sensory attributes of wines during alcohol content reduction. *J. Sci. Food Agric.* **2017**, *97*, 8–16. [CrossRef] [PubMed]

33. Longo, R.; Blackman, J.W.; Antalick, G.; Torley, P.J.; Rogiers, S.Y.; Schmidtke, L. A comparative study of partial dealcoholisation versus early harvest: Effects on wine volatile and sensory profiles. *Food Chem.* **2018**, *261*, 21–29. [CrossRef]

34. Longo, R.; Blackman, J.W.; Antalick, G.; Torley, P.J.; Rogiers, S.Y.; Schmidtke, L.M. Harvesting and blending options for lower alcohol wines: A sensory and chemical investigation. *J. Sci. Food Agric.* **2018**, *98*, 33–42. [CrossRef]

35. Varela, J.; Varela, C. Microbiological strategies to produce beer and wine with reduced ethanol concentration. *Curr. Opin. Biotechnol.* **2019**, *56*, 88–96. [CrossRef]

36. Laoretani, D.S.; Sánchez, R.J.; Paredes, D.A.F.; Iribarren, O.A.; Espinosa, J. On the conceptual modeling, economic analysis and life cycle assessment of partial dealcoholization alternatives of bitter extracts. *Sep. Purif. Technol.* **2020**, *251*, 117331. [CrossRef]

37. Paredes, D.A.F.; Laoretani, D.S.; Morero, B.; Sánchez, R.; Iribarren, O.A.; Espinosa, J. Screening of membrane technologies in concentration of bitter extracts with simultaneous alcohol recovery: An approach including both economic and environmental issues. *Sep. Purif. Technol.* **2020**, *237*, 116339. [CrossRef]

38. Margallo, M.; Aldaco, R.; Barceló, A.; Diban, N.; Ortiz, I.; Irabien, A. Life cycle assessment of technologies for partial dealcoholisa-tion of wines. *Sustain. Prod. Consum.* **2015**, *2*, 29–39. [CrossRef]

39. Baschali, A.; Tsakalidou, E.; Kyriacou, A.; Karavasiloglou, N.; Matalas, A.-L. Traditional low-alcoholic and non-alcoholic fermented beverages consumed in European countries: A neglected food group. *Nutr. Res. Rev.* **2017**, *30*, 1–24. [CrossRef]

40. Zendeboodi, F.; Jannat, B.; Sohrabvandi, S.; Khanniri, E.; Mortazavian, A.M.; Khosravi, K.; Sarmadi, B.; Asadzadeh, S.; Amirabadi, P.E.; Esmaeili, S. Monitoring of ethanol content in non-alcoholic beer stored in different packages under different storage temperatures. *Biointerface Res. Appl. Chem.* **2019**, *9*, 4624–4628. [CrossRef]

41. Anderson, P.; Llopis, E.J.; O'Donnell, A.; Manthey, J.; Rehm, J. Impact of low and no alcohol beers on purchases of alcohol: Interrupted time series analysis of British household shopping data, 2015–2018. *BMJ Open* **2020**, *10*, e036371. [CrossRef] [PubMed]

42. Anderson, P.; Llopis, E.J.; Rehm, J. Evaluation of Alcohol Industry Action to Reduce the Harmful Use of Alcohol: Case Study from Great Britain. *Alcohol Alcohol.* **2020**, *55*, 424–432. [CrossRef] [PubMed]

43. Kypri, K.; Harrison, S.; McCambridge, J. Ethanol Content in Australian and New Zealand Beer Markets: Exploratory Study Examining Public Health Implications of Official Data and Market Intelligence Reports. *J. Stud. Alcohol Drugs* **2020**, *81*, 320–330. [CrossRef]

44. Franco, L.; Galán, C.; Bravo, R.; Bejarano, I.; Peñas-Lledo, E.; Rodríguez, A.B.; Barriga, C.; Cubero, J. Effect of non-alcohol beer on anxiety: Relationship of 5-HIAA. *Neurochem. J.* **2015**, *9*, 149–152. [CrossRef]

45. Franco, L.; Bravo, R.; Galan, C.; Rodríguez, A.; Barriga, C.; Cubero, J. Effect of non-alcoholic beer on Subjective Sleep Quality in a university stressed population. *Acta Physiol. Hung.* **2014**, *101*, 353–361. [CrossRef]

46. Franco, L.; Sanchez, C.L.; Bravo, R.; Rodríguez, A.B.; Barriga, C.; Romero, E.; Cubero, J. The Sedative Effect of Non-Alcoholic Beer in Healthy Female Nurses. *PLoS ONE* **2012**, *7*, e37290. [CrossRef]

47. Mateo-Gallego, R.; Pérez-Calahorra, S.; Lamiquiz-Moneo, I.; Marco-Benedí, V.; Bea, A.M.; Fumanal, A.J.; Prieto-Martín, A.; Laclaustra, M.; Cenarro, A.; Civeira, F. Effect of an alcohol-free beer enriched with isomaltulose and a resistant dextrin on insulin resistance in diabetic patients with overweight or obesity. *Clin. Nutr.* **2020**, *39*, 475–483. [CrossRef] [PubMed]

48. Macías-Rodríguez, R.U.; Ruiz-Margáin, A.; Román-Calleja, B.M.; Espin-Nasser, E.M.; Flores-García, N.C.; Torre, A.; Galicia-Hernández, G.; Rios-Torres, S.L.; Fernández-Del-Rivero, G.; Orea-Tejeda, A.; et al. Clinical Trials Study: Effect of non-alcoholic beer, diet and exercise on endothelial function, nutrition and quality of life in patients with cirrhosis. *World J. Hepatol.* **2020**, *12*, 1299–1313. [CrossRef] [PubMed]

49. Daimiel, L.; Micó, V.; Díez-Ricote, L.; Ruiz-Valderrey, P.; Istas, G.; Rodríguez-Mateos, A.; Ordovás, J.M. Alcoholic and Non-Alcoholic Beer Modulate Plasma and Macrophage microRNAs Differently in a Pilot Intervention in Humans with Cardiovascular Risk. *Nutrients* **2021**, *13*, 69. [CrossRef] [PubMed]

50. Chiva-Blanch, G.; Condines, X.; Magraner, E.; Roth, I.; Valderas-Martínez, P.; Arranz, S.; Casas, R.; Martínez-Huélamo, M.; Vallverdú-Queralt, A.; Rada, P.Q.; et al. The non-alcoholic fraction of beer increases stromal cell derived factor 1 and the number of circulating endothelial progenitor cells in high cardiovascular risk subjects: A randomized clinical trial. *Atherosclerosis* **2014**, *233*, 518–524. [CrossRef] [PubMed]

51. Chiva-Blanch, G.; Magraner, E.; Condines, X.; Valderas-Martínez, P.; Roth, I.; Arranz, S.; Casas, R.; Navarro, M.; Hervas, A.; Sisó, A.; et al. Effects of alcohol and polyphenols from beer on atherosclerotic biomarkers in high cardiovascular risk men: A randomized feeding trial. *Nutr. Metab. Cardiovasc. Dis.* **2015**, *25*, 36–45. [CrossRef]

52. Noguer, M.; Cerezo, A.B.; Navarro, E.D.; Garcia-Parrilla, M. Intake of alcohol-free red wine modulates antioxidant enzyme activities in a human intervention study. *Pharmacol. Res.* **2012**, *65*, 609–614. [CrossRef]

53. Codoñer-Franch, P.; Hernandez-Aguilar, M.-T.; Navarro-Ruiz, A.; López-Jaén, A.B.; Borja-Herrero, C.; Valls-Bellés, V. Diet Supplementation During Early Lactation with Non-alcoholic Beer Increases the Antioxidant Properties of Breastmilk and Decreases the Oxidative Damage in Breastfeeding Mothers. *Breastfeed. Med.* **2013**, *8*, 164–169. [CrossRef]

54. Schneider, C.; Thierauf, A.; Kempf, J.; Auwärter, V. Ethanol Concentration in Breastmilk After the Consumption of Non-alcoholic Beer. *Breastfeed. Med.* **2013**, *8*, 291–293. [CrossRef]

55. Yin, J.; Winzenberg, T.; Quinn, S.; Giles, G.; Jones, G. Beverage-specific alcohol intake and bone loss in older men and women: A longitudinal study. *Eur. J. Clin. Nutr.* **2011**, *65*, 526–532. [CrossRef]

56. Scherr, J.; Nieman, D.C.; Schuster, T.; Habermann, J.; Rank, M.; Braun, S.; Pressler, A.; Wolfarth, B.; Halle, M. Nonalcoholic Beer Reduces Inflammation and Incidence of Respiratory Tract Illness. *Med. Sci. Sports Exerc.* **2012**, *44*, 18–26. [CrossRef]

57. Castro-Sepulveda, M.; Johannsen, N.; Astudillo, S.; Jorquera, C.; Álvarez, C.; Zbinden-Foncea, H.; Ramírez-Campillo, R. Effects of Beer, Non-Alcoholic Beer and Water Consumption before Exercise on Fluid and Electrolyte Homeostasis in Athletes. *Nutrients* **2016**, *8*, 345. [CrossRef] [PubMed]

58. Desbrow, B.; Murray, D.; Leveritt, M. Beer as a Sports Drink? Manipulating Beer's Ingredients to Replace Lost Fluid. *Int. J. Sport Nutr. Exerc. Metab.* **2013**, *23*, 593–600. [CrossRef] [PubMed]

59. Desbrow, B.; Cecchin, D.; Jones, A.; Grant, G.; Irwin, C.; Leveritt, M. Manipulations to the Alcohol and Sodium Content of Beer for Postexercise Rehydration. *Int. J. Sport Nutr. Exerc. Metab.* **2015**, *25*, 262–270. [CrossRef] [PubMed]

60. Wijnen, A.H.C.; Steennis, J.; Catoire, M.; Wardenaar, F.C.; Mensink, M. Post-Exercise Rehydration: Effect of Consumption of Beer with Varying Alcohol Content on Fluid Balance after Mild Dehydration. *Front. Nutr.* **2016**, *3*, 45. [CrossRef] [PubMed]

61. Krennhuber, K.; Kahr, H.; Jäger, A. Suitability of Beer as An Alternative to Classical Fitness Drinks. *Curr. Res. Nutr. Food Sci. J.* **2016**, *4*, 26–31. [CrossRef]

62. Blackmore, H.; Hidrio, C.; Godineau, P.; Yeomans, M.R. The effect of implicit and explicit extrinsic cues on hedonic and sensory expectations in the context of beer. *Food Qual. Prefer.* **2019**, *81*, 103855. [CrossRef]

63. Fukuda, M. The effects of non-alcoholic beer on response inhibition: An open-label study. *Learn. Motiv.* **2019**, *66*, 46–54. [CrossRef]

64. Bucher, T.; Frey, E.; Wilczynska, M.; Deroover, K.; Dohle, S. Consumer perception and behaviour related to low-alcohol wine: Do people overcompensate? *Public Health Nutr.* **2020**, *23*, 1939–1947. [CrossRef]

65. Smeets, P.A.M.; De Graaf, C. Brain Responses to Anticipation and Consumption of Beer with and without Alcohol. *Chem. Senses* **2018**, *44*, 51–60. [CrossRef]

66. Frost, R.; Quiñones, I.; Veldhuizen, M.; Alava, J.-I.; Small, D.; Carreiras, M. What Can the Brain Teach Us about Winemaking? An fMRI Study of Alcohol Level Preferences. *PLoS ONE* **2015**, *10*, e0119220. [CrossRef]

67. Naspetti, S.; Alberti, F.; Mozzon, M.; Zingaretti, S.; Zanoli, R. Effect of information on consumer preferences and willingness-to-pay for sparkling mock wines. *Br. Food J.* **2020**, *122*, 2621–2638. [CrossRef]

68. Lachenmeier, D.W.; Pflaum, T.; Nieborowsky, A.; Mayer, S.; Rehm, J. Alcohol-free spirits as novel alcohol placebo—A viable approach to reduce alcohol-related harms? *Int. J. Drug Policy* **2016**, *32*, 1–2. [CrossRef] [PubMed]

69. Saliba, A.J.; Ovington, L.A.; Moran, C.C. Consumer demand for low-alcohol wine in an Australian sample. *Int. J. Wine Res.* **2013**, *5*, 1–8. [CrossRef]

70. Stanco, M.; Lerro, M.; Marotta, G. Consumers' Preferences for Wine Attributes: A Best-Worst Scaling Analysis. *Sustainability* **2020**, *12*, 2819. [CrossRef]

71. Paixão, J.A.; Filho, E.T.; Bolini, H.M.A. Investigation of Alcohol Factor Influence in Quantitative Descriptive Analysis and in the Time-Intensity Profile of Alcoholic and Non-Alcoholic Commercial Pilsen Beers Samples. *Beverages* **2020**, *6*, 73. [CrossRef]

72. Chrysochou, P. Drink to get drunk or stay healthy? Exploring consumers' perceptions, motives and preferences for light beer. *Food Qual. Prefer.* **2014**, *31*, 156–163. [CrossRef]

73. Vasiljevic, M.; Couturier, D.-L.; Marteau, T.M. What are the perceived target groups and occasions for wines and beers labelled with verbal and numerical descriptors of lower alcohol strength? An experimental study. *BMJ Open* **2019**, *9*, e024412. [CrossRef]

74. Silva, A.P.; Jager, G.; van Bommel, R.; van Zyl, H.; Voss, H.-P.; Hogg, T.; Pintado, M.M.; de Graaf, C. Functional or emotional? How Dutch and Portuguese conceptualise beer, wine and non-alcoholic beer consumption. *Food Qual. Prefer.* **2016**, *49*, 54–65. [CrossRef]

75. Silva, A.P.; Jager, G.; Voss, H.-P.; van Zyl, H.; Hogg, T.; Pintado, M.; de Graaf, C. What's in a name? The effect of congruent and incongruent product names on liking and emotions when consuming beer or non-alcoholic beer in a bar. *Food Qual. Prefer.* **2016**, *55*, 58–66. [CrossRef]

76. Vasiljevic, M.; Coulter, L.; Petticrew, M.; Marteau, T.M. Marketing messages accompanying online selling of low/er and regular strength wine and beer products in the UK: A content analysis. *BMC Public Health* **2018**, *18*, 147. [CrossRef]

77. Vasiljevic, M.; Couturier, D.L.; Marteau, T.M. Impact of low alcohol verbal descriptors on perceived strength: An experimental study. *Br. J. Health Psychol.* **2018**, *23*, 38–67. [CrossRef]

78. Vasiljevic, M.; Couturier, D.-L.; Frings, D.; Moss, A.; Albery, I.; Marteau, T.M. Impact of lower strength alcohol labeling on consumption: A randomized controlled trial. *Health Psychol.* **2018**, *37*, 658–667. [CrossRef] [PubMed]

79. Vasiljevic, M.; Couturier, D.; Marteau, T.M. Supplemental Material for Impact on Product Appeal of Labeling Wine and Beer With (a) Lower Strength Alcohol Verbal Descriptors and (b) Percent Alcohol by Volume (%ABV): An Experimental Study. *Psychol. Addict. Behav.* **2018**, *32*, 779. [CrossRef] [PubMed]

80. Llopis, E.J.; O'Donnell, A.; Anderson, P. Impact of price promotion, price, and minimum unit price on household purchases of low and no alcohol beers and ciders: Descriptive analyses and interrupted time series analysis of purchase data from 70, 303 British households, 2015–2018 and first half of 2020. *Soc. Sci. Med.* **2021**, *270*, 113690. [CrossRef] [PubMed]

81. Srivastava, P.; McLaren, K.R.; Wohlgenant, M.; Zhao, X. Disaggregated econometric estimation of consumer demand response by alcoholic beverage types. *Aust. J. Agric. Resour. Econ.* **2014**, *59*, 412–432. [CrossRef]

82. Vandenberg, B.; Jiang, H.; Livingston, M. Effects of changes to the taxation of beer on alcohol consumption and government revenue in Australia. *Int. J. Drug Policy* **2019**, *70*, 1–7. [CrossRef]

83. European Commission. PEFCR for Beer. Available online: https://ec.europa.eu/environment/eussd/smgp/pdf/Beer%20PEFCR%20June%202018%20final.pdf (accessed on 1 September 2021).

Article

Nature and Potential Impact of Alcohol Health Warning Labels: A Scoping Review

Daša Kokole [1,*], Peter Anderson [1,2] and Eva Jané-Llopis [1,3,4]

[1] Department of Health Promotion, CAPHRI Care and Public Health Research Institute, Maastricht University, POB 616, 6200 MD Maastricht, The Netherlands; peteranderson.mail@gmail.com (P.A.); eva.jane@esade.edu (E.J.-L.)
[2] Population Health Sciences Institute, Newcastle University, Baddiley-Clark Building, Richardson Road, Newcastle upon Tyne NE2 4AX, UK
[3] ESADE Business School, University Ramon Llull, Avenida de Pedralbes, 60-62, 08034 Barcelona, Spain
[4] Institute for Mental Health Policy Research, CAMH, 33 Russell Street, Toronto, ON M5S 2S1, Canada
* Correspondence: d.kokole@maastrichtuniversity.nl

Abstract: Alcohol is toxic to human health. In addition to providing nutritional information, labels on alcohol products can be used to communicate warnings on alcohol-related harms to consumers. This scoping review examined novel or enhanced health warning labels to assess the current state of the research and the key studied characteristics of labels, along with their impact on the studied outcomes. Four databases (Web of Science, MEDLINE, PsycInfo, CINAHL) were searched between January 2010 and April 2021, and 27 papers were included in the review. The results found that most studies were undertaken in English-speaking populations, with the majority conducted online or in the laboratory setting as opposed to the real world. Seventy percent of the papers included at least one cancer-related message, in most instances referring either to cancer in general or to bowel cancer. Evidence from the only real-world long-term labelling intervention demonstrated that alcohol health warning labels designed to be visible and contain novel and specific information have the potential to be part of an effective labelling strategy. Alcohol health warning labels should be seen as tools to raise awareness on alcohol-related risks, being part of wider alcohol policy approaches.

Keywords: alcohol; labelling; health warning labels; effectiveness; implementation

Citation: Kokole, D.; Anderson, P.; Jané-Llopis, E. Nature and Potential Impact of Alcohol Health Warning Labels: A Scoping Review. *Nutrients* **2021**, *13*, 3065. https://doi.org/10.3390/nu13093065

Academic Editor: Joerg Koenigstorfer

Received: 21 July 2021
Accepted: 27 August 2021
Published: 31 August 2021

Publisher's Note: MDPI stays neutral with regard to jurisdictional claims in published maps and institutional affiliations.

1. Introduction

Globally, 5.3% of all deaths and 5.0% of all disability adjusted life years are caused by ethanol [1], a toxic substance [2]. Alcohol consumption is causally related to more than 40 ICD-10 three-digit categories [3], including non-communicable diseases such as liver disease [4], cancers of the colon, liver and breast [5–7], alcohol use disorders [8], non-ischaemic cardiovascular diseases [9]; communicable diseases such as tuberculosis [10]; and intentional and unintentional injuries [3]. Alcohol can also impair mental health (particularly depression [11,12]) and cause harm to others, such as in foetal alcohol spectrum disorders [13,14].

Despite alcohol's detrimental effect on a variety of health outcomes, public awareness about some health risks associated with alcohol consumption remains relatively low. A 2018 review of 32 studies found that while awareness of alcohol as a risk factor for cancer varies by country, in most studies, less than half of the respondents correctly identified the alcohol–cancer link [15]. This finding is confirmed in more recent studies [16–18], which additionally found that awareness of the link between alcohol and cancer differed by type of cancer (e.g., the link between alcohol and breast cancer tended to be the least well known).

One of the approaches to raise awareness of the health risks associated with alcohol consumption can be product labelling. A review from 2013 on enhanced labelling [19]

identified five elements that could be useful to consumers: a list of ingredients, nutritional information, serving size and servings per container, a definition of moderate intake (low risk drinking guidelines) and a health warning label (HWL). While the majority of countries have mandated alcohol volume content on the labels [20], this is not the case with nutritional or health labelling. In many of the high consumption regions, such as EU countries, UK, US, Australia, New Zealand and Canada, alcoholic beverages are exempt from regulations to include nutritional information (for example, nutritional values and calorie content) on the product label [21–24], as opposed to non-alcoholic drinks and food products. Nutritional labelling policy thus leaves much room for improvement, with clear guidelines available on how information should be presented based on the requirements for food and non-alcoholic beverages should nutritional labels be mandated for alcoholic beverages. Whilst there is scarce alcohol-specific research on nutrition information [25,26], much is known from the nutrition field on enhanced presentation of nutritional information [27,28], which could be applied to alcoholic beverages.

There is little available evidence on the effectiveness of health warning labels that would inform consumers about the risks associated with alcohol consumption. Most older studies examining the effectiveness of implemented health warning messages report on the US health warning label, which focuses on risks surrounding drinking during pregnancy, and has not changed since its introduction in 1989 [29,30]. More recent studies on existing health warning labelling schemes, either mandatory [31,32] or voluntary [33–35], show that the existing labels (e.g., pregnancy logo or responsibility message) are suboptimal and are either not being noticed or not being understood. However, evidence from tobacco shows that health warning labels can be a very important part of the broader package of interventions [36,37], and there have been calls to apply tobacco-style health warning labels to alcohol products [38]. Given the evidence for the less effective existing alcohol HWL, this review focuses on examining evidence with regard to novel (not previously implemented in practice) or enhanced labels (improved versions of existing ones).

The importance of labelling is recognised by the World Health Organisation, whose global strategy to reduce the harmful use of alcohol [39] calls for providing consumer information about the harm related to alcohol and labelling alcoholic beverages to indicate such risks. Awareness of such information is, in turn, linked to increased public support for more stringent policies, such as taxation [40,41], and can therefore be an important part of effective alcohol policy. There has been increased interest among policymakers to provide health information to consumers through labels. Ireland passed the Public Health Alcohol Act in 2018, mandating cancer warnings on labels [42] and becoming only the second country in the world after South Korea so to do [20], but nothing else has been implemented as of 2021. There has also been increased research activity, and a number of labelling-related reviews have been published recently. Previous reviews focusing on alcohol HWL operated with a relative scarcity of studies [43], focused only on one aspect of the label (image vs. text) [44] or made a brief narrative overview of health labelling studies as part of broader alcohol labelling [45]. Whilst our work partially overlaps with other reviews in examining the impact of labels, our review adds a thorough and systematic examination of the scope of (quasi-)experimental research on new or enhanced health warning labels, including an overview of the labels used in the studies and the explanatory variables studied. As there is not a single definition of "alcohol HWL", labels can differ in how they are developed, implemented and evaluated, and the current review aims to contextualise impacts in light of this information. As health warning labels, we consider labels that are containing information on the relationship between alcohol and health outcomes. We do not focus on drinking guidelines and standard drinks labelling, as this has been carried out elsewhere [46].

Thus, the aim of this scoping review is to examine the peer-reviewed literature studying novel or enhanced health warning labels and to answer two research questions: (1) what is the scope of the current alcohol HWL research (in terms of research aims, studied independent variables, the outcomes of interest, as well as label content and format), and (2) what is the impact of the labels and key label characteristics on the studied outcomes?

2. Methods

2.1. Design

To answer the research questions, we used scoping review methodology, guided by the methodological framework of Arksey and O'Malley [47]. Scoping reviews are used to map the main concepts in the research area and present a broad overview of the existing evidence, including identification of the gaps, regardless of the study quality [48]. The protocol was not published in advance, and the results are reported in accordance with PRISMA-ScR checklist [49].

2.2. Eligibility Criteria

To be included in the review, studies had to evaluate new or enhanced alcohol health warning labels in any population. By "new", we considered labels developed for the purpose of the research and not previously implemented in practice, and by "enhanced", we considered improved versions of existing labels already implemented in practice. Studies had to include experimental or quasi-experimental methodology with clearly delineated independent and dependent variables, in order to be able to synthesise results based on the manipulated variables and to identify the key characteristics of effective labels. Only peer-reviewed articles in the English language published in 2010 or after were included, as in the previous reviews, there were no older experimental studies of new messages. Studies that did not include health warning messages, that focused on qualitative data or were only cross-sectional evaluations were excluded.

2.3. Information Sources and Search Strategy

Four databases (Web of Science, MEDLINE, PsycInfo, CINAHL) were searched with the abovementioned restrictions on language and dates (English language, January 2010–April 2021). Additionally, we complemented the database searches with inspection of reference lists of relevant articles and reviews, and we undertook a Google Scholar search. The search strategy contained a combination of terms related to alcohol labelling: [alcohol AND (label OR label*) AND (message OR information OR warning) AND (experiment* OR eval* OR effect)].

2.4. Study Selection and Summary

Study selection was performed by one researcher according to the abovementioned inclusion and exclusion criteria, with ambiguous cases being resolved through consultation with the second researcher. After the removal of duplicates, the studies were first screened by title and abstract, followed by full text examination and final article selection. Next, the data were charted to answer the main research questions. Next to general information on the papers' aim and methodology, information on label development and implementation was charted, as well as the independent studied variables and the results for each of the outcomes. Papers rather than studies were taken as the units of interest, as the focus of the review was to appraise the scope of the published literature and to identify results for all the studied outcomes.

3. Results

In total, we identified 27 papers that fit our inclusion criteria (see Appendix A for PRISMA diagram) [50–76]. One paper reported on two studies using different methodologies to study the same variables [50], and four papers reported on different aspects of one intervention [51–54]. The remaining papers were individual studies (with a complete or partial focus on alcohol HWL labelling), although some of them were part of the same research line of label development and testing (e.g., [55–60]) or studied the same labels (e.g., [61,62]).

Table 1 provides an overview of the papers with their aim and methodology. Two-thirds (18) of the included papers were published in 2018 or later, indicating an increasing research interest in the topic. All of the research was undertaken in Western countries: most papers reported on research conducted in the UK (nine), Canada (six) and Australia (five). The remaining papers reported on research conducted in the US (three), Germany (two) or were multi-country (twice Luxembourg/Germany, and once Italy/France). Only four papers described real-world interventions. The remainder of the studies were either conducted online (sixteen) or in a laboratory (eight, of which one was in a naturalistic shopping laboratory). Six studies had under 100 participants, eleven studies had above 1000 participants and one study relied on sales data. The majority (80%) of the studies with a lower number of participants were conducted in the laboratory and before 2018, indicating that in recent years, larger sample sizes were accessible in an easier manner through online panels. Two-thirds of the papers focused on the general adult (of legal drinking age in the country) drinking population, although the exact inclusion criteria differed and ranged from people drinking at least once in the past 12 months, to people consuming alcohol weekly. Eight of the papers focused on the younger population/students (some by convenience rather than by inclusion criteria) and one paper focused on secondary school students.

3.1. Research Scope

To answer the first research question, we looked at the aim of the papers and the variables studied (both independent variables and the outcomes), as well as at the content and format of the studied labels.

In terms of study aims and studied independent variables, five papers (19%) compared only new or enhanced labels with no or regular labels [51–54,63]. Focusing on label content, eight papers (30%) compared text-only labels with text and image labels [55,56,61,62,64–67], twelve studies (44%) compared text message characteristics (e.g., content, specificity, framing, source, use of causal language) [57–60,67–74] and three (11%) compared image characteristics [57,66,75]. Four (15%) papers compared label formats, such as size, position, colour or branding [50,64,70,76]. Additionally, nine papers (33%) studied other variables, such as alcohol content, alcohol/substance type, efficacy information, expectancy and self-affirmation [64–66,68,69,73–76].

The majority of papers investigated more than one outcome (Table 2). The most commonly studied outcomes were attitude (41% of the papers), intentions (30%), emotion-related outcomes (e.g., fear, worry) (26%) and acceptability/support for the labels (26%). Behaviour was studied in six papers (22%)—in three of them through self-report [51,54,67], but also through purchasing [53,56] or consuming alcohol [62].

The tested labels were predominantly developed by the researchers themselves, based either on other research (e.g., related to tobacco and alcohol labels, alcohol harm, effective messages) or on own development and piloting (e.g., through qualitative studies), or both. In the four papers related to a real-world intervention [51–54], the labels were additionally (next to own and other research) developed in consultation with local stakeholders. In two cases, the labels used in the research were developed by a non-governmental organisation [67,75], and in one case [50], existing voluntary labels were enhanced based on theory.

Table 1. Studies overview—aim and methodology.

Authors and Year Published	Country	Study Aim (According to Authors)	Setting	Methodology/Study Design	Sample Characteristics	Population (P) or Inclusion Criteria (IC)	Sampling and Recruitment	Year Data Collected
Al-Hamdani and Smith (2015) [64]	Canada	To apply the lessons learned from the tobacco health warnings and plain packaging literature to an alcohol packaging study and test whether labelling alters consumer perceptions.	Online (survey)	Experiment: 3 × 4 mixed design	N = 92 $M(SD)_{age}$ = 36.4 (13.3) 66.2% female	P: 60.2% students, 39.8% hospital employees 91.1% participants drinking at least weekly	Convenience sample, recruited through posters and ads in the organisations	Not reported
Al-Hamdani and Smith (2017) [76]	Canada	To examine whether increasing the size of HWL and plain packaging lowers ratings of alcohol products and the consumers who use them, increases ratings of bottle "boringness" and enhances warning recognition compared with branded packaging.	Online (survey)	Experiment: 3 × 2 × 3 mixed design	N = 440 initially/241 finally $M(SD)_{age}$ = 26 (7.1) 51.7% female	IC: Adults of legal age who consumed alcohol in the past 12 months P: 91% participants drinking at least weekly	Convenience sample, recruited online	Not reported
Annunziata et al. (2019) [70]	Italy and France	To analyse Generation Y consumers' preferences for, interest in and attitudes towards different formats of health warnings on wine labels in two countries with different legal approaches: France and Italy.	Online (survey)	Discrete choice experiment	N = 500 (250 per country) $M(SD)_{age}$ = 23.3 (3.4)—FR; 25.2 (4.5)—IT 54% females—FR; 60% females—IT	IC: Generation Y (1978–2000)	Convenience sample, recruited online	2018
Blackwell et al. (2018) [71]	UK	To examine the influence of unit labels and health warnings on drinkers' understanding, attitudes and behavioural intentions regarding drinking and examine optimal methods of delivering this information on labels.	Online (survey)	Between subjects experimental study	N = 1184 $M(SD)_{age}$ = 35 (12) 50% female	IC: Adult (18+) drinkers only	Recruitment through crowdsourcing platform/panel	Not reported
Clarke, Pechey et al. (2021a) [55]	UK	To obtain a preliminary assessment of the possible impact of (i) image-and-text, (ii) text only and (iii) image-only HWLs on selection of alcoholic versus non-alcoholic drinks.	Online (survey)	Between-subjects experimental study, 2 × 2 factorial design	N = 6024 (completed the study) $M(SD)_{age}$ = 49.5 (15.5)50% female	IC: Adults (18+) who consumed beer or wine regularly (i.e., at least once a week)	Recruited via market research agency	2019
Clarke, Blackwell et al. (2021b) [56]	UK	To estimate the impact of HWLs describing adverse health consequences of excessive alcohol consumption on selection of alcoholic drinks.	Naturalistic shopping laboratory	Between-subjects design	N = 399 $M(SD)_{age}$ = 39.9 (13.7)55% female	IC: Adults (18+) who purchased beer or wine weekly to drink at home	Recruited via market research agency	2020
Glock and Krolak-Schwerdt (2013) [72]	Luxembourg and Germany	Compared the effectiveness of warning labels that contradicted positive outcome expectancies with health-related warning labels.	Laboratory	Between-subjects experimental study, two factorial mixed design	N = 40 $M(SD)_{age}$ = 24.0 (3.2) 60% female	P: Undergraduates, native German speakers 95% drinkers	Recruited through university courses	Not reported
Gold et al. (2020) [63]	UK	To examine whether showing people a health warning alongside our label designs would have a further effect on our secondary outcomes, increasing the perceived risk of alcohol consumption, decreasing the motivation to drink and lowering the level of drinking which people believe to be health-damaging.	Online (survey)	Parallel randomised-controlled trial	N = 7516 total/500 for HWL $M(SD)_{age}$ = 44.2 (16.5)50.5% female	IC: English adults (18+) reporting drinking alcohol	Representative sample of the adult population of Englandin terms of age, gender and region, recruited through online panel platform	2019
Hall et al. (2019) [73]	US	To examine US adults' reactions to health warnings with strong versus weak causal language.	Online (survey)	Between-subjects experimental study	N = 1360 $M(SD)_{age}$ = 37.4 (11.6) 47.3% female	IC: US residents, (18+)	Convenience sample, recruitment via online crowdsourcing platform	2018

Table 1. *Cont.*

Authors and Year Published	Country	Study Aim (According to Authors)	Setting	Methodology/Study Design	Sample Characteristics	Population (P) or Inclusion Criteria (IC)	Sampling and Recruitment	Year Data Collected
Hall et al. (2020) [65]	US	To examine reactions to graphic versus text-only warnings for cigarettes, SSBs and alcohol.	Online (survey)	$2 \times 2 \times 3$ Between-subjects experimental study	N= 1352 M_{age} = 37 47% female	IC: US residents, (18+)	Convenience sample, recruitment via online crowdsourcing platform	2018
Hobin, Schoueri-Mychasiw et al. (2020a) [51]	Canada	To examine the effects of strengthening alcohol labels on consumer attention and message processing, and a self-reported reduction in drinking due to the labels.	Real-world	Quasi experimental design	N = 2049 unique cohort participants $M(SD)_{age}$ = intervention 47.4 (14.6); comparison 41.2 (13.7) 50.7% female intervention, 45.1% female control	IC: Adult (19+), current drinkers (at least 1 drink in past 30 days), living in the intervention or comparison cities, bought alcohol at the liquor store and did not self-report being pregnant or breast-feeding	Systematic recruitment -standard intercept technique of approaching every person that passed a pre-identified landmark	2017/2018
Hobin, Weerasinghe et al. (2020b) [52]	Canada	To test the initial and continued effects of cancer warning labels on drinkers' recall and knowledge that alcohol can cause cancer.	Real-world	Quasi experimental design	N = 2049 unique cohort participants $M(SD)_{age}$ = intervention 47.4 (14.6); comparison 41.2 (13.7) 50.7% female intervention, 45.1% female control	IC: Adult (19+) current drinkers (at least 1 drink in past 30 days), living in the intervention or comparison cities, bought alcohol at the liquor store, did not self-report being pregnant or breast-feeding	Systematic recruitment—standard intercept technique of approaching every person that passed a pre-identified landmark	2017/2018
Hobin, Shokar et al. (2020c) [54]	Canada	To investigate the impact of alcohol labels on (i) unprompted recall of label messages, (ii) depth of cognitive processing of label messages and (iii) self-reported impact on alcohol consumption.	Real-world	Quasi experimental design	N= 1647 unique cohort participants Age: Intervention: 6.8% 19–24, 34.3% 25–44, 58.8% 45 + Control: 12.3% 19–24, 45.5% 25–44, 42.2% 45 + 51.5% female intervention, 44.1% female control	IC: Adult (19+) current drinkers (at least 1 drink in past 30 days), living in the intervention or comparison cities, bought alcohol at the liquor store, did not self-report being pregnant or breast-feeding	Systematic recruitment—standard intercept technique of approaching every person that passed a pre-identified landmark	2017/2018
Jarvis and Pettigrew (2013) [74]	Australia	To assess the effects of warning statements on youth's alcohol purchase decisions in the context of information relating to proprietary brands and alcohol content.	Online (survey)	Discrete choice experiment	N = 300 Age and gender not reported	IC: 18–25 year old current drinkers of pre-mixed alcoholic beverages, purchased a packaged pre-mixed alcoholic beverage in the past the 4 weeks	Through web panel provider, including equal proportional representation of two age categories	Not reported
Jongenelis, Pettigrew et al. (2018a) [58]	Australia	To examine the effectiveness of messages when delivered by single versus multiple sources.	Online (survey)	Between-subjects experimental study	N = 2087 $M(SD)_{age}$= 36.05 (12.67) 50.1% female	IC: Adult drinkers consuming alcohol at least twice per month	Web panel provider	Not reported
Jongenelis, Pratt et al. (2018b) [59]	Australia	To examine whether exposing at-risk drinkers to warning statements relating to specific chronic diseases increases the extent to which alcohol is believed to be a risk factor for those diseases and influences consumption intentions.	Online (survey)	Between-subjects experimental study	N = 364 18–34 years 29% 35–65 years 71% 28% female	IC: Adults 18–65 years who reported drinking at levels associated with long-term risk of harm (more than 2 SD per day)	Web panel provider	Not reported
Krischler and Glock (2015) [68]	Luxembourg and Germany	To investigate the effectiveness of alcohol warning labels tailored toward young adults' positive outcome expectancies.	Laboratory	Experiment: 3×2 mixed design	N = 122 $M(SD)_{age}$ = 23.5 (3.5) 68.9% female	P: 91.7% Undergraduates	Recruited on campus	2014

Table 1. *Cont.*

Authors and Year Published	Country	Study Aim (According to Authors)	Setting	Methodology/Study Design	Sample Characteristics	Population (P) or Inclusion Criteria (IC)	Sampling and Recruitment	Year Data Collected
Ma (2021) [69]	US	To determine the impact of pictorial warning labels featuring narrative content on risk perceptions and behavioural intentions.	Online (survey)	Between-subjects experimental study	$N = 169$ $M(SD)_{age} = 43.2$ (11.5) 37.9% female	IC: Alcohol consumers	Recruited through web panel provider	Not reported
Monk et al. (2017) [66]	UK	To investigate the amount of time spent looking at the different elements of alcohol-related health warnings.	Laboratory	Experiment: $2 \times 2 \times 2$ mixed factorial design	$N = 22$ $M(SD)_{age} = 21.3$ (1.7) 68.2% female	University students	Opportunity sampling	Not reported
Morgenstern et al. (2021) [67]	Germany	To investigate impact of alcohol warning labels on knowledge and negative emotions.	Online (survey)	Three factorial experiment	$N = 9260$ $M(SD)_{age} = 12.9$ (1.8) 48.6% female	IC: Secondary school students (10–17)	Recruited through schools from randomly selected sub-regions	2017–2018
Pechey et al. (2020) [57]	UK	to describe the potential effectiveness and acceptability of image-and-text (also known as pictorial or graphic) HWLs applied to alcoholic drinks.	Online (survey)	Between-subjects experimental study	$N = 5528$ $M(SD)_{age} = 47.5$ (15.8) 50.9% female	IC: 18+, self-reported consuming either beer or wine at least once a week	Purposeful sampling to include range of age, gender and social grades (recruited via market research agency)	2018
Pettigrew et al. (2016) [60]	Australia	To investigate the potential effectiveness of alcohol warning statements designed to increase awareness of the alcohol–cancer link.	Online (survey)	Between-subjects experimental study	$N = 1680$ < 31 years 49.5 31–45 years 25.2 46–65 years 25.3 49.9% female	IC: Adult (18–65) drinkers, consuming alcohol at least two days per month	Web panel provider	Not reported
Pham et al. (2018) * [50]	Australia	To investigate attention of current in market alcohol warning labels and examine whether attention can be enhanced through theoretically informed design.	Study 1: Online (survey) Study 2: Laboratory	Between-subjects experimental study	Study 1: $N = 559$ $M(SD)_{age} = 31.9$ (7.8) Gender not reported Study 2: $N = 87$ $M(SD)_{age} = 26.6$ (10.5)Gender not reported	Study 1: P: 72% employed, 6.6% full time students Study 2: P 49.4% full time students, 41.4% employed	Study 1: Snowball recruitment online Study 2: Face to face on campus	2015
Sillero-Rejon et al. (2018) [75]	UK	To examine whether enhancing self-affirmation among a population of drinkers, prior to viewing threatening alcohol pictorial health warning labels, would reduce defensive reactions and promote reactions related to behaviour change, and whether there is an interaction between self-affirmation and severity of warning.	Laboratory	Between-subjects experimental study	$N = 128$ $M(SD)_{age} = 22$ (4) Gender not reported	IC: Adult (18+) regular alcohol consumers who have consumed over 14 units per week during the preceding week	Opportunity sampling, recruited via e-mail, posters and websites	Not reported
Stafford and Salmon (2017) [62]	UK	To test whether the speed of alcohol consumption is influenced by the type of alcohol health warning contained on the beverage.	Laboratory	Between-subjects experimental study	$N = 45$ $M(SD)_{age} = 18.9$ (1.1) 100% female	IC: Female (18–25) regular alcohol consumers	Recruited using an online system and via social media	Not reported
Wigg and Stafford (2016) [61]	UK	To test the effectiveness of a range of alcohol health warnings, comparing no health warning, text only and pictorial warning.	Laboratory	Between-subjects experimental study	$N = 60$ $M(SD)_{age} = 19.4$ (3.1) 71.7% female	IC: Alcohol consumers	Recruited using an online system, received credit points	Not reported
Zhao et al. (2020) [53]	Canada	To test if the labelling intervention was associated with reduced alcohol consumption.	Real-world	Quasi experimental design	Monthly retail sales data, no individual participants	Population 15+ in the research areas	/	2017/2018

* two studies considered as one, as they are studying the same outcome.

Table 2. Studied outcomes.

	N	%	Papers
Attitude	11	41%	Al-Hamdani & Smith (2015) [64]; Al-Hamdani & Smith (2017) [76]; Annunziata et al. (2019) [70]; Blackwell et al. (2018) [71]; Glock & Krolak-Schwerdt (2013) [72]; Hall et al. (2020) [65]; Jarvis & Pettigrew (2013) [74]; Jongenelis et al. (2018a) [58]; Krischler & Glock (2015) [68]; Pettigrew et al. (2016) [60]; Stafford & Salmon (2017) [62]
Intentions	8	30%	Clarke et al. (2021a) [55]; Glock & Krolak-Schwerdt (2013) [72]; Jongenelis et al. (2018a) [58]; Jongenelis et al. (2018b) [59]; Krischler & Glock (2015) [68]; Ma (2021) [69]; Pettigrew et al. (2016) [60]; Wigg & Stafford (2016) [61]
Emotion	7	26%	Clarke et al. (2021a) [55]; Clarke et al. (2021b) [56]; Hall et al. (2020) [65]; Ma (2021) [69]; Morgenstern et al. (2021) [67]; Pechey et al. (2020) [57]; Wigg & Stafford (2016) [61]
Acceptability/support	7	26%	Blackwell et al. (2018) [71]; Clarke et al. (2021a) [55]; Clarke et al. (2021b) [56]; Hall et al. (2020) [65]; Hall et al. (2019) [73]; Hobin et al. (2020b) [52]; Pechey et al. (2020) [57]
Behaviour	6	22%	Clarke et al. (2021b) [56]; Hobin et al. (2020a) [51]; Hobin et al. (2020c) [54]; Morgenstern et al. (2021) [67]; Stafford & Salmon (2017) [62]; Zhao et al. (2020) [53]
Awareness/recognition	5	19%	Al-Hamdani & Smith (2015) [64]; Al-Hamdani & Smith (2017) [76]; Hobin et al. (2020a) [51]; Hobin et al. (2020b) [52]; Hobin et al. (2020c) [54]
Risk perception	5	19%	Clarke et al. (2021a) [55]; Gold et al. (2020) [63]; Ma (2021) [69]; Sillero-Rejon et al. (2018) [75]; Wigg & Stafford (2016) [61]
Motivation	4	15%	Blackwell et al. (2018) [71]; Gold et al. (2020) [63]; Pechey et al. (2020) [57]; Sillero-Rejon et al. (2018) [75]
Reactance	4	15%	Blackwell et al. (2018) [71]; Clarke et al. (2021a) [55]; Hall et al. (2020) [65]; Sillero-Rejon et al. (2018) [75]
Knowledge	3	11%	Hobin et al. (2020b) [52]; Jongenelis et al. (2018b) [59]; Morgenstern et al. (2021) [67]
Perceived effectiveness	3	11%	Hall et al. (2020) [65]; Hall et al. (2019) [73]; Sillero-Rejon et al. (2018) [75]
Attention	3	11%	Monk et al. (2017) [66]; Pham et al. (2018) [50]; Sillero-Rejon et al. (2018) [75]
Cognitive processing	3	11%	Hall et al. (2020) [65]; Hobin et al. (2020a) [51]; Hobin et al. (2020c) [54]
Avoidance	3	11%	Blackwell et al. (2018) [71]; Clarke et al. (2021a) [55]; Sillero-Rejon et al. (2018) [75]
Efficacy	2	7%	Blackwell et al. (2018) [71]; Hall et al. (2020) [65]

In terms of label content, eight studies studied a single message; the remaining were either comparing different labels (or groups of labels) or used several messages as a single intervention. Most commonly, the messages were referring either to cancer in general (e.g., "Warning: Alcohol causes cancer") [51–54,57–60,63,67,71,75], or specifically to bowel cancer (e.g., "Alcohol causes bowel cancer") [51–60,69,71], both in 41% of all papers. Overall, 70% of papers included at least one cancer-related message, with other mentioned cancers being also breast (33%) [51–58,60] and liver cancer (30%) [55–57,61,62,64,69,76]. Liver cirrhosis or liver disease message was used in nine (33%) papers (e.g., "Warning: Alcohol causes liver disease") ([56,57,59,65–67,72,73,75]. Heart disease [55,57,59,72], mental illness [59,67,71,75], brain damage [70,72,74,75] and drinking and driving-related messages [67,70,74,75] were used in four papers each (15%), and pregnancy messages in three papers (11%) [50,67,75]. Four papers (15%; [59,67,68,72]), included other topics of messages such as diabetes, positive expectancies related to alcohol, warning against operating machinery or serving alcohol to minors. Fifteen papers (56%) included images in the labels—ten of them included only graphic images (e.g., images of diseased body parts) [55–57,61,62,64,65,68,69,76], two included both graphic and neutral images [66,75], one included only neutral images [67] and two included pictograms [50,70].

In terms of label format, the majority of the papers formatted the labels to be clearly visible. This meant putting labels on the front of the pack, with a predetermined size and clear contrast between text and background and separation from the rest of the label (e.g., black text on white background with black or red border, such as [55–57,61,63] or black text on a bright yellow background and red border, such as [51–54]). A complete description of the studied labels content and format is available in Supplementary Table S1.

3.2. Key Label Characteristics—Impact on the Outcomes

Results for all the studied outcomes from each of the papers are presented in Table 3. Below, the results are summarised based on the study aims identified through the two research questions: (1) studying overall label effectiveness by comparing image and text labels and (2) studying other label message characteristics and label formats.

3.2.1. Label Effectiveness

A large online UK study [63] did not find that adding a text warning to alcohol drink label changed perceived personal risk, motivation to drink less or perception of damaging drinking among adult drinkers. On the other hand, the Canadian real-world study, which included putting three types of labels (one of them a cancer warning) on the majority of alcohol drinks in the participating liquor stores in the studied region, found that these enhanced labels led to increased recall, cognitive processing and self-reported impact of drinking [51,54], as well as decreased alcohol sales [53] in the intervention municipality compared to the control municipality. While the whole intervention included both cancer warnings as well as low-risk guidelines/standard drink labels, it is not possible to completely disentangle the impact of solely the cancer warning labels from the results, although some results pointed to effects being driven predominantly by the cancer label. The one paper that looked specifically at the cancer labels [52] found that cancer labels increased knowledge of alcohol as a carcinogen, and that knowledge increased with time. Additionally, the authors found that knowledge about alcohol causing cancer was associated with a greater likelihood of supporting health warning labels.

3.2.2. Label Content—Image vs. Text

Among the eight papers that compared image and text labels, six had control group comparison [55,61,62,64,67] and in two [65,66], only text and image were compared. Six of the papers used graphic images (e.g., diseased body part); in one, the images were neutral, and one contained both graphic and neutral images.

In the studies where graphic images were used, comparison between text-only and image-and-text labels showed higher negative emotional arousal [55,56], fear [55,56,61,65] and reactance [65] for the image-and-text labels among adults compared to text only labels, as well as lower product appeal [64,65], label acceptability [55,56] and believability [65]. Moreover, in the study using neutral images [67], there were higher negative emotions for image-and-text labels compared to text-only were found among the secondary school students, but only for some messages. The same study also found that both types of labels are better than no label for increasing knowledge about alcohol-related risks when knowledge is low at baseline. While an online experiment showed a selection of alcoholic drinks in the given scenario to be lower after seeing image-and-text labels compared to text-only labels [55], this finding did not replicate in a naturalistic lab shopping setting [56] among adult regular drinkers, where none of the labels affected purchasing behaviour. In another study, both text-only and image-and-text labels decreased the speed of consumption compared to no label, with no significant difference between them [62].

Table 3. Results of the studies.

Authors	Studied Groups/Variables (Independent Variables)	Outcome (Dependent Variable)	Category	Significant Results * (< Lower Than, > Higher Than)	Mechanism of Impact (if Tested)
Al-Hamdani & Smith (2015) [64]	Between subjects: • standard • text warning • combined text and image warning • combined text and image warning on plain packaged bottles Within subjects: • alcohol type: beer, wine and spirits	Product-based perception	Attitude	Plain packaging < Standard (all three alcohol types) Text and image < Standard (all three alcohol types) Text < Image (spirits only)	
		Consumer-based perception	Attitude	Plain packaging < Standard (all three alcohol types) Text and image < Standard (wine and spirits only) No difference text and plain packaging	
		Warning recognition	Awareness/recognition	Higher odds of recognition of warning on plain packaged bottle of wine (but not beer and spirits)	
Al-Hamdani & Smith (2017) [76]	Between subjects: • branding type: branded or plain packaged • warning size: medium, large, or extra-large Within subjects: • alcohol type: beer, wine and spirits	Product-based perception	Attitude	Plain packaging < Branded, no difference in size Interaction between alcohol type and branding and warning size (Extra large < Medium (wine and spirits only)	
		Consumer-based perception	Attitude	Plain packaging < Branded, no difference in size	
		Boringness of the bottle	Attitude	No difference in size or branding Interaction between alcohol type and branding	
		Warning recognition	Awareness/recognition	Higher odds of recognition of warning on plain packaged spirit bottle (but not beer and wine)	
Annunziata et al. (2019) [70]	Attributes: • alcohol content • framing of warning statement • warning size • warning position	Utility	Attitude	By attribute importance: No pictorial > Pictorial Back label > Front label No message > Neutral framed Message > Negative framed message	
Blackwell et al. (2018) [71]	Between subjects: • message specificity (general vs. specific) • message framing (positive vs. negative) • message content (cancer vs. mental health)	Support for alcohol labelling	Acceptability/support	Support HWL < Support calorie information and strength information	
		Motivation to drink less	Motivation	Cancer message > Mental health message Negative framing > Positive framing No difference in specificity	
		Reactance	Reactance	Specific message < General message Negative framing > Positive framing No difference in content	
		Avoidance	Avoidance	Cancer message > Mental health message Negative framing > Positive framing No difference in specificity	
		Believability	Attitude	Specific message > General message No difference in framing and content	
		Self-efficacy	Efficacy	No difference in any characteristic	
		Response efficacy	Efficacy	Specific message > General message No difference in framing and content	

Table 3. *Cont.*

Authors	Studied Groups/Variables (Independent Variables)	Outcome (Dependent Variable)	Category	Significant Results * (< Lower Than, > Higher Than)	Mechanism of Impact (if Tested)
Clarke, Pechey et al. (2021a) [55]	Between subjects: • image: present versus absent • text: present versus absent	Negative emotional arousal	Emotions	Any HWL > No HWL Image and text > Text only	(T) Tested model suggested negative emotional arousal possibly mediates the effect of HWL on alcohol selection
		Acceptability of labels	Acceptability/support	Text > Image and text > Image only	
		Reactance	Reactance	Any HWL > No HWLImage and text > Text only	
		Avoidance	Avoidance	Any HWL > No HWL Image and text > Text only Image > Image and text	
		Perceived disease risk	Risk perception	Any HWL > No HWL No difference between HWLs	
		Proportion of participants selecting an alcoholic beverage to be consumed either immediately or later on that day	Intentions	Any HWL < No label Image or Image and text < Text alone	
Clarke, Blackwell et al. (2021b) [56]	Between subjects: • image and text HWL • text HWL • no HWL (control)	Negative emotional arousal	Emotions	Image and text > text	
		Acceptability of labels	Acceptability/support	Text > Image and text	
		Proportion of total alcoholic drinks among all selected	Behaviour	No difference between the three groups	
Glock & Krolak-Schwerdt (2013) [72]	Between subjects: • labels (health-related vs. positive-related) Within subjects: • time (before vs. after presentation of warning labels)	Implicit attitudes	Attitude	Interaction between warning and time: Health labels: before < after Positive expectancies: before > after	
		Explicit attitudes	Attitude	No difference	
		Intention to drink	Intentions	Health-related > Positive-related	
Gold et al. (2020) [63]	Between subjects: • label: present vs. absent	Perceived personal risk	Risk perception	No difference	
		Motivation to drink less	Motivation	No difference	
		Perception of damaging drinking	Risk perception	No difference	
Hall et al. (2019) [73]	Between subjects: • substance: cigarette, alcohol, sugar sweetened beverages • causal language: "causes", "contributes to", "can contribute to" and "may contribute to".	Perceived message effectiveness	Perceived effectiveness	Causes > Others (overall, for alcohol less than for cigarettes)	
		Perceived message ineffectiveness	Perceived effectiveness	May contribute to > Others (overall, and for alcohol)	
		Public support	Acceptability/support	No difference in support in alcohol group	

Table 3. *Cont.*

Authors	Studied Groups/Variables (Independent Variables)	Outcome (Dependent Variable)	Category	Significant Results * (< Lower Than, > Higher Than)	Mechanism of Impact (if Tested)
Hall et al. (2020) [65]	Between subjects: • label: text vs. image and text • efficacy information: present vs. absent Within subjects: • substance: cigarette, alcohol, sugar sweetened beverages	Perceived message effectiveness	Perceived effectiveness	Image and text > Text (overall, and for alcohol)	
		Believability	Attitude	Text > Image and text (overall, and for alcohol)	
		Reactance	Reactance	Image and text > Text (overall, and for alcohol)	
		Fear	Emotions	Image and text > Text (overall, and for alcohol)	
		Thinking about harms	Cognitive processing	Image and text > Text (overall, and for alcohol)	
		Product appeal	Attitude	Text > Image and text (overall, and for alcohol)	
		Policy support	Acceptability/support	Text > Image and text (overall, and for alcohol)	
		Self-efficacy	Efficacy	No difference in efficacy information presence	
Hobin, Schoueri-Mychasiw et al. (2020a) [51]	Between subjects: • labels: enhanced vs. regular	Recognition of labels	Awareness/recognition	Enhanced label > regular label	(T) Consumer attention to and processing of label messages partially mediated the relationship between the enhanced alcohol warning labels and self-reported drinking less → suggests that strengthening HWL will increase their effectiveness because they draw attention to and increase processing of the labels
		Cognitive processing	Cognitive processing	Enhanced label > Regular label	
		Self-reported impact on drinking (reduction)	Behaviour	Enhanced labels > Regular label	
Hobin, Weerasinghe et al. (2020b) [52]	Between subjects: • labels: enhanced vs. regular	Prompted and unprompted recall	Awareness/recognition	Enhanced label > Regular label (largest difference between groups at T2)	(T) Knowledge about alcohol causing cancer associated with greater likelihood of supporting health warning labels in the study
		Knowledge of alcohol as carcinogen	Knowledge	Enhanced label > Regular label (largest difference between groups at T3)	
		Support for health warning labels on alcohol containers	Acceptability/support	Intervention: Agree (Wave 1 = 57.4%; Wave 2 = 57.3%; Wave 3 = 61.3%) Comparison: Agree (Wave 1 = 53.7%; Wave 2 = 51.6%; Wave 3 = 53.7%)	
Hobin, Shokar et al. (2020c) [54]	Between subjects: • labels: enhanced vs. regular	Unprompted recall	Awareness/recognition	Enhanced label > Regular label (cancer label)	
		Cognitive processing	Cognitive processing	Enhanced label > Regular label	
		Self-reported impact on drinking (reduction)	Behaviour	Enhanced labels > Regular label	
Jarvis & Pettigrew (2013) [74]	Attributes: • brand • alcohol content • warning message (framed positive or negative)	Utility	Attitude	Brand > Alcohol content > Warning statement Two warning statements significant: one positive and one negative	

Table 3. *Cont.*

Authors	Studied Groups/Variables (Independent Variables)	Outcome (Dependent Variable)	Category	Significant Results * (< Lower Than, > Higher Than)	Mechanism of Impact (if Tested)
Jongenelis, Pettigrew et al. (2018a) [58]	Between subjects: • Source: single-source or multiple-source of message	Intentions to reduce alcohol consumption	Intentions	Multiple source > Single source	
		Message believability	Attitude	Multiple source > Single source	
		Message convincingness	Attitude	Multiple source > Single source	
		Personal relevance of the message	Attitude	Multiple source > Single source	
Jongenelis, Pratt et al. (2018b) [59]	Between subjects: • Label content	Alcohol as risk factor beliefs	Knowledge	Present message > Absent message (for all messages except for liver damage)	
		Intentions to reduce alcohol consumption	Intentions	Present message > Absent message (for cancer, diabetes and mental illness message, but not for liver damage and heart disease)	
Krischler & Glock (2015) [68]	Between subjects: • labels (questions vs. statements vs. control) Within subjects: • expectancy category (positive vs. negative, tension reduction vs. negative sedating vs. socially related)	Drinking intentions	Intentions	No difference	
		Individual outcome expectancies	Attitude	No difference	
		General outcome expectancies	Attitude	No difference	
Ma (2021) [69]	Between subjects: • narrative PWLs • non-narrative PWLs • control (no stimulus)	Worry about developing alcohol-related cancer	Emotions	Narrative labels > no labels	
		Feelings of risk of developing alcohol-related cancer	Risk perception	Narrative labels > no labels	
		Comparative likelihood of developing alcohol-related cancer	Risk perception	No difference	(T) Mediation analysis found that narrative PWLs (vs. control) indirectly influenced intentions through worry, but not through feelings of risk, comparative likelihood, or perceived severity.
		Perceived severity of harm of developing alcohol-related cancer	Risk perception	Narrative labels > no labels	
		Intentions to reduce alcohol use	Intentions	No difference	
Monk et al. (2017) [66]	Between subjects: • image: graphic or neutral • area of interest: text or image • positive expectancy change: increase or decrease/no change	Dwell time	Attention	Image > Text (overall and in the positive expectancies increase group)No difference between graphic or neutral, and between increase and decrease in positive expectancies	
Morgenstern et al. (2021) [67]	Between subjects: • position of the HWL on the questionnaire (before vs. after alcohol items) • type of HWL (text only vs. image and text) • content of the HWL (one out of a pool of ten)	Knowledge about alcohol-related risks	Knowledge	Any label >No label (for cancer and liver cirrhosis)	
		Self-report of alcohol use	Behaviour	No difference	
		Negative emotions	Emotions	Text and image > Text only (only for some messages—driving, liver cirrhosis, pharmaceuticals, minors)	

Table 3. *Cont.*

Authors	Studied Groups/Variables (Independent Variables)		Outcome (Dependent Variable)	Category	Significant Results * (< Lower Than, > Higher Than)	Mechanism of Impact (if Tested)
Pechey et al. (2020) [57]	Between subjects:		Negative emotional arousal	Emotions	Bowel cancer > Others (> liver cancer > liver cirrhosis > heart disease > liver disease > 7 types of cancer > breast cancer)	
	• message type (7 messages of different health consequences)		Desire to consume the labelled product	Motivation	Bowel cancer < Liver cirrhosis < Breast cancer < Liver cancer < Heart disease < Liver disease < 7 types of cancer	
	• image type (3 different graphic images per health consequence)		Acceptability of the label	Acceptability/support	Low overall acceptability, Bowel cancer < Others (Breast cancer < Liver disease < 7 types of cancer < Liver cancer < Liver disease < Liver cirrhosis)	
Pettigrew et al. (2016) [60]	Between subjects:		Intention to reduce alcohol consumption	Intentions	Pre < Post No difference between messages	
	• content of the message (6 messages)		Believability of message	Attitude	Bowel cancer > Other messages	
	Within subjects: • before and after (only for intention)		Convincingness of the message	Attitude	Bowel cancer > Other messages	
			Personal relevance of the message	Attitude	Bowel cancer > Other messages	
Pham et al. (2018) [50]	Between subjects:		Attention	Attention	Enhanced colour and size > Other groups	
	• colour change (use of red colouring instead of black) • size of the HWL (increase by 50%) • changes in both colour and size • control (existing label)		Number and length of visual fixations	Attention	No difference between groups	
Sillero-Rejon et al. (2018) [75]	Between subjects:		Visual attention	Attention	No difference in severity or self-affirmation	
	• self-affirmation (self-affirmed vs. control)		Avoidance	Avoidance	Highly severe > Moderately severe No difference in self-affirmation	
	Within-subject:		Reactance	Reactance	Highly severe > Moderately severe No difference in self-affirmation	
	• warning severity (moderately-severe vs. highly-severe)		Perceived susceptibility to health risks	Risk perception	No difference in severity or self-affirmation	
			Perceived warning effectiveness	Perceived effectiveness	Highly severe > Moderately severe No difference in self-affirmation	
			Motivation to drink less	Motivation	Highly severe > Moderate severe No difference in self-affirmation	
Stafford & Salmon (2017) [62]	Between subjects:		Product design evaluation	Attitude	Image and text HWL < No HWL No difference between text only and no HWL	(H) The mechanism responsible for slower consumption is theorised to be due to higher levels of fear arousal in the two health warning conditions
	• no HWL • text only HWL • image and text HWL		The duration to consume the test beverage	Behaviour	No HWL < Text HWL No HWL < Image and text HWL No difference between text HWL and image and text HWL	

Table 3. *Cont.*

Authors	Studied Groups/Variables (Independent Variables)	Outcome (Dependent Variable)	Category	Significant Results * (< Lower Than, > Higher Than)	Mechanism of Impact (if Tested)
Wigg & Stafford (2016) [61]	Between subjects: • no HWL • text only HWL • image and text HWL	Fear arousal	Emotions	Image and text HWL > Text HWL Image and text HWL > No HWL No difference between text HWL and no HWL	
		Risk perception	Risk perception	Image and text HWL > No HWL No difference between text HWL and no HWL No difference between image and text HWL and text HWL	
		Intention to quit/reduce alcohol consumption	Intentions	Image and text HWL > No HWL No difference between text HWL and no HWL No difference between image and text HWL and text HWL	
Zhao et al. (2020) [53]	Between subjects: • label: enhanced vs. regular	Alcohol sales	Behaviour	Enhanced labels < Regular labels	

* Significant results column describes the impact of independent variables (tested groups) on the dependent variable (outcome) and should be read together with the outcome column, for example, "plain packaging group had significantly lower product based perception compared to standard packaging group".

The two studies actively looking at the differences between graphic and neutral images did not find any difference between the two in attention [66,75], although the latter found higher avoidance, reactance, perceived warning effectiveness and motivation to drink less in graphic compared to neutral images.

3.2.3. Label Content—Message Characteristics

Several papers also compared different messages or their characteristics. In terms of message *content*, in Australia, one study among adult drinkers found increased intention to reduce drinking for cancer, diabetes and mental illness messages, but not for heart disease and liver damage-related messages [59]. Another study found that a bowel cancer message had the highest believability, convincingness and personal relevance [60]. A bowel cancer message also had the highest negative emotional arousal, lowest desire to consume the labelled product and lowest acceptability in a large-scale UK online study of adult drinkers [57], but in this study, all the labels included graphic photos, which might have influenced the outcome. Another large scale online experiment in the UK among adult drinkers [71] compared a cancer message to a mental health message and found the former led to a higher motivation to drink less and higher avoidance, but there was no difference in reactance, believability, self-efficacy and response efficacy. Another smaller study found that positive expectancy-related messages lowered drinking intentions compared to health-related messages among students from Germany and Luxembourg [72]. In a German sample of secondary school students, cancer and liver cirrhosis messages, but not a pregnancy label, increased knowledge about alcohol-related risks [67].

In terms of *framing*, the discrete choice experiments gave mixed results [70,74]; in the first study, no message at all was more preferred than any message in a sample of Italian and French young wine drinkers, and in the latter, the negative message was less preferred than the neutral message among Australian students. In an online study of British adult drinkers [71], a negative framing led to a higher motivation to drink less, with increased reactance and avoidance compared to a positive framing, but did not influence believability, self-efficacy and response efficacy. This study also looked at *specificity* and found that specific messages led to lower reactance, higher response efficacy and higher believability compared to general messages, but there was no difference in motivation to drink less, avoidance and self-efficacy.

One study looked at whether framing labels as questions or statements impacts attitude and intentions, and found no difference [68]. Another study compared narrative and non-narrative labels and found narrative labels led to higher worry about developing alcohol-related cancer and higher feelings of risk of developing alcohol-related cancer, but no difference in intentions to reduce alcohol use [69]. The study examining causal language [73] found that using the word "causes" had higher perceived message effectiveness, and "may contribute to" had the highest perceived ineffectiveness, but there was no difference in support for any of the labels. Finally, a study of Australian adult drinkers found that receiving the message from multiple sources is better than receiving a message from one single source [58].

3.2.4. Label Format

Finally, four studies examined label format: branding/plain packaging [64,76], warning size [50,70,76] and warning position [70]. Two smaller-scale Canadian studies found that plain packaging (removing any branding) decreased product-based and consumer-based perception [64,76]. The latter study [76] also found that after a certain size of the warning (50% of the bottle), there is no further increased difference in outcomes. In a discrete choice experiment, young Italian and French wine drinkers preferred to have warning sizes as little as possible, and on the back of the label rather than on the front of the label [70]. Finally, one study [50] found that an increase in size and change in prominent colour was effective in self-reported measures of attention among the Australian drinkers, but not when measured directly with eye-tracking.

4. Discussion

This review aimed to examine the scope of the (quasi-)experimental research on novel and enhanced alcohol health warning labels and to identify the key characteristics of the effective warning labels. Twenty-seven papers were included, the majority of which were published in 2018 or later, indicating an uptake in interest surrounding the topic from the research community (recognised also in other broader reviews [45]). This is aligned with the increase in policy-related labelling discussions, such as in Codex Alimentarius, at the EU or national level (e.g., [77,78]). All of the research has been undertaken in Western countries, where alcohol labelling has either not been updated since its implementation decades ago (e.g., Canada, US), carried out on a voluntary basis (e.g., Australia, UK) or not mandated at all (e.g., many of the EU countries). Given high alcohol consumption in those regions [79], this may reflect the recognition of labelling of potentially effective means of reducing alcohol-related harm—although it is debatable if this focus adds to or replaces focus on "best buys" policies [80], such as taxation, pricing and advertising restrictions. In addition, the majority of the studies focused on general population, which reflects a shift from findings of previous review [43], where less than a third of studies were targeting the broader population, and the remaining were targeting younger populations and (pregnant) women.

In terms of content, most commonly, the studied messages were referring to cancer—70% of papers included at least one cancer-related message, out of which the most commonly mentioned cancers were bowel, breast and liver cancer. This was followed by liver cirrhosis or liver disease messages, used in a third of papers. Other topics of messages were heart disease, mental illness, brain damage, pregnancy, drinking and driving and other (e.g., positive expectancies surrounding alcohol). This brings us to the question on how to select the content of the message to be included on the labels. In the papers included in this review, researchers have commonly either developed messages based on "evidence on alcohol-related harm" or have taken messages used in other studies. We would argue that the message first has to reflect evidence-based risk (e.g., that causal relationship between alcohol and the outcome has been clearly demonstrated [3]), and then the message should be tailored to the context in which the labels will be displayed, based on the burden of disease. Another factor that has been found to be important for the selection of the message is also its novelty for the given audience [16,53]. For example, in the EU, 29.4% of all alcohol-attributable deaths and 19.4% of all attributable DALYs are due to cancers [81], while awareness of the risk between alcohol and cancer is low [15]. Based on these guidelines we can conclude that cancer warnings should be a priority when mandating health warning labels in this region.

Next, the focus of research implies the researcher's theory of change—the mechanisms by which they consider labelling to work [82]—and this will influence the targeted outcomes and how effectiveness is measured. This is relevant because the policy debate will inevitably discuss the merit of the health warning labels based on their "effectiveness"—previously, one of the arguments in the World Trade Organisation's discussions was that labels are considered unnecessary interference with international trade because there is no scientific evidence to support such warnings as an effective public health measure [83]. Based on the scope of the existing research, we propose that two possible theories of how labelling brings about change are currently being investigated. The first one is focused on persuasion and short-term individual behaviour change, and optimises using health warning labels that aim to elicit reactance, avoidance, fear and other negative emotional reactions as means of changing behaviour, drawing heavily on the experience from tobacco warning labels. For this line of research, there is no real-world evidence from the alcohol field, and mixed evidence from lab studies regarding (short-term) behaviour [56,62]. The (online and lab) studies that focused on studying the graphic images on labels found that they elicited higher negative emotional arousal, fear and reactance, and were not well accepted by the participants. These results are also corroborated by the metanalysis investigating the impact of HWL in both alcohol and food products, and found that while health warning labels reduce product selection compared to no labels, the difference between image-and-text HWLs and text-only HWLs was not statistically significant (although in the former, the effect was slightly larger) [44].

The second theory of change sees labelling as a tool to raise awareness of alcohol-related risks as a part of a comprehensive strategy to reduce alcohol-related harm (which should include other alcohol policies and interventions), and emphasises optimising labels to increase attention to and cognitive processing of the information on the labels, leading to increased knowledge (e.g., on alcohol health risks). This approach is more in line with system perspectives [84] and its focus is to use health warning labels to arrive at a better informed population that could in turn lead to more support for other alcohol policies. Individual short-term behaviour change might be a welcome by-product of this approach, but in the long term, labels can be seen as a tool to help with changing societal conversation and norms surrounding alcohol (as suggested also by O'Brien [85]). In terms of evidence, the best current evidence points to the second theory of change as having greater potential to be explored in future research and practice. The most comprehensive study to date and the only one measuring long-term effects in the real world [86] showed that simple, specific and well-visible messages led to an increase in knowledge (with associated support for stronger alcohol policies [87]) and change in the amount of alcohol consumed, partially mediated by the consumer attention to and processing of label messages [54]. The labels were also well accepted and supported by the population, as well as the drinkers [51].

It is important to note that this study [86] was implemented in context with existing relatively strict other policy measures—e.g., national monopoly on the alcohol stores (alcohol being sold separately in the government-owned liquor stores), and enhanced label replacing existing label (rather than no label). Nevertheless, a key approach to label development and implementation in this study is a good practice example and should be used in any future attempts to study alcohol HWLs in different contexts. This means developing labels in consultation with local stakeholders, including a range of different messages (not only health but also related to standard drinks and lower-risk guidelines) and including evidence-based messages for which there was low awareness at baseline and are relevant to the population. This study [83] also points to the unintended consequences that can be anticipated—originally, the study was planned to run for eight months with all three labels, but the alcohol industry intervened and threatened with a lawsuit if cancer warning labels were to continue being applied on the products [83]. Thus, cancer labels were only applied to the alcohol products for one instead of eight months. This shows that in any attempts to mandate evidence-based alcohol health warning labels, strong opposition from the industry can be expected, similar to what happened in the tobacco field [88].

Thus, while there should be some lessons from tobacco taken for alcohol health warning labels, we must be careful to still consider alcohol in its own context, and note that the evolution of alcohol legislation lags behind the tobacco field by a couple of decades. There is evidence for the effectiveness of pictorial labels for tobacco [36], but even here, the argument can be made that labels have been part of a broader approach that involved international collaboration and changed societal norms around tobacco [89], and their purpose was perhaps less to deter smokers from smoking than to indirectly prevent younger populations from taking up smoking, as tobacco warning labels replaced packaging as the last venue of marketing after other forms of advertising had been prohibited [38]. The value of adding images on alcohol labels might be to help with attracting attention and take away space from the branding of the product, but not enough studies with only neutral images or symbols have been conducted to make any firm conclusions.

The main implications for research and policy based on the results of this review are summarised in Table 4. We would like to note that the same theory of change useful for health warnings could also be useful to develop and test for optimal nutrition information on alcohol labels (not direct focus on changing behaviour, but focus on providing the information), although as previously mentioned, the issue with nutritional labels is to first align requirements for alcoholic beverages with existing labelling requirements for food and non-alcoholic beverages.

Table 4. Research and policy implications.

Future Labelling Research

- Aim to be explicit about the theory of change underpinning the research and adjust the studied outcomes accordingly.
- Develop labels to be tested with own qualitative research and preferably co-created with local stakeholders/populations—focus on both content and format.
- The information provided in the labels should be clear and reflect the up-to-date literature on alcohol-attributable burden of disease.
- Online surveys with large samples and experimental design can be a good lower-cost approach to help with tailoring the messages to new contexts before testing them in the real world.
- Real-world studies with longer-term outcomes are preferable in the context a lot of existing research (e.g., online and lab surveys).
- When reporting, describe the context in which the intervention was conducted; frameworks such as Medical Research Councils' process evaluation guidance [82] can be applied to support design and reporting.

Development of mandatory labels

As currently implemented mandatory and voluntary labels are often suboptimal, existing research on new health warning labels points to the following characteristics:

- In terms of format:
- good visibility of the label is essential: this is achieved by large size, contrasting colours, large text, prominent position on the product
- In terms of content:
- it is important to have messages that convey novel information to the target population
- messages should be evidence-based—for example, take into account alcohol-attributable risk, morbidity and mortality
- wording should be simple, messages should be specific rather than general
- current evidence suggests that graphic images are not well accepted, and does not point to a superiority of their inclusion
- use rotating health warning labels presented in combination with other messages
- information on health warning labels should be presented also through other sources of information (e.g., information campaigns), and labelling should preferably also be introduced as part of a broader package of alcohol policy measures

4.1. Limitations

Our review is subject to general limitations of scoping reviews: we did not appraise studies for quality, and the conclusions are still somewhat broad and qualitative. In terms of this specific review, we narrowed the scope of our investigation to health warning labels and overall outcomes and we did not focus on other labels, such as lower-risk drinking guidelines or standard drinks, even though some research suggests it might be worth taking them into account simultaneously. We also did not investigate differential impact of labels on different subgroups. Evidence from Canada points to some subgroup differences in cognitive processing of the labels, although concludes that vulnerable populations still attended to the labels [52]. Our search was also focused only on peer-reviewed English language papers; therefore, we might have missed peer-reviewed papers in other languages and non-peer-reviewed research.

4.2. Conclusions

The majority of the current evidence regarding alcohol health warning labels comes from very short interventions in the online or the laboratory setting as opposed to the real world, and there is rather large diversity in the content and format of labels being evaluated. Nevertheless, evidence from real-world, long-term interventions shows that alcohol health warning labels designed to be visible and contain novel and specific information have the potential to be part of an effective labelling strategy. Labelling should be seen as tool to raise awareness and a small part of wider alcohol prevention and policy approach. Future research could focus on importance of labels in a broader societal context, and go beyond individual behaviour change, recognising that change in social norms can be a complex, multifaceted and long-term process.

Supplementary Materials: The following are available online at https://www.mdpi.com/article/10.3390/nu13093065/s1, Table S1. Label development, content and format.

Author Contributions: Conceptualisation: D.K., P.A., E.J.-L.; Methodology: D.K., P.A., E.J.-L.; Investigation: D.K.; Data analysis: D.K., Supervision; E.J.-L.; Writing—original draft: D.K.; Writing—Review and Editing: D.K., P.A., E.J.-L. All authors have read and agreed to the published version of the manuscript.

Funding: Part of this research was produced under the service contract AlHaMBRA Project (Alcohol Harm—Measuring and Building Capacity for Policy Response and Action, Contract No. 20197105).

Institutional Review Board Statement: Not applicable.

Informed Consent Statement: Not applicable.

Data Availability Statement: Extracted data are presented in the main and supplementary tables.

Acknowledgments: Part of this research was produced under the service contract AlHaMBRA Project (Alcohol Harm—Measuring and Building Capacity for Policy Response and Action, Contract No. 20197105). The information and views presented are those of the authors, and hence represent their sole responsibility. Accordingly, the information and views presented cannot be considered to reflect the views of all AlHaMBRA Project consortium members, the European Commission and/or the Health and Digital Executive Agency or any other body of the European Union. The European Commission and the Agency do not accept any responsibility for use that may be made of the information contained herein.

Conflicts of Interest: The authors declare no conflict of interest.

Appendix A

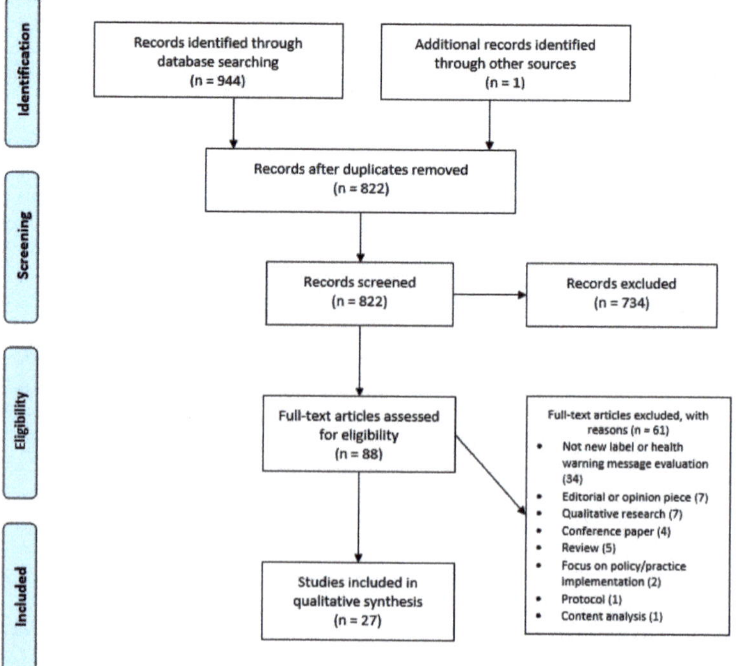

Figure A1. PRISMA Diagram.

References

1. Shield, K.; Manthey, J.; Rylett, M.; Probst, C.; Wettlaufer, A.; Parry, C.D.H.; Rehm, J. National, regional, and global burdens of disease from 2000 to 2016 attributable to alcohol use: A comparative risk assessment study. *Lancet Public Health* **2020**, *5*, e51–e61. [CrossRef]
2. Okaru, A.O.; Lachenmeier, D.W. Margin of exposure analyses and overall toxic effects of alcohol with special consideration of carcinogenicity. *Nutrients* **2021**, in press.
3. Rehm, J.; Gmel Sr, G.E.; Gmel, G.; Hasan, O.S.M.; Imtiaz, S.; Popova, S.; Probst, C.; Roerecke, M.; Room, R.; Samokhvalov, A.V.; et al. The relationship between different dimensions of alcohol use and the burden of disease—An update. *Addiction* **2017**, *112*, 968–1001. [CrossRef]
4. Rehm, J.; Taylor, B.; Mohapatra, S.; Irving, H.; Baliunas, D.; Patra, J.; Roerecke, M. Alcohol as a risk factor for liver cirrhosis: A systematic review and meta-analysis. *Drug Alcohol Rev.* **2010**, *29*, 437–445. [CrossRef]
5. Bagnardi, V.; Rota, M.; Botteri, E.; Tramacere, I.; Islami, F.; Fedirko, V.; Scotti, L.; Jenab, M.; Turati, F.; Pasquali, E.; et al. Alcohol consumption and site-specific cancer risk: A comprehensive dose–response meta-analysis. *Br. J. Cancer* **2015**, *112*, 580–593. [CrossRef] [PubMed]
6. International Agency for Research on Cancer IARC. *IARC Monographs on the Evaluation of Carcinogenic Risks to Humans: Volume 96: Alcohol Consumption and Ethyl Carbamate*; International Agency for Research on Cancer IARC: Lion, France, 2010; Volume 96.
7. International Agency for Research on Cancer. *Monographs on the Evaluation of Carcinogenic Risks to Humans: Alcohol Drinking*; International Agency for Research on Cancer IARC: Lion, France, 1988; Volume 44.
8. Grant, B.F.; Goldstein, R.; Saha, T.D.; Chou, S.P.; Jung, J.; Zhang, H.; Pickering, R.P.; Ruan, W.J.; Smith, S.M.; Huang, B.; et al. Epidemiology of DSM-5 Alcohol Use Disorder results from the national epidemiologic survey on alcohol and related conditions III. *JAMA Psychiatry* **2015**, *72*, 757–766. [CrossRef] [PubMed]
9. Rehm, J.; Roerecke, M. Cardiovascular effects of alcohol consumption. *Trends Cardiovasc. Med.* **2017**, *27*, 534–538. [CrossRef]
10. Imtiaz, S.; Shield, K.D.; Roerecke, M.; Samokhvalov, A.V.; Lonnroth, K.; Rehm, J. Alcohol consumption as a risk factor for tuberculosis: Meta-analyses and burden of disease. *Eur. Respir. J.* **2017**, *50*, 1700216. [CrossRef]
11. Bellos, S.; Skapinakis, P.; Rai, D.; Zitko, P.; Araya, R.; Lewis, G.; Lionis, C.; Mavreas, V. Longitudinal association between different levels of alcohol consumption and a new onset of depression and generalized anxiety disorder: Results from an international study in primary care. *Psychiatry Res.* **2016**, *243*, 30–34. [CrossRef]
12. Boden, J.M.; Fergusson, D.M. Alcohol and depression. *Addiction* **2011**, *106*, 906–914. [CrossRef]
13. Popova, S.; Lange, S.; Shield, K.; Mihic, A.; Chudley, A.E.; Mukherjee, R.A.S.; Bekmuradov, D.; Rehm, J. Comorbidity of fetal alcohol spectrum disorder: A systematic review and meta-analysis. *Lancet* **2016**, *387*, 978–987. [CrossRef]
14. Testa, M.; Quigley, B.M.; Eiden, R. The Effects of Prenatal Alcohol Exposure on Infant Mental Development: A Meta-Analytical Review. *Alcohol Alcohol.* **2003**, *38*, 295–304. [CrossRef]
15. Scheideler, J.K.; Klein, W.M.P. Awareness of the Link between Alcohol Consumption and Cancer across the World: A Review. *Cancer Epidemiol. Biomark. Prev.* **2018**, *27*, 429–437. [CrossRef]
16. Vallance, K.; Stockwell, T.; Zhao, J.; Shokar, S.; Schoueri-Mychasiw, N.; Hammond, D.; Greenfield, T.K.; McGavock, J.; Weerasinghe, A.; Hobin, E. Baseline Assessment of Alcohol-Related Knowledge of and Support for Alcohol Warning Labels Among Alcohol Consumers in Northern Canada and Associations With Key Sociodemographic Characteristics. *J. Stud. Alcohol Drugs* **2020**, *81*, 238–248. [CrossRef] [PubMed]
17. Calvert, C.M.; Toomey, T.; Jones-Webb, R. Are people aware of the link between alcohol and different types of Cancer? *BMC Public Health* **2021**, *21*, 1–10. [CrossRef] [PubMed]
18. Thomsen, K.L.; Christensen, A.S.P.; Meyer, M.K.H. Awareness of alcohol as a risk factor for cancer: A population-based cross-sectional study among 3000 Danish men and women. *Prev. Med. Rep.* **2020**, *19*, 101156. [CrossRef]
19. Martin-Moreno, J.M.; Harris, M.E.; Breda, J.; Møller, L.; Alfonso-Sanchez, J.L.; Gorgojo, L. Enhanced labelling on alcoholic drinks: Reviewing the evidence to guide alcohol policy. *Eur. J. Public Health* **2013**, *23*, 1082–1087. [CrossRef] [PubMed]
20. World Health Organisation Global Information System on Alcohol and Health (GISAH). Available online: http://apps.who.int/gho/data/node.gisah.A1191?lang=en&showonly=GISAH (accessed on 27 August 2021).
21. Government of Canada Labelling Requirements for Alcoholic Beverages. Available online: https://inspection.canada.ca/food-label-requirements/labelling/industry/alcohol/eng/1392909001375/1392909133296 (accessed on 11 June 2021).
22. European Commission Regulation (EU) No 1169/2011 of the European Parliament and of the Council of 25 October 2011 on the Provision of Food Information to Consumers. Available online: https://eur-lex.europa.eu/legal-content/EN/ALL/?uri=CELEX%3A32011R1169 (accessed on 27 August 2021).
23. Food Standards Australia and New Zealand Labelling of Alcoholic Beverages. Available online: https://www.foodstandards.gov.au/consumer/labelling/Pages/Labelling-of-alcoholic-beverages.aspx (accessed on 11 June 2021).
24. Alcohol and Tobacco Tax and Trade Bureau Labeling Resources. Available online: https://www.ttb.gov/labeling/labeling-resources (accessed on 11 June 2021).
25. Maynard, O.M.; Langfield, T.; Attwood, A.S.; Allen, E.; Drew, I.; Votier, A.; Munafò, M.R. No Impact of Calorie or Unit Information on Ad Libitum Alcohol Consumption (Special Issue: Communicating messages about drinking). *Alcohol Alcohol.* **2018**, *53*, 12–19. [CrossRef] [PubMed]

26. Maynard, O.; Blackwell, A.; Munafò, M.; Attwood, A. *Know Your Limits: Labelling Interventions to Reduce Alcohol Consumption*; Alcohol Research UK: London, UK, 2018.

27. Croker, H.; Packer, J.; Russell, S.J.; Stansfield, C.; Viner, R.M. Front of pack nutritional labelling schemes: A systematic review and meta-analysis of recent evidence relating to objectively measured consumption and purchasing. *J. Hum. Nutr. Diet.* **2020**, *33*, 518–537. [CrossRef]

28. Kelly, B.; Jewell, J. *What Is the Evidence on the Policy Specifications, Development Processes and Effectiveness of Existing Front-of-Pack Food Labelling Policies in the WHO European Region?* WHO Regional Office for Europe: Copenhagen, Denmark, 2018.

29. Wilkinson, C.; Allsop, S.; Cail, D.; Chikritzhs, T.; Daube, M.; Kirby, G.; Mattick, R. *Alcohol Warning Labels: Evidence of Effectiveness on Risky Alcohol Consumption and Short Term Outcomes*; National Drug Research Institute: Bentley, WA, Australia, 2009.

30. Wilkinson, C.; Room, R. Warnings on alcohol containers and advertisements: International experience and evidence on effects. *Drug Alcohol Rev.* **2009**, *28*, 426–435. [CrossRef]

31. Dossou, G.; Gallopel-Morvan, K.; Diouf, J.-F. The effectiveness of current French health warnings displayed on alcohol advertisements and alcoholic beverages. *Eur. J. Public Health* **2017**, *27*, 699–704. [CrossRef]

32. Dumas, A.; Toutain, S.; Hill, C.; Simmat-Durand, L. Warning about drinking during pregnancy: Lessons from the French experience. *Reprod. Health* **2018**, *15*, 20. [CrossRef] [PubMed]

33. Coomber, K.; Martino, F.; Barbour, I.R.; Mayshak, R.; Miller, P.G. Do consumers "Get the facts"? A survey of alcohol warning label recognition in Australia. *BMC Public Health* **2015**, *15*, 816. [CrossRef] [PubMed]

34. Critchlow, N.; Jones, D.; Moodie, C.; Mackintosh, A.M.; Fitzgerald, N.; Hooper, L.; Thomas, C.; Vohra, J. Awareness of product-related information, health messages and warnings on alcohol packaging among adolescents: A cross-sectional survey in the United Kingdom. *J. Public Health* **2020**, *42*, e223–e230. [CrossRef] [PubMed]

35. Kersbergen, I.; Field, M. Alcohol consumers' attention to warning labels and brand information on alcohol packaging: Findings from cross-sectional and experimental studies. *BMC Public Health* **2017**, *17*, 1–11. [CrossRef] [PubMed]

36. Noar, S.M.; Francis, D.B.; Bridges, C.; Sontag, J.M.; Ribisl, K.M.; Brewer, N.T. The impact of strengthening cigarette pack warnings: Systematic review of longitudinal observational studies. *Soc. Sci. Med.* **2016**, *164*, 118–129. [CrossRef]

37. Noar, S.M.; Francis, D.B.; Bridges, C.; Sontag, J.M.; Brewer, N.T.; Ribisl, K.M. Effects of Strengthening Cigarette Pack Warnings on Attention and Message Processing: A Systematic Review. *J. Mass Commun. Q.* **2017**, *94*, 416–442. [CrossRef]

38. Al-Hamdani, M. The case for stringent alcohol warning labels: Lessons from the tobacco control experience. *J. Public Health Policy* **2013**, *35*, 65–74. [CrossRef]

39. World Health Organization. *Global Strategy to Reduce the Harmful Use of Alcohol*; World Health Organization: Geneva, Switzerland, 2010.

40. Bates, S.; Holmes, J.; Gavens, L.; De Matos, E.G.; Li, J.; Ward, B.; Hooper, L.; Dixon, S.; Buykx, P. Awareness of alcohol as a risk factor for cancer is associated with public support for alcohol policies. *BMC Public Health* **2018**, *18*, 1–11. [CrossRef]

41. Buykx, P.; Gilligan, C.; Ward, B.; Kippen, R.; Chapman, K. Public support for alcohol policies associated with knowledge of cancer risk. *Int. J. Drug Policy* **2015**, *26*, 371–379. [CrossRef]

42. Alcohol Action Ireland What is Public Health Alcohol Act? Available online: https://alcoholireland.ie/what-is-the-public-health-alcohol-bill/ (accessed on 11 June 2021).

43. Hassan, L.M.; Shiu, E. A systematic review of the efficacy of alcohol warning labels: Insights from qualitative and quantitative research in the new millennium. *J. Soc. Mark.* **2018**, *8*, 333–352. [CrossRef]

44. Clarke, N.; Pechey, E.; Kosīte, D.; König, L.M.; Mantzari, E.; Blackwell, A.K.M.; Marteau, T.M.; Hollands, G.J. Impact of health warning labels on selection and consumption of food and alcohol products: Systematic review with meta-analysis. *Health Psychol. Rev.* **2020**, 1–24. [CrossRef] [PubMed]

45. Dimova, E.D.; Mitchell, D. Rapid literature review on the impact of health messaging and product information on alcohol labelling. *Drugs: Educ. Prev. Policy* **2021**, 1–13. [CrossRef]

46. Wettlaufer, A. Can a Label Help me Drink in Moderation? A Review of the Evidence on Standard Drink Labelling. *Subst. Use Misuse* **2018**, *53*, 585–595. [CrossRef] [PubMed]

47. Arksey, H.; O'Malley, L. Scoping studies: Towards a methodological framework. *Int. J. Soc. Res. Methodol.* **2005**, *8*, 19–32. [CrossRef]

48. Tricco, A.C.; Lillie, E.; Zarin, W.; O'Brien, K.; Colquhoun, H.; Kastner, M.; Levac, D.; Ng, C.; Sharpe, J.P.; Wilson, K.; et al. A scoping review on the conduct and reporting of scoping reviews. *BMC Med. Res. Methodol.* **2016**, *16*, 1–10. [CrossRef]

49. Tricco, A.C.; Lillie, E.; Zarin, W.; O'Brien, K.K.; Colquhoun, H.; Levac, D.; Moher, D.; Peters, M.D.J.; Horsley, T.; Weeks, L.; et al. PRISMA extension for scoping reviews (PRISMA-ScR): Checklist and explanation. *Ann. Intern. Med.* **2018**, *169*, 467–473. [CrossRef]

50. Pham, C.; Rundle-Thiele, S.; Parkinson, J.; Li, S. Alcohol Warning Label Awareness and Attention: A Multi-method Study. *Alcohol Alcohol.* **2017**, *53*, 39–45. [CrossRef]

51. Hobin, E.; Schoueri-Mychasiw, N.; Weerasinghe, A.; Vallance, K.; Hammond, D.; Greenfield, T.K.; McGavock, J.; Paradis, C.; Stockwell, T. Effects of strengthening alcohol labels on attention, message processing, and perceived effectiveness: A quasi-experimental study in Yukon, Canada. *Int. J. Drug Policy* **2020**, *77*, 102666. [CrossRef]

52. Hobin, E.; Weerasinghe, A.; Vallance, K.; Hammond, D.; McGavock, J.; Greenfield, T.K.; Schoueri-Mychasiw, N.; Paradis, C.; Stockwell, T. Testing Alcohol Labels as a Tool to Communicate Cancer Risk to Drinkers: A Real-World Quasi-Experimental Study. *J. Stud. Alcohol Drugs* **2020**, *81*, 249–261. [CrossRef]

53. Zhao, J.; Stockwell, T.; Vallance, K.; Hobin, E. The Effects of Alcohol Warning Labels on Population Alcohol Consumption: An Interrupted Time Series Analysis of Alcohol Sales in Yukon, Canada. *J. Stud. Alcohol Drugs* **2020**, *81*, 225–237. [CrossRef] [PubMed]

54. Hobin, E.; Shokar, S.; Vallance, K.; Hammond, D.; McGavock, J.; Greenfield, T.K.; Schoueri-Mychasiw, N.; Paradis, C.; Stockwell, T. Communicating risks to drinkers: Testing alcohol labels with a cancer warning and national drinking guidelines in Canada. *Can. J. Public Health* **2020**, *111*, 716–725. [CrossRef] [PubMed]

55. Clarke, N.; Pechey, E.; Mantzari, E.; Blackwell, A.K.M.; De Loyde, K.; Morris, R.W.; Munafò, M.R.; Marteau, T.M.; Hollands, G.J. Impact of health warning labels communicating the risk of cancer on alcohol selection: An online experimental study. *Addiction* **2021**, *116*, 41–52. [CrossRef]

56. Clarke, N.; Blackwell, A.K.M.; De Loyde, K.; Pechey, E.; Hobson, A.; Pilling, M.; Morris, R.W.; Marteau, T.M.; Hollands, G.J. Health warning labels and alcohol selection: A randomised controlled experiment in a naturalistic shopping laboratory. *Addiction* **2021**, 3. [CrossRef]

57. Pechey, E.; Clarke, N.; Mantzari, E.; Blackwell, A.K.M.; Deloyde, K.; Morris, R.W.; Marteau, T.M.; Hollands, G.J. Image-and-text health warning labels on alcohol and food: Potential effectiveness and acceptability. *BMC Public Health* **2020**, *20*, 1–14. [CrossRef]

58. Jongenelis, M.; Pettigrew, S.; Wakefield, M.; Slevin, T.; Pratt, I.S.; Chikritzhs, T.; Liang, W. Investigating Single-Versus Multiple-Source Approaches to Communicating Health Messages Via an Online Simulation. *Am. J. Health Promot.* **2018**, *32*, 979–988. [CrossRef] [PubMed]

59. Jongenelis, M.I.; Pratt, I.S.; Slevin, T.; Chikritzhs, T.; Liang, W.; Pettigrew, S. The effect of chronic disease warning statements on alcohol-related health beliefs and consumption intentions among at-risk drinkers. *Health Educ. Res.* **2018**, *33*, 351–360. [CrossRef] [PubMed]

60. Pettigrew, S.; Jongenelis, M.I.; Glance, D.; Chikritzhs, T.; Pratt, I.S.; Slevin, T.; Liang, W.; Wakefield, M. The effect of cancer warning statements on alcohol consumption intentions. *Health Educ. Res.* **2016**, *31*, 60–69. [CrossRef]

61. Wigg, S.; Stafford, L.D. Health Warnings on Alcoholic Beverages: Perceptions of the Health Risks and Intentions towards Alcohol Consumption. *PLoS ONE* **2016**, *11*, e0153027. [CrossRef]

62. Stafford, L.D.; Salmon, J. Alcohol health warnings can influence the speed of consumption. *J. Public Health* **2017**, *25*, 147–154. [CrossRef] [PubMed]

63. Gold, N.; Egan, M.; Londakova, K.; Mottershaw, A.; Harper, H.; Burton, R.; Henn, C.; Smolar, M.; Walmsley, M.; Arambepola, R.; et al. Effect of alcohol label designs with different pictorial representations of alcohol content and health warnings on knowledge and understanding of low-risk drinking guidelines: A randomized controlled trial. *Addiction* **2021**, *116*, 1443–1459. [CrossRef]

64. Al-Hamdani, M.; Smith, S. Alcohol warning label perceptions: Emerging evidence for alcohol policy. *Can. J. Public Health* **2015**, *106*, e395–e400. [CrossRef]

65. Hall, M.G.; Grummon, A.H.; Lazard, A.J.; Maynard, O.M.; Taillie, L.S. Reactions to graphic and text health warnings for cigarettes, sugar-sweetened beverages, and alcohol: An online randomized experiment of US adults. *Prev. Med.* **2020**, *137*, 106120. [CrossRef] [PubMed]

66. Monk, R.L.; Westwood, J.; Heim, D.; Qureshi, A.W. The effect of pictorial content on attention levels and alcohol-related beliefs: An eye-tracking study. *J. Appl. Soc. Psychol.* **2017**, *47*, 158–164. [CrossRef]

67. Morgenstern, M.; Dumbili, E.W.; Hansen, J.; Hanewinkel, R. Effects of alcohol warning labels on alcohol-related cognitions among German adolescents: A factorial experiment. *Addict. Behav.* **2021**, *117*, 106868. [CrossRef] [PubMed]

68. Krischler, M.; Glock, S. Alcohol warning labels formulated as questions change alcohol-related outcome expectancies: A pilot study. *Addict. Res. Theory* **2015**, *23*, 343–349. [CrossRef]

69. Ma, Z. The Role of Narrative Pictorial Warning Labels in Communicating Alcohol-Related Cancer Risks. *Health Commun.* **2021**, 1–9. [CrossRef]

70. Annunziata, A.; Agnoli, L.; Vecchio, R.; Charters, S.; Mariani, A. Health warnings on wine labels: A discrete choice analysis of Italian and French Generation Y consumers. *Wine Econ. Policy* **2019**, *8*, 81–90. [CrossRef]

71. Blackwell, A.K.M.; Drax, K.; Attwood, A.S.; Munafò, M.R.; Maynard, O.M. Informing drinkers: Can current UK alcohol labels be improved? *Drug Alcohol Depend.* **2018**, *192*, 163–170. [CrossRef]

72. Glock, S.; Krolak-Schwerdt, S. Changing Outcome Expectancies, Drinking Intentions, and Implicit Attitudes toward Alcohol: A Comparison of Positive Expectancy-Related and Health-Related Alcohol Warning Labels. *Appl. Psychol. Health Well-Being* **2013**, *5*, 332–347. [CrossRef]

73. Hall, M.G.; Grummon, A.H.; Maynard, O.M.; Kameny, M.R.; Jenson, D.; Popkin, B.M. Causal Language in Health Warning Labels and US Adults' Perception: A Randomized Experiment. *Am. J. Public Health* **2019**, *109*, 1429–1433. [CrossRef]

74. Jarvis, W.; Pettigrew, S. The relative influence of alcohol warning statement type on young drinkers' stated choices. *Food Qual. Prefer.* **2013**, *28*, 244–252. [CrossRef]

75. Sillero-Rejon, C.; Attwood, A.S.; Blackwell, A.K.M.; Ibáñez-Zapata, J.-A.; Munafò, M.R.; Maynard, O.M. Alcohol pictorial health warning labels: The impact of self-affirmation and health warning severity. *BMC Public Health* **2018**, *18*, 1403. [CrossRef] [PubMed]

76. Al-Hamdani, M.; Smith, S.M. Alcohol Warning Label Perceptions: Do Warning Sizes and Plain Packaging Matter? *J. Stud. Alcohol Drugs* **2017**, *78*, 79–87. [CrossRef]

77. Jané-Llopis, E.; Kokole, D.; Neufeld, M.; Hasan, O.S.M.; Rehm, J. What Is the Current Alcohol Labelling Practice in the WHO European Region and What Are Barriers and Facilitators to Development and Implementation of Alcohol Labelling Policy? Health Evidence Network Synthesis Report, No. 68. Available online: https://www.ncbi.nlm.nih.gov/books/NBK558550/ (accessed on 11 June 2021).

78. Neufeld, M.; Ferreira-Borges, C.; Rehm, J. Implementing Health Warnings on Alcoholic Beverages: On the Leading Role of Countries of the Commonwealth of Independent States. *Int. J. Environ. Res. Public Health* **2020**, *17*, 8205. [CrossRef] [PubMed]

79. World Health Organisation. *Global Status Report on Alcohol and Health 2018*; World Health Organization: Geneva, Switzerland, 2018; Volume 65. Available online: http://www.who.int/iris/handle/10665/112736 (accessed on 27 August 2021).

80. Chisholm, D.; Moro, D.; Bertram, M.; Pretorius, C.; Gmel, G.; Shield, K.; Rehm, J. Are the "Best Buys" for Alcohol Control Still Valid? An Update on the Comparative Cost-Effectiveness of Alcohol Control Strategies at the Global Level. *J. Stud. Alcohol Drugs* **2018**, *79*, 514–522. [CrossRef] [PubMed]

81. WHO. *Status Report on Alcohol Consumption, Harm and Policy Responses in 30 European Countries 2019*; WHO: Copenhagen, Denmark, 2019.

82. Moore, G.F.; Audrey, S.; Barker, M.; Bond, L.; Bonell, C.; Hardeman, W.; Moore, L.; O'Cathain, A.; Tinati, T.; Wight, D.; et al. Process evaluation of complex interventions: Medical Research Council guidance. *BMJ* **2015**, *350*. [CrossRef]

83. Stockwell, T.; Solomon, R.; O'Brien, P.; Vallance, K.; Hobin, E. Cancer Warning Labels on Alcohol Containers: A Consumer's Right to Know, a Government's Responsibility to Inform, and an Industry's Power to Thwart. *J. Stud. Alcohol Drugs* **2020**, *81*, 284–292. [CrossRef] [PubMed]

84. Petticrew, M.; Shemilt, I.; Lorenc, T.M.; Marteau, T.; Melendez-Torres, G.J.; O'Mara-Eves, A.; Stautz, K.; Thomas, J. Alcohol advertising and public health: Systems perspectives versus narrow perspectives. *J. Epidemiol. Community Health* **2017**, *71*, 308–312. [CrossRef] [PubMed]

85. O'Brien, P.; Mitchell, A.D. On the Bottle: Health Information, Alcohol Labelling and the WTO Technical Barriers to Trade Agreement. *QUT Law Rev.* **2018**, *18*, 124. [CrossRef]

86. Vallance, K.; Stockwell, T.; Hammond, D.; Shokar, S.; Schoueri-Mychasiw, N.; Greenfield, T.; McGavock, J.; Zhao, J.; Weerasinghe, A.; Hobin, E. Testing the Effectiveness of Enhanced Alcohol Warning Labels and Modifications Resulting From Alcohol Industry Interference in Yukon, Canada: Protocol for a Quasi-Experimental Study. *JMIR Res. Protoc.* **2020**, *9*, e16320. [CrossRef] [PubMed]

87. Weerasinghe, A.; Schoueri-Mychasiw, N.; Vallance, K.; Stockwell, T.; Hammond, D.; McGavock, J.; Greenfield, T.K.; Paradis, C.; Hobin, E. Improving Knowledge that Alcohol Can Cause Cancer is Associated with Consumer Support for Alcohol Policies: Findings from a Real-World Alcohol Labelling Study. *Int. J. Environ. Res. Public Health* **2020**, *17*, 398. [CrossRef] [PubMed]

88. Hiilamo, H.; Crosbie, E.; A Glantz, S. The evolution of health warning labels on cigarette packs: The role of precedents, and tobacco industry strategies to block diffusion. *Tob. Control* **2014**, *23*, e2. [CrossRef] [PubMed]

89. Myers, M.L. The FCTC's evidence-based policies remain a key to ending the tobacco epidemic. *Tob. Control* **2013**, *22*, i45–i46. [CrossRef]

Article

Government Options to Reduce the Impact of Alcohol on Human Health: Obstacles to Effective Policy Implementation

Tim Stockwell [1,*], Norman Giesbrecht [2], Kate Vallance [1] and Ashley Wettlaufer [2]

1   Canadian Institute for Substance Use Research, University of Victoria, Victoria, BC V8P 5C2, Canada; vallance@uvic.ca
2   Centre for Addiction and Mental Health, Toronto, ON M5V 2B4, Canada; Norman.Giesbrecht@camh.ca (N.G.); Ashley.Wettlaufer@camh.ca (A.W.)
*   Correspondence: timstock@uvic.ca

Abstract: Evidence for effective government policies to reduce exposure to alcohol's carcinogenic and hepatoxic effects has strengthened in recent decades. Policies with the strongest evidence involve reducing the affordability, availability and cultural acceptability of alcohol. However, policies that reduce population consumption compete with powerful commercial vested interests. This paper draws on the Canadian Alcohol Policy Evaluation (CAPE), a formal assessment of effective government action on alcohol across Canadian jurisdictions. It also draws on alcohol policy case studies elsewhere involving attempts to introduce minimum unit pricing and cancer warning labels on alcohol containers. Canadian governments collectively received a failing grade (F) for alcohol policy implementation during the most recent CAPE assessment in 2017. However, had the best practices observed in any one jurisdiction been implemented consistently, Canada would have received an A grade. Resistance to effective alcohol policies is due to (1) lack of public awareness of both need and effectiveness, (2) a lack of government regulatory mechanisms to implement effective policies, (3) alcohol industry lobbying, and (4) a failure from the public health community to promote specific and feasible actions as opposed to general principles, e.g., 'increased prices' or 'reduced affordability'. There is enormous untapped potential in most countries for the implementation of proven strategies to reduce alcohol-related harm. While alcohol policies have weakened in many countries during the COVID-19 pandemic, societies may now also be more accepting of public health-inspired policies with proven effectiveness and potential economic benefits.

Keywords: alcohol; policy; public health; alcohol industry; Canada

Citation: Stockwell, T.; Giesbrecht, N.; Vallance, K.; Wettlaufer, A. Government Options to Reduce the Impact of Alcohol on Human Health: Obstacles to Effective Policy Implementation. *Nutrients* **2021**, *13*, 2846. https://doi.org/10.3390/nu13082846

Academic Editor: Peter Anderson

Received: 2 July 2021
Accepted: 11 August 2021
Published: 19 August 2021

Publisher's Note: MDPI stays neutral with regard to jurisdictional claims in published maps and institutional affiliations.

1. Introduction

Alcohol and public health policy as an organised field of academic inquiry is relatively new. In 1975, the assertion by Kettil Bruun, Griffith Edwards, Robin Room and others that alcohol was a public health issue was seen as new and controversial [1]. Using case studies mostly from Scandinavia and other parts of Europe, they provided evidence that the total consumption of alcohol was a reliable predictor of the extent of serious alcohol-related harms in any particular population, e.g., liver disease, injury and alcohol use disorders. Perhaps their most revolutionary proposal was that governments could take action to reduce alcohol-related harm, principally by reducing its affordability (e.g., through higher taxes) and its availability (e.g., by reduced days and hours of sale).

In the decades since this landmark publication, often referred to as the "Purple Book", thousands of studies have been published on the intersections between population consumption of alcohol, the extent of population harm from alcohol and the relative effectiveness of alternative government policies. Learnings from these research mountains are now distilled in systematic reviews and meta-analyses and are further informed by advances in theory and methodology. For example, the International Model of Alcohol Harms and Policies (InterMAHP) [2] provides an online tool to accurately estimate the extent of alcohol-related harm in any population (be it a city, region, or country) from available data on per capita alcohol consumption and the recorded prevalence of various causes of death, injury, and illness. InterMAHP also helps to estimate how these harms change when policies are introduced to increase or decrease alcohol consumption (e.g., [3]). The tool is built on some solid foundations including (a) a demonstrated mathematical relationship between the average consumption of a population and the distribution of drinkers according to their typical daily consumption [4] and (b) the latest systematic reviews and meta-analyses of published studies estimating risks of different types of disease and injury as a function of how much people drink.

As a result of such advances, it is now possible to provide very specific and detailed advice to governments regarding the public health consequences of policy decisions in such concrete terms as how many people will become ill, injured or die prematurely from alcohol-related reasons if policy X or Y is introduced—and what might be the economic costs and benefits. Furthermore, the Sheffield Alcohol Policy Model (SAPM, e.g., [5]) has been effectively applied over the last decade to inform legal and policy processes culminating in the introduction of a Minimum Unit Price (MUP) for alcohol in Scotland in May 2018 [6], a successful policy which is rapidly being emulated in other countries (e.g., [7–9]). Both InterMAHP and SAPM have been used to estimate policy impacts of MUP on such sensitive outcomes as government revenues and consumer expenditure (e.g., [3,10]).

In this paper, we take the view that while there are considerable obstacles and challenges to the implementation of effective alcohol policies, theories and methodologies to inform effective implementation are available to policy decision makers, the public health field and the broader community of concerned citizens. While authors of the most highly cited and comprehensive evidence review on effective alcohol policies [11] concluded, rather depressingly, that effective policy is inevitably unpopular, we take a more optimistic approach. We highlight "circuit breaker" strategies which can create a more favourable climate for governments being prepared to implement policies with demonstrated effectiveness in reducing alcohol-related harms.

2. What Is Effective Alcohol Policy?

The World Health Organization (WHO) recently launched the WHO-SAFER international initiative to reduce alcohol-related harm in member countries [12]. Built around a growing evidence base and the pre-existing WHO Global Strategy on Alcohol [13], this initiative promotes the following objectives:

(1)   Strengthening restrictions on alcohol availability
(2)   Advancing and enforcing drink-driving countermeasures
(3)   Facilitating access to screening, brief interventions and treatment
(4)   Enforcing bans or comprehensive restrictions on alcohol advertising, sponsorship and promotion
(5)   Raising prices on alcohol through excise taxes and pricing policies.

A number of research-led initiatives have sought to conduct systematic assessments of the extent to which such evidence-based strategies are being implemented in jurisdictions and, in some cases, relate the extent of policy implementation to outcomes (e.g., [14,15]). The co-authors of this manuscript developed the Canadian Alcohol Policy Evaluation (CAPE) project [16–19] in which 250 indicators of alcohol policy implementation were evaluated across 11 domains of effective government action in each of Canada's 10 provinces and three territories. Table 1 below identifies seven domains for which there is evidence of direct effectiveness for the reduction in alcohol consumption and harms and also indicates the strength of this evidence and the likely scope of impacts on population level harms. These domains include each of the five WHO SAFER domains listed above and, in addition, two others with long-standing evidence to support them: Liquor Law Enforcement (e.g., targeting enforcement to high-risk premises and enforcing laws prohibiting service to intoxicated persons) and Minimum Legal Drinking Age Laws.

Table 1. CAPE provincial and territorial alcohol policy domains and weights.

Direct Policy Domains	Effectiveness (out of 5)	Scope (out of 5)	Weight [1] (out of 25)
1. Pricing and Taxation	5	5	25
2. Physical Availability of Alcohol	4	4	16
3. Impaired Driving Countermeasures	5	3	15
4. Marketing and Advertising Controls	3	5	15
5. Minimum Legal Drinking Age	4	3	12
6. Screening, Brief Intervention and Referral	3	3	9
7. Liquor Law Enforcement:	3	3	9
Indirect Policy Domains	Facilitation (out of 5)	Scope (out of 5)	Weight [2] (out of 25)
8. Alcohol Control System	5	5	25
9. Alcohol Strategy	4	5	20
10. Monitoring and Reporting	4	4	16
11. Health and Safety Messaging	3	4	12

[1] Weight (direct) = Effectiveness × Scope. [2] Weight (indirect) = Facilitation × Scope. Weights are boldfaced to emphasise these were the final estimates of potential policy impacts.

Uniquely, CAPE also identified and assessed four "indirect policy domains" with the capacity to facilitate the implementation of the other seven directly effective policies. In short, the extent of government ownership of alcohol distribution and retail systems (i.e., Alcohol Control System) facilitates control of direct policy levers such as pricing and availability; a comprehensive Alcohol Strategy outlining multiple policy objectives with funding for implementation, clear leadership independent of commercial vested interests and an evaluation plan will guide multiple arms of government towards effective action; comprehensive, transparent and regular public Monitoring and Reporting of alcohol harms and policies will help to sustain and continue to direct effective policies; Health and Safety Messaging, particularly at the point of purchase and consumption through container labelling, will motivate effective policy action by highlighting risks and harms from alcohol consumption, both among citizens and their elected representatives [17].

3. What Are the Obstacles to Effective Policy Implementation?

There are some recurring themes that characterise resistance to the implementation of the types of effective alcohol policies discussed above. We will restrict our discussion to four key themes: (1) a lack of awareness about the extent of alcohol-related harm and the effectiveness of alcohol policies; (2) a lack of government regulatory and legislative structures focused on reducing harm from alcohol; (3) effective lobbying by alcohol industry groups to foster such skepticism and to propose less-effective policies; (4) absent or ineffective lobbying by public health advocacy groups. We will discuss each of these in turn and suggest ways such impediments can be overcome or at least be rendered less obstructive.

3.1. Low Public Awareness of the Extent of Alcohol Related Harm

Alcohol use is causally implicated in several hundred diagnostic categories of illness and injury [2] including cancer and cardiovascular diseases, the two most common causes of death in developed countries. However, alcohol's contribution to the illnesses and injuries leading to emergency department presentation, hospitalisation and/or premature death are usually not recorded on death certificates or diagnostic records. Special studies are required to estimate the extent to which these premature deaths and presentations might be attributable to alcohol. The methods of such studies are complex and technical, well beyond the grasp of many health professionals, let alone decision-makers or members of the public. There is also a widespread belief evidenced by national surveys and perpetuated by industry groups that moderate alcohol consumption can protect people from a wide array of harms, cardiovascular diseases in particular. For example, one survey found that 57% of the Canadian population believed alcohol in moderation was good for their health (e.g., [20]). The prevalence of this belief may have waned in recent years, perhaps reflecting more open debate and critique of this idea, with a 2020 survey of Quebecers finding this proportion to now be lower, at 40% [21]. Opinion surveys in different countries have also frequently found that when people are asked to estimate whether alcohol or illicit drugs contribute the most problems to society, usually alcohol comes in a poor second [22].

Contrary to these popular and governmental biases, the hard evidence is that alcohol use contributes more health harms than other psychoactive drugs that are prohibited in many countries (i.e., cannabis, opioids, stimulants, etc.). This is borne out of WHO Global Burden of Disease estimates which most recently indicate that 2.4 million premature deaths globally are attributable to alcohol, compared with 0.49 million for illicit drugs in 2019 [23]. In Canada, the ongoing Canadian Substance Use Costs and Harms [24] project reports annually on estimated hospitalisations, deaths and economic costs from alcohol, cannabis, nicotine, opioids and other psychoactive substances [24]. The overall economic cost of substance use across healthcare, lost productivity, criminal justice and other direct-cost domains was estimated at $46 billion in 2017. Accounting for more than three-quarters of the total cost are Canada's three legal drugs: alcohol ($16.6 billion), tobacco ($12.3 billion) and cannabis ($3.2 billion). Among illicit substances, only opioids ($5.9 billion) and cocaine ($3.7 billion) make up more than 5% of the total cost [24]. This disparity persists even at a time when there is an opioid overdose crisis. Typically, public debate focuses on the "crises" even if the harm and death rates are not substantially higher than those from alcohol.

Part of the reason for the lack of awareness of alcohol-related harms may be the extent to which these are not wholly attributable to the use of alcohol. Alcohol's contributions to serious illnesses, injuries and premature deaths are usually contributory and often not even recorded. Sherk et al. [25] estimated that over 90% of alcohol-attributable deaths in a Canadian jurisdiction were partially alcohol attributable, and less than 10% wholly or 100% attributable. A good example is alcohol-related breast cancer. It has been estimated that approximately 7% of these cancers can be attributed to alcohol [26]. Nonetheless, despite this relatively small contribution, because of the high prevalence of cancer in general, about 30% of alcohol-attributable deaths in Canada are from alcohol-related cancers [24] (Note that these estimates sum up the fractions of lives lost or hospitalisations caused to make up each partially attributable case).

Not surprisingly, there is evidence worldwide that few members of the public are aware that alcohol is a carcinogen, despite the WHO's International Agency for Research on Cancer having confirmed a causal association more than 30 years ago [27]. For example, a national UK survey found that only 13% of respondents could identify alcohol as a cause of cancer without prompting, 33% could do so if prompted and 54% were completely unaware of any alcohol–cancer connection [28].

The lack of public awareness of the hidden, partially alcohol-attributable illnesses and injuries is also reflected in the widespread perception that the only serious harms from alcohol are alcohol dependence, liver cirrhosis and crashes from impaired driving. In fact, these are just tips of much larger icebergs. At the outset of the COVID-19 pandemic, there were several North American examples of perverse rationales being given by governments for classifying alcohol as an "essential commodity", which reflect this restricted view of alcohol-related harm. In California [29] and at least two Canadian jurisdictions [30], this essential-commodity status was justified as a way of preventing healthcare services from being overwhelmed by people going into alcohol withdrawal. In Canada, at most only 5% of alcohol-attributable hospitalisations are associated with alcohol dependence or withdrawal [30]. When India imposed a strict lockdown with a complete alcohol prohibition, there was indeed observed a spike in presentations for alcohol withdrawal, but this was transient and soon demand dropped to zero [31]. It was also documented during a Nordic strike of government alcohol monopoly workers that demand for alcohol withdrawal treatment dropped to less than 40% of normal levels [32]. It has been shown elsewhere [30] that when partial as well as wholly alcohol-attributable hospitalisations are considered, maintaining alcohol use imposes a greater burden on healthcare services in many countries than has COVID-19, e.g., 105,000 alcohol-attributable hospital admissions were estimated for 2017 [24] compared with 40,000 COVID-19-related admissions in the first full year of the pandemic [33].

3.2. Low Public Awareness of the Effectiveness of Alcohol Policies

Babor et al. [11] advanced the influential narrative that effective alcohol policies (principally restrictions on pricing and availability) are unpopular with the public and their elected representatives, while less effective educational strategies receive the strongest support in opinion surveys. This position is well based on opinion research spanning multiple countries which consistently suggests that across-the-board price increases in particular are the most unpopular of all alcohol policy options. For example, a careful analysis of public opinion on alcohol policy in the UK found 84% support for public information campaigns, whereas only 33% supported increasing the price of alcohol, 41% for reducing the number of liquor outlets and 39% for reducing their hours of trading [34]. Critically, Li et al. [34] were able to identify skepticism about the effectiveness of different policies as a determining factor in whether they would be supported.

There is also a widespread belief that "alcoholics" will always access alcohol regardless of price or availability, a viewpoint that anyone who has discussed alcohol control policies with decision-makers or members of the public will have encountered. This is contrary to evidence from a thorough systematic review that heavy drinkers do reduce their drinking when prices increase [35] and from the accounts of alcohol-dependent drinkers themselves [36].

3.3. Absent or Inadequate Government Regulatory and Legislative Structures

Governments in both the developed and developing worlds rarely give high level consideration to alcohol policy. Responsibility for the implementation of the kinds of evidence-based policies listed above is usually scattered across multiple ministries and departments. Health ministries are mostly responsible for providing healthcare to the sick and injured rather than addressing underlying causes and reducing risk factors, thus they do not have direct access to the most important policy levers, namely, price, availability and marketing restrictions. Finance departments are responsible for taxation and pricing but are not mandated to reduce adverse health and safety impacts of alcohol and often are unaware of the economic costs of these. They see their major responsibilities as raising government revenues and maintaining free and fair markets for all commodities. Direct regulation of the sale and distribution of liquor is mostly delegated to local licensing and civic authorities who perhaps are more inclined to see their roles as limiting public nuisance and overseeing a fair market rather than protecting public health. It is rare for public health considerations to be represented in the local, regional and national regulation of alcohol.

The low priority given to alcohol by governments is often reflected in the extent of national and international government funding for research, prevention and treatment. A prime example is the substantially larger budget for the US National Institute for Drug Abuse than for the US National Institute for Alcoholism and Alcohol Abuse, i.e., USD1.3 billion versus USD0.45 billion in 2020 [37,38]. This balance of effort and resourcing is also frequently reflected in national government departmental budgets and numbers of dedicated staff. For example, Health Canada's Alcohol Policy Unit (APU) had about 1/6th of the staff positions allocated to the Office of Controlled Substances [39]. The APU has since been downsized and absorbed into another public health unit.

At various times and places, alcohol-related problems have been deemed so severe that they have become a top government priority. Total prohibitions on alcohol sales have occasionally been attempted, notably in the US and parts of Canada in the early part of the last century and, most recently, in India and South Africa as part of government measures to reduce the spread of COVID-19 (e.g., [40]). These tend to be short lived and associated with increased crime and corruption even if health and safety outcomes improve, e.g., reduced deaths from liver cirrhosis during the US Prohibition [41] and reduced violence in the South African COVID-19 lockdowns [40]. More sustainable policies to control alcohol consumption and reduce harms were introduced in the 19th and 20th centuries, principally in North America and Scandinavia, namely, direct alcohol distribution and retail systems otherwise known as "government alcohol monopolies". Remnants of these monopolies still operate in some 17 US states, mostly restricted to sales of spirits and wine, and to some degree in all Canadian jurisdictions (even Alberta maintains government control over the distribution of alcohol) and some Scandinavian countries. The Finnish and Swedish alcohol monopolies (Alko and Systembolaget, respectively) are notable for both reporting to Ministries for Health and Social Affairs rather than following the North American practice of placing such operations within finance ministries focused on revenue collection rather than health and social outcomes [42,43]. Government alcohol control systems can facilitate ready access to the key policy levers of pricing and availability, but they also need a special mandate to do so in the interests of public health and safety rather than simple free-market economics. For this reason, the CAPE project documents both the extent of government ownership of alcohol distribution and retail systems in a jurisdiction and also to what ministry they report as an indication of the public purpose they are intended to fulfil [18].

Perhaps the most telling indication of the low priority afforded to alcohol and public health policy globally is the lack of alcohol-specific national and international laws. Room et al. [44] have made the case for an international treaty on alcohol noting such treaties exist for tobacco and illicit drug trade. In Canada, we have a federal Tobacco Act and a federal Cannabis Act but no Alcohol Act [19]. The legalisation of cannabis in Canada casts the lack of legislation and regulation of alcohol in sharp relief, especially when considering the substantially lower harms and costs of cannabis in contrast to alcohol [24].

3.4. Effectiveness of Alcohol Industry Influence

Globally, nationally and sub-nationally, alcohol industry groups and companies apply their immense resources to influence public debates on alcohol-related harm and policies to address this harm so as to protect their marketplace. Babor [45] described some of the ways alcohol industry-funded groups have sought to influence research agendas and shape public discussions around alcohol-related harm and prevention policies. He highlighted how these activities help to confuse public discussion of health issues and policy options and are a convenient way to demonstrate 'corporate responsibility' in their attempts to avoid taxation and regulation. McCambridge et al. [46] have further analysed how industry-funded social aspects bodies such as DrinkAware in the UK attempt to work closely with government agencies to influence policy agendas away from evidence-based policies that might restrict alcohol markets. Needless to say, industry groups can use their financial clout to directly influence the agendas of political parties through donations and direct lobbying in ways that are not available to the public health community.

Petticrew and colleagues published a thematic analysis of industry-sponsored publications on the role of alcohol as a risk factor for cancer. They showed how the evidence for alcohol's causal role in cancer is downplayed, ignored or even denied with heavy emphasis on the strength of other rival risk factors [47]. There are also many examples of industry-sponsored organisations extolling various health benefits from using their product with little discussion of the growing critical literature of this interpretation (e.g., [48]). A comprehensive analysis of direct alcohol industry funding for alcohol research reported that this had increased by 56% in recent decades [49].

The CAPE project has worked to identify the extent of alcohol industry involvement in the development and implementation of alcohol policies in Canada, provincially and federally [49,50]. A key indicator used is whether alcohol industry representatives are directly involved in the development of government alcohol strategies. In 2017, it was noted that Canada's National Alcohol Strategy Advisory Council had several Canadian alcohol manufacturers represented, a situation that has been subsequently remedied. These efforts, and many others besides, work towards creating a positive climate of opinion around alcohol and discourage effective government action to restrict its consumption.

Industry players have also shown they are not afraid to directly counter efforts to educate the public about alcohol's role as a contributing cause of cancer. The phenomenon of "pink washing" has been widely documented whereby industry groups support cancer awareness activities [51]. A prime Canadian example some years ago was the creation of Mike's Hard Pink Lemonade in association with the Canadian Cancer Associations breast cancer awareness week [52], i.e., an alcohol producer associating itself with a cancer prevention activity while marketing a product which contributes directly to cancer [26]. More recently, an initiative taken by the Yukon government in Canada to place cancer warning labels on all products sold in a major liquor store in its capital Whitehorse as part of a Health Canada-funded research project was shut down as a result of thinly veiled legal threats from major drinks producers [53].

3.5. Ineffectiveness of Public Health Advocacy

Perhaps surprisingly, even agencies concerned with the promotion of public health have often worked in opposition to or ignored the potential of evidence-based alcohol policies to improve public health. The "Pink-Washing" of breast cancer noted above is just one example of perverse strategies from the public health field. Amin et al. [54] conducted a formal review of the websites of the Organisation for Economic Co-operation and Development cancer control agencies in 2017 as to the extent to which they identified alcohol as a risk factor for cancer. Websites in all countries except the USA and Canada at that time identified alcohol as a Class 1 carcinogen. While the US cancer agency recommended using taxation as a means of reducing tobacco-related cancers, no such action was recommended in relation to alcohol taxes. Canada is also an example of an OECD country with no national advocacy agency dedicated to promoting effective alcohol policy.

3.6. Summary and Ways Forward

There could not be a starker contrast between government responses to the COVID-19 pandemic versus those to the health and safety problems associated with alcohol use, despite the related harms often being similar in scale [30]. Naturally, precautions to avoid contracting a potentially deadly virus that might make you seriously ill or kill you within weeks are more likely to be supported than measures to reduce the availability and affordability of an enjoyable intoxicant, despite its short- and long-term risks. However, while the harms from COVID-19 have been counted and reported daily through multimedia throughout the course of the pandemic crisis, the majority of people in most countries are unaware of the serious harms (e.g., cancer) of alcohol consumption and of the full scale of its burden on health and well-being. Since greater than 90% of associated mortality is partially alcohol attributable, this important contribution often flies under the radar, and is not recorded on death certificates or routinely reported by health authorities. It should not be surprising, therefore, that effective government action to reduce our ability to access and purchase a favourite recreational drug, widely believed to be innocuous or even beneficial to health, is often not taken. If one adds to this the existence of powerful commercial vested interest groups who are prepared to threaten litigation against governments that exercise their legal right to warn consumers of alcohol's health risks by placing cancer warning labels [53], it is a wonder that any government action at all is ever taken to inform citizens of alcohol's health risks, let alone implement effective policies to reduce its consumption.

The evidence that the minority of people who understand that alcohol is a carcinogen, for example, are more likely to support evidence-based alcohol strategies (e.g., [50,55]) should be a wake-up call to the public health community to increase awareness of alcohol-related harms. While awareness rising and educational strategies in general have been decried as lacking evidence for changing population behaviours (e.g., [11]), they could likely have a critical role as a means of creating a more favourable environment for the implementation of more directly effective policies [18].

The CSUCH project in Canada has begun the difficult task of enumerating both the partially and wholly alcohol-attributable deaths and adverse health outcomes for all its jurisdictions, for all years and in comparison with other popular psychoactive substances [24]. The estimates are presented to the public and to policymakers in a variety of formats and through a variety of media, including a web-based data visualisation tool (see: https://csuch.ca (access date 11 August 2021)) for individuals to explore and download figures and data tables. Such a resource on its own is completely insufficient to substantially impact national policies on such weighty matters as taxation and pricing, but, if communicated effectively to stakeholders with an interest in, for example, preventing cancer, heart disease, violence and/or substance use disorders, it could help turn the tide of public opinion.

Perhaps the single most effective tool to raise awareness of alcohol's potentially harmful effects would be mandated warning labels of the variety trialled by the Yukon Liquor Corporation (see Figure 1). The Health Canada-funded evaluations of this intervention demonstrated increased awareness of key messages (i.e., cancer risk, national drinking guidelines and standard drink contents) among those who recalled seeing the messages (e.g., [56]), greater intentions to reduce alcohol intake [57] and significantly reduced per capita consumption at the intervention site [58]. These labels were likely effective as they were carefully developed on the basis of published research regarding what makes a warning label noticeable, relevant and memorable (unlike the static, small, black and white US labels). A unique feature of warning labels as a means of raising awareness of alcohol-related issues is that they are more likely to be seen and recalled by those who would perhaps most benefit from the information they contain, i.e., those who consume alcohol most frequently [59].

Cancer Warning   National Drinking Guidelines   Standard Drink Labels (beer example)

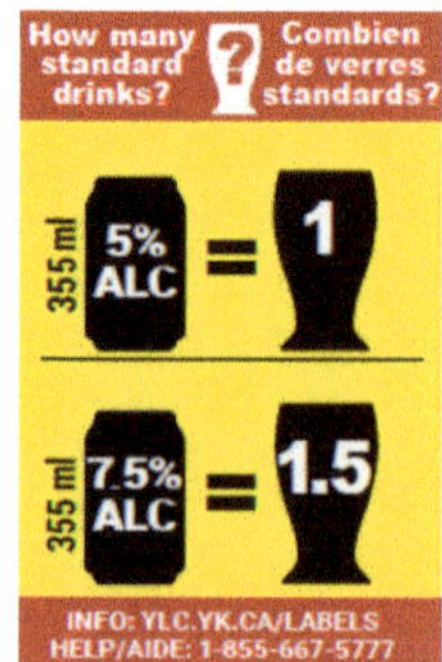

Figure 1. Warning labels trialled in the Yukon Territory, Canada.

What else might the public health community attempt by way of increasing the likelihood that effective alcohol policies might be implemented by governments? Most critically, putting pressure on their governments to give public health rather than industry groups a place at the policy-making table. At the very least, they can attempt to hold governments accountable by regularly monitoring and reporting the extent to which present policies match up with gold-standard best practices for reducing alcohol-related harm. The CAPE project, and other similar projects in Europe and the USA, draw attention to the disparities between actual implementation and best practice evidence-based policies. The CAPE project has extensively engaged with stakeholders and relevant government sectors (e.g., health, liquor regulators and government retailers, finance) to communicate clearly using publicly available and transparent indicators of effective practice [60]. This has involved going beyond lofty ideals and exhortations to "reduce alcohol's availability" or "increase the price of alcohol" and instead reported on tangible, verifiable performance indicators such as the density of different types of liquor outlets, the extent of private versus government ownership, whether alcohol prices are keeping up with the cost of living and the level of minimum prices [18]. We have discovered that, contrary to what might be expected, across Canada's 13 jurisdictions, examples of best practice could be found somewhere in every one of the 11 policy domains identified earlier. In fact, if identified best practices were implemented uniformly across Canada, the national score would have been 87%, i.e., an A grade. It may well be that government failures to implement effective alcohol policies are at least partly due to a lack of awareness of what is effective policy rather than a lack of willingness to implement this. However, then again, Canada may be an exception in this regard given that it still has significant, if eroding, government controls over alcohol's distribution and retail sale(e.g., [61]).

A final thought. This essay was written in the middle of 2021 as the world is struggling to emerge from the COVID-19 pandemic. It has been widely documented that alcohol controls have weakened in many countries and that the alcohol industry has lobbied effectively to persuade governments to remove restrictions on their trade that have been in place for generations (e.g., [62]). However, there is also a significant opportunity at this moment for governments to take effective actions on alcohol to achieve a variety of both public health and fiscal outcomes. The population at large may now be more sensitised to the importance of public health measures to protect the population's health and safety. There is also a substantial shortfall in government revenues as a consequence of economic downturns during the pandemic crisis. Alcohol pricing, taxation and availability policies could be a key part of the process of recovery. A combination of MUP and increased alcohol taxes has been shown to be a means of delivering such diverse benefits as improved public health outcomes, increased government revenues and greater industry profits [3]. Restrictions on outlet densities or at least moratoria on the issuing of new liquor licences could also help the financial sustainability of existing businesses. Such restrictions on the alcohol market could enable producers and retailers to make greater profits from selling less alcohol while also helping to shore up much-needed government revenues and reducing related harms.

Author Contributions: Writing—original draft, T.S.; Writing—review and editing, T.S., N.G., K.V. and A.W. All authors have read and agreed to the published version of the manuscript.

Funding: This research received no external funding.

Conflicts of Interest: I have received travel expenses, research funds and limited personal fees from two Nordic government alcohol monopolies for research on alcohol and public health in the past 5 years.

References

1. Bruun, K.; Edwards, G.; Lumui, M.; Mäkelä, K.; Pan, L.; Popham, R.; Room, R.; Schmidt, W.; Skog, O.-J.; Sulkunen, P.; et al. *Alcohol Control Policies in Public Health Perspective*; Finnish Foundation for Alcohol Studies: Helsinki, Finland, 1975.
2. Sherk, A.; Stockwell, T.; Rehm, J.; Dorocicz, J.; Shield, K.D. *InterMAHP: The International Model of Alcohol Harms and Policies: A Comprehensive Guide to the Estimation of Alcohol-Attributable Morbidity and Mortality*; Version 1.0: December 2017; Canadian Institute for Substance Use Research, University of Victoria: Victoria, BC, Canada, 2017. Available online: www.intermahp.cisur.ca (accessed on 13 August 2021).
3. Stockwell, T.; Churchill, S.; Sherk, A.; Sorge, J.; Gruenewald, P. How many alcohol attributable deaths and hospital admissions could be prevented by alternative pricing and taxation policies? Modelling impacts on alcohol consumption, revenues and related harms in Canada. *Health Promot. Chronic Dis. Prev. Can.* **2020**, *40*, 153–164. [CrossRef]
4. Kehoe, T.; Gmel, G.; Shield, K.; Rehm, J. Determining the best population-level alcohol consumption model and its impact on estimates of alcohol-attributable harms. *Popul. Health Metr.* **2012**, *10*, 6. [CrossRef] [PubMed]
5. Holmes, J.; Meng, Y.; Meier, P.; Brennan, A.; Angus, C.; Campbell-Burton, A.; Guo, Y.; Hill-McManus, D.; Purshouse, R. Effects of minimum unit pricing for alcohol on different income and socioeconomic groups: A modelling study. *Lancet* **2014**, *383*, 1655–1664. [CrossRef]
6. Scottish Government. Alcohol and Drugs: Minimum Unit Pricing. 2018. Available online: https://www.gov.scot/policies/alcohol-and-drugs/minimum-unit-pricing/ (accessed on 13 August 2021).

7.   KTVZ News. Oregon High-Proof, Low-Cost Spirits to Rise in Price; Bend Distiller Sees Little Impact. 2021. Available online: https://ktvz.com/news/central-oregon/2021/04/12/olcc-raises-pricimg-on-high-proof-low-cost-spirits-to-increase-in-cost/ (accessed on 13 August 2021).
8.   Taylor, N.; Miller, P.; Coomber, K.; Livingston, M.; Scott, D.; Buykx, P.; Chikritzhs, T. The impact of a minimum unit price on wholesale alcohol supply trends in the Northern Territory, Australia. *Aust. N. Z. J. Public Health* **2021**, *45*, 26–33. [CrossRef]
9.   Welsh Government. *Guidance on the Implementation of Minimum Pricing for Alcohol in Wales*; Welsh Government: Cardiff, UK, 2020.
10.   Sherk, A.; Stockwell, T.; April, N.; Churchill, S.; Sorge, J.; Gamache, P. The potential health impact of an alcohol minimum unit price in Québec: An application of the International Model of Alcohol Harms and Policies. *J. Stud. Alcohol Drugs* **2020**, *81*, 631–640. [CrossRef] [PubMed]
11.   Babor, T.; Caetano, R.; Casswell, S.; Edwards, G.; Giesbrecht, N.; Grube, J.W.; Graham, K. *Alcohol: No Ordinary Commodity. Research and Public Policy*, 2nd ed.; Oxford University Press: Oxford, UK, 2010.
12.   WHO. *The SAFER Technical Package: Five Areas of Intervention at National and Subnational Levels*; WHO Technical Document; WHO: Geneva, Switzerland, 2019.
13.   WHO. *Global Strategy to Reduce the Harmful Use of Alcohol*; WHO Department of Mental Health and Substance Abuse: Geneva, Switzerland, 2016. Available online: https://www.who.int/publications/i/item/9789241599931 (accessed on 13 August 2021).
14.   Canadian Institute for Health Information. Alcohol Harm in Canada: Examining Hospitalizations Entirely Caused by Alcohol and Strategies to Reduce Alcohol Harm. 2019. Available online: https://www.cihi.ca/sites/default/files/document/report-alcohol-hospitalizations-en-web.pdf (accessed on 13 August 2021).
15.   Naimi, T.S.; Blanchette, J.; Nelson, T.F.; Nguyen, T.; Oussayef, N.; Heeren, T.C.; Gruenewald, P.; Mosher, J.; Xuan, Z. A new scale of the U.S. alcohol policy environment and its relationship to binge drinking. *Am. J. Prev. Med.* **2014**, *46*, 10–16. [CrossRef]
16.   Giesbrecht, N.; Wettlaufer, A.; Simpson, S.; April, N.; Asbridge, M.; Cukier, S.; Mann, R.E.; McAllister, J.; Murie, A.; Pauley, C.; et al. Strategies to reduce alcohol-related harms and costs in Canada: A comparison of provincial policies. *Int. J. Alcohol Drug Res.* **2016**, *5*, 33–45. [CrossRef]
17.   Stockwell, T.; Wettlaufer, A.; Vallance, K.; Chow, C.; Giesbrecht, N.; April, N.; Asbridge, M.; Callaghan, R.; Cukier, S.; Davis-MacNevin, P.; et al. *Strategies to Reduce Alcohol-Related Harms and Costs in Canada: An Evaluation of Provincial and Territorial Policies*; Canadian Institute for Substance Use Research, University of Victoria: Victoria, BC, Canada, 2019. Available online: https://www.uvic.ca/research/centres/cisur/assets/docs/report-cape-pt-en.pdf (accessed on 13 August 2021).
18.   Vallance, K.; Stockwell, T.; Wettlaufer, A.; Giesbrecht, N.; Chow, C.; Card, K.; Farrell-Low, A.; The Canadian Alcohol Policy Evaluation (CAPE) project: Findings from a review of provincial and territorial alcohol policies. Drug Alcohol Rev. 2021. Available online: https://onlinelibrary.wiley.com/doi/epdf/10.1111/dar.13313 (accessed on 13 August 2021).
19.   Wettlaufer, A.; Vallance, K.; Chow, C.; Stockwell, T.; Giesbrecht, N.; April, N.; Asbridge, M.; Callaghan, R.; Cukier, S.; Hynes, G.; et al. *Strategies to Reduce Alcohol-Related Harms and Costs in Canada: A Review of Federal Policies*; Canadian Institute for Substance Use Research, University of Victoria: Victoria, BC, Canada, 2019. Available online: https://www.uvic.ca/research/centres/cisur/assets/docs/report-cape-fed-en.pdf (accessed on 13 August 2021).
20.   Ogborne, A.C.; Smart, R.G. Public opinion on the health benefits of moderate drinking: Results from a Canadian National Population Health Survey. *Addiction* **2001**, *96*, 641–649. [CrossRef]
21.   Institut National de Santé Publique du Québec. Alcohol Consumption: What do the People of Quebec Think? Research Report, Montreal, Canada. 6 May 2021. Available online: https://www.inspq.qc.ca/publications/2772-consommation-alcool-population-quebec (accessed on 13 August 2021).
22.   Australian Institute of Health and Welfare. *National Drug Strategy Household Survey 2019*; Drug Statistics Series No. 32. PHE 270; AIHW: Canberra, Australia, 2020.
23.   Institute for Health Metrics and Evaluation. Global Health Data Exchange (GHMx) Data Tool. 2021. Available online: http://ghdx.healthdata.org/gbd-results-tool (accessed on 9 June 2021).
24.   *Canadian Substance Use Costs and Harms Scientific Working Group Canadian Substance Use Costs and Harms (2007–2017)*; Canadian Centre on Substance Use and Addiction: Ottawa, ON, Canada, 2020.
25.   Sherk, A.; Thomas, G.; Churchill, S.; Stockwell, T. Does drinking within low-risk guidelines prevent harm? Implications for high-income countries using the International Model of Alcohol Harms and Policies. *J. Stud. Alcohol Drugs* **2020**, *81*, 352–361. [CrossRef] [PubMed]
26.   Shield, K.; Soerjomataram, I.; Rehm, R. Alcohol Use and Breast Cancer: A Critical Review. *Alcohol. Clin. Exp. Res.* **2016**, *40*, 1166–1181. [CrossRef] [PubMed]
27.   International Agency for Research on Cancer. *Alcohol Consumption and Ethyl Carbamate*; IARC Monographs on the Evaluation of Carcinogenic Risks to Humans; IARC: Lyon, France, 2010.
28.   Bates, S.; Holmes, J.; Gavens, L.; De Matos, E.G.; Li, J.; Ward, B.; Hooper, L.; Dixon, S.; Bukyx, P. Awareness of alcohol as a risk factor for cancer is associated with public support for alcohol policies. *BMC Public Health* **2018**, *18*, 688. [CrossRef] [PubMed]
29.   Opp, S.M.; Mosier, S.L. Liquor, marijuana, and guns: Essential services or political tools during the COVID-19 pandemic? *Policy Des. Pract.* **2020**, *3*, 297–311. [CrossRef]
30.   Stockwell, T.; Andreasson, S.; Cherpitel, C.; Chikritzhs, T.; Danghardt, F.; Holder, H.; Naimi, T.; Sherk, A. The burden of alcohol on health care during COVID-19. *Drug Alcohol Rev.* **2020**, *40*, 3–7. [CrossRef] [PubMed]

31. Narasimha, V.L.; Shukla, L.; Mukherjee, D.; Menon, J.; Huddar, S.; Panda, U.K.; Mahadevan, J.; Kandasamy, A.; Chand, P.K.; Benegal, V.; et al. Complicated Alcohol Withdrawal-An Unintended Consequence of COVID-19 Lockdown. *Alcohol Alcohol.* **2020**, *55*, 350–353. [CrossRef] [PubMed]

32. Rossow, I. The strike hits: The 1982 wine and liquor monopoly strike in Norway and its impact on various harm indicators. In *The Effects of Nordic Alcohol Policies: What Happens to Drinking and Harm When Alcohol Controls Change?* Room, R., Ed.; Nordic Council for Alcohol and Drug Research: Helsinki, Finland, 2002; pp. 133–144, Publication No. 42.

33. Government of Canada. Epidemiological Summary of COVID-19 Cases in Canada. 2021. Available online: https://health-infobase.canada.ca/covid-19/epidemiological-summary-covid-19-cases.html (accessed on 13 August 2021).

34. Li, J.; Lovatt, M.; Eadie, D.; Dobbie, F.; Meier, P.; Holmes, J.; Hastings, G.; MacKintosh, A. Public attitudes towards alcohol control policies in Scotland and England: Results from a mixed-methods study. *Soc. Sci. Med.* **2017**, *177*, 177–189. Available online: https://www.sciencedirect.com/science/article/pii/S0277953617300448 (accessed on 13 August 2021). [CrossRef] [PubMed]

35. Wagenaar, A.C.; Salois, M.J.; Komro, K.A. Effects of beverage alcohol price and tax levels on drinking: A meta-analysis of 1003 estimates from 112 studies. *Addiction* **2009**, *104*, 179–190. [CrossRef]

36. Erickson, R.; Stockwell, T.; Pauly, B.; Chow, C.; Roemer, A.; Zhao, J.; Vallance, K.; Wettlaufer, A. How do people with homelessness and alcohol dependence cope when alcohol is unaffordable? A comparison of residents of Canadian Managed Alcohol Programs (MAPs) and locally recruited controls. *Drug Alcohol Rev.* **2018**, *37* (Suppl. S1), S174–S183. Available online: https://onlinelibrary.wiley.com/doi/abs/10.1111/dar.12649 (accessed on 13 August 2021). [CrossRef]

37. US National Institute on Drug Abuse. *Fiscal Year 2020 Budget Information—Congressional Justification for National Institute on Drug Abuse*; NIDA: Bethesda, MD, USA, 2020. Available online: https://www.drugabuse.gov/about-nida/legislative-activities/budget-information/fiscal-year-2020-budget-information-congressional-justification-national (accessed on 13 August 2021).

38. US National Institute on Alcohol Abuse and Alcoholism. *Fiscal Year 2020 Budget Information—Congressional Justification for National Institute on Alcohol Abuse and Alcoholism*; NIDA: Bethesda, MD, USA, 2020. Available online: https://www.niaaa.nih.gov/about-niaaa/our-funding/congressional-budget-justification/fy-2020-congressional-budget-justification (accessed on 13 August 2021).

39. Government Electronic Directory Services. 2021. Available online: https://www.canada.ca/en/shared-services/corporate/transparency/access-information-privacy/publications/privacy-impact-assessments-summary-government-electronic-directory-services.html (accessed on 13 August 2021).

40. Parry, C. *The Temporary Ban on Alcohol Sales During Levels 4 & 5 of COVID-19 Lockdown in South Africa: Lessons Learnt*; South African Medical Research Council: Cape Town, South Africa, 2020. Available online: https://www.wcla.gov.za/sites/default/files/2020-07/C%20Parry%20Alcohol%20%20Covid19%20-%203%20July%20Webinar%20Edinburgh.pdf (accessed on 13 August 2021).

41. Hall, W. What are the policy lessons of National Alcohol Prohibition in the United States, 1920–1933? *Addiction* **2010**, *105*, 1164–1173. [CrossRef]

42. Stockwell, T.; Sherk, A.; Norstrom, T.; Angus, C.; Ramstedt, M.; Andreasson, S.; Chikritzhs, T.; Gripenberg, J.; Holder, H.; Holmes, J.; et al. Estimating the public health impact of disbanding a government alcohol monopoly: Application of new methods to the case of Sweden. *BMC Public Health* **2018**, *18*, 1400. [CrossRef]

43. Stockwell, T.; Sherk, A.; Sorge, J.; Norström, T.; Angus, C.; Chikritzhs, T.; Churchill, S.; Holmes, J.; Meier, P.; Naimi, T.; et al. *Finnish Alcohol Policy at the Crossroads: The Health, Safety and Economic Consequences of Alternative Systems to Manage the Retail Sale of Alcohol*; A Report Prepared for the Finnish Alcohol Monopoly, Alko; Canadian Institute for Substance Use Research, University of Victoria: Victoria, BC, Canada, 2019. Available online: https://www.uvic.ca/research/centres/cisur/assets/docs/report-alko.pdf (accessed on 13 August 2021).

44. Room, R.; Schmidt, L.; Rehm, J.; Mäkelä, P. International regulation of alcohol. *Br. Med. J.* **2008**, *337*, a2364. [CrossRef]

45. Babor, T.F. Alcohol research and the alcoholic beverage industry: Issues, concerns and conflicts of interest. *Addiction* **2009**, *104*, 34–47. [CrossRef]

46. McCambridge, J.; Kypri, K.; Miller, P.; Hawkins, B.; Hastings, G. Be aware of Drinkaware. *Addiction* **2014**, *109*, 519–524. [CrossRef] [PubMed]

47. Petticrew, M.; Maani Hessari, N.; Knai, C.; Weiderpass, E. How alcohol industry organisations mislead the public about alcohol and cancer. *Drug Alcohol Rev.* **2018**, *37*, 293–303. [CrossRef] [PubMed]

48. Éduc'alcool. Alcohol and Heart Health. Montreal: Canada. PDF Copy. 2019. Available online: https://www.educalcool.qc.ca/wp-content/uploads/2021/03/Alcohol-and-heart-health.pdf (accessed on 13 August 2021).

49. Golder, S.; Garry, J.; McCambridge, J. Declared funding and authorship by alcohol industry actors in the scientific literature: A bibliometric study. *Eur. J. Public Health* **2020**, *30*, 1193–1200. [CrossRef] [PubMed]

50. Buykx, P.; Li, J.; Gavens, L.; Hooper, L.; Lovatt, M.; Gomes de Matos, E.; Holmes, J. Public awareness of the link between alcohol and cancer in England in 2015: A population-based survey. *BMC Public Health* **2016**, *16*, 1194. [CrossRef]

51. Weerasinghe, A.; Schoueri-Mychasiw, N.; Vallance, K.; Stockwell, T.; Hammond, D.; McGavock, J.; Greenfield, T.K.; Paradis, C.; Hobin, E. Improving Knowledge that Alcohol Can Cause Cancer is Associated with Consumer Support for Alcohol Policies: Findings from a Real-World Alcohol Labelling Study. *Int. J. Environ. Res. Public Health* **2020**, *17*, 398. [CrossRef]

52. Mart, S.; Giesbrecht, N. Red flags on pink-washed drinks: Contradictions and dangers in marketing alcohol to prevent cancer. *Addiction* **2015**, *110*, 1541–1548. [CrossRef]

53. Buteau, E. Drink Pink for Breast Cancer Research with Mike's Hard Pink Lemonade. 2014. Available online: http://ericabuteau.com/2014/09/06/hard-pink-lemonade/ (accessed on 13 August 2021).

54. Stockwell, T.; Solomon, R.; O'Brien, P.; Vallance, K.; Hobin, E. Cancer Warning Labels on Alcohol Containers: A Consumer's Right to Know, a Government's Responsibility to Inform, and an Industry's Power to Thwart. *J. Stud. Alcohol Drugs* **2020**, *81*, 284–292. [CrossRef]
55. Amin, G.; Siegel, M.; Naimi, T. National Cancer Societies and their public statements on alcohol consumption and cancer risk. *Addiction* **2018**, *113*, 1802–1808. [CrossRef]
56. Hobin, E.; Weerasinghe, A.; Vallance, K.; Hammond, D.; McGavock, J.; Greenfield, T.K.; Schoueri-Mychasiw, N.; Paradis, C.; Stockwell, T. Testing Alcohol Labels as a Tool to Communicate Cancer Risk to Drinkers: A Real-World Quasi-Experimental Study. *J. Stud. Alcohol Drugs* **2020**, *81*, 249–261. [CrossRef]
57. Hobin, E.; Shokar, S.; Vallance, K.; Hammond, D.; McGavock, J.; Greenfield, T.K.; Schoueri-Mychasiw, N.; Paradis, C.; Stockwell, T. Communicating risks to drinkers: Testing alcohol labels with a cancer warning and national drinking guidelines in Canada. *Can. J. Public Health* **2020**, *111*, 716–725. [CrossRef] [PubMed]
58. Zhao, J.H.; Stockwell, T.; Vallance, K.; Hobin, E. The effects of alcohol warning labels on population alcohol consumption: An interrupted time-series analysis of alcohol sales in Yukon, Canada. *J. Stud. Alcohol Drugs* **2020**, *81*, 225–237. [CrossRef] [PubMed]
59. Greenfield, T. Warning labels: Evidence on harm reduction fromlong-term American surveys. In *Alcohol: Minimizing the Harm*; Plant, M., Single, E., Stockwell, T., Eds.; Free Association Books: London, UK, 1997; pp. 105–125.
60. Vallance, K.; Stockwell, T.; Wettlaufer, A.; Giesbrecht, N.; Chow, C.; Card, K.; Farrell-Low, A.; Strategies for engaging policy stakeholders to translate research knowledge into practice more effectively: Lessons learned from the Canadian Alcohol Policy Evaluation project. *Drug Alcohol Rev.* **2021**. Available online: https://onlinelibrary.wiley.com/doi/epdf/10.1111/dar.13313 (accessed on 13 August 2021).
61. Giesbrecht, N.; Wettlaufer, A.; Stockwell, T.; Vallance, K.; Chow, C.; April, N.; Asbridge, M.; Callaghan, R.; Cukier, S.; Hynes, G.; et al. Alcohol retail privatization in Canadian provinces between 2012 and 2017. Is decision-making oriented to harm reduction? *Drug Alcohol Rev.* **2020**, *40*, 459–467. [CrossRef] [PubMed]
62. Andreasson, S.; Chikritzhs, T.; Dangardt, F.; Holder, H.; Naimi, T.; Stockwell, T. *Alcohol and the Coronavirus Pandemic: Individual, Societal and Policy Perspectives*; Alcohol and Society 2021; Swedish Society of Nursing: Stockholm, Sweden; SFAM: Stockholm, Sweden; SAFF: Stockholm, Sweden; CERA: Stockholm, Sweden; The Swedish Society of Addiction Medicine: Stockholm, Sweden; SIGHT: Stockholm, Sweden; Movendi International: Stockholm, Sweden; IOGT-NTO: Stockholm, Sweden, 2021.

MDPI
St. Alban-Anlage 66
4052 Basel
Switzerland
Tel. +41 61 683 77 34
Fax +41 61 302 89 18
www.mdpi.com

Nutrients Editorial Office
E-mail: nutrients@mdpi.com
www.mdpi.com/journal/nutrients